SELECTED PAPERS ON

NOISE AND STOCHASTIC PROCESSES

Edited by NELSON WAX

PROFESSOR OF ELECTRICAL ENGINEERING
UNIVERSITY OF ILLINOIS

DOVER PUBLICATIONS, INC.
New York

Published in Canada by General Publishing Company, Ltd.,
30 Lesmill Road, Don Mills, Toronto, Ontario.

Published in the United Kingdom by Constable and Company, Ltd.,
10 Orange Street, London WC 2.

This Dover edition, first published in 1954, is a selection of papers published for the first time in collected form. The editor and publisher are indebted to the authors and original publishers of all the papers for permission to reproduce these articles.

International Standard Book Number: 0-486-60262-1
Library of Congress Catalog Card Number: 54-4062

Manufactured in the United States of America

DOVER PUBLICATIONS, INC.
180 Varick Street
New York, N. Y. 10014

PREFACE

This is the third volume in a series which Dover Publications is issuing. Each volume in the series is to be a collection of significant papers in a given branch of science. The previous volumes ('Foundations of Nuclear Physics' edited by R. T. Beyer, and 'Foundations of High Speed Aerodynamics,' edited by G. F. Carrier) covered the fields of nuclear physics and the flow of compressible fluids. Random processes, and their applications to physical questions, are treated in the present volume.

The papers have been chosen with certain assumed interests and needs of American physicists and engineers in mind. It was felt, first, that a handy reference work containing a number of well known reprints would be of value to people concerned with 'noise,' with Brownian motion, and with statistical mechanics. Secondly, it seemed that a collection of some diversified introductions to noise theory and to the theory of fluctuation phenomena would be useful for workers new to these subjects. This collection, therefore, consists almost entirely of reprints of 'review' papers.

The references cited in each paper form a working bibliography of the literature up to about 1940-45. The following publications describe more recent work, and contain supplementary references to earlier investigations:

J. L. Doob, 'Stochastic Processes,' John Wiley & Sons, Inc. New York, 1953

An advanced mathematical treatise. Extensive bibliography of the mathematical literature.

W. Feller, 'An Introduction to Probability Theory and its Applications,' Volume One, John Wiley & Sons, Inc., New York, 1950.

A number of applications to physics and engineering are given. Considers holding and waiting time problems.

W. Jackson, 'Communication Theory,' Butterworths Scientific Publications, London, 1953. (U.S.A. Edition published by Academic Press, Inc., New York).

A symposium on applications of communication theory. One section is concerned with signal discrimination in the presence of noise.

Journal of the Royal Statistical Society, Series B, Volume II, No. 2, 1949, pp. 150-282.

A symposium on stochastic processes. Applications to physics, in particular J. E. Moyal's paper, pp. 110-210

Massachusetts Institute of Technology, Radiation Laboratory Series, Volumes 1-28, McGraw Hill Book Company, New York, 1947-53.

Discussions of noise and of fluctuation phenomena in a wide variety of devices, are scattered through the series. The index, vol. 28, 1953, gives specific references.

University of Illinois
Urbana, Illinois

Nelson Wax

CONTENTS

PAPERS

REVIEWS OF

MODERN PHYSICS

VOLUME 15, NUMBER 1 JANUARY, 1943

Stochastic Problems in Physics and Astronomy

S. CHANDRASEKHAR
Yerkes Observatory, The University of Chicago, Williams Bay, Wisconsin

CONTENTS

APPENDIXES

INTRODUCTION

IN this review we shall consider certain fundamental probability methods which are finding applications increasingly in a wide variety of problems and in fields as different as colloid chemistry and stellar dynamics. However, a common characteristic of all these problems is that interest is focused on a property which is the result of superposition of a large number of variables, the values which these variables take being governed by certain probability laws. We may cite as illustrations two examples:

(i) The first example is provided by the *problem of random flights*. In this problem, a particle undergoes a sequence of displacements $\mathbf{r}_1, \mathbf{r}_2, \cdots, \mathbf{r}_i, \cdots$, the magnitude and direction of each displacement being independent of all the preceding ones. But the probability that the displacement $\mathbf{r}_i$ lies between $\mathbf{r}_i$ and $\mathbf{r}_i+d\mathbf{r}_i$ is governed by a distribution function $\tau_i(\mathbf{r}_i)$ assigned *a priori*. We ask: What is the probability $W(\mathbf{R})d\mathbf{R}$ that after N displacements the coordinates of the particle lie in the interval $\mathbf{R}(=[x, y, z])$ and $\mathbf{R}+d\mathbf{R}$. It is seen that in this problem the position $\mathbf{R}$ of the particle is the resultant of N vectors, $\mathbf{r}_i$, $(i=1, \cdots, N)$ the position and direction of each vector being governed by the probability distributions $\tau_i(\mathbf{r}_i)$. As we shall see the solution to this problem provides us with one of the principal weapons of the theory.[1]

(ii) We shall take our second illustration from stellar dynamics. The gravitational force acting on a star (per unit mass) is given by

$$\mathbf{F}=G\Sigma M_i\mathbf{r}_i/|\mathbf{r}_i|^3 \tag{1}$$

where M_i denotes the mass of a typical "field" star and $\mathbf{r}_i$ its position vector relative to the star under consideration and G the constant of gravitation. Further in Eq. (1) the summation is extended over all the neighboring stars. We now suppose that the distribution of stars in the neighborhood of a given one is subject to fluctuations and that stars of different masses occur in the stellar system according to some well defined empirically established law. However, the fluctuations in density are assumed to be subject to the restriction of a constant average density of n stars per unit volume. We ask: What is the probability that $\mathbf{F}$ lies between $\mathbf{F}$ and $\mathbf{F}+d\mathbf{F}$? Again, the force acting on a star is the resultant of the forces due to all the neighboring stars while the spatial distribution of these stars and their masses are subject to well-defined laws of fluctuations.

From the foregoing two examples it is clear that one of the principal problems under the circumstances envisaged is the specification of the distribution function $W(\mathbf{\Phi})$ of a quantity $\mathbf{\Phi}$ (in general a vector in hyper-space) which is the resultant of a large number of other quantities having assigned distributions over a range of values. A second fundamental problem in the theories we shall consider concerns questions relating to *probability after-effects*[2]—a notion first introduced by Smoluchowski. We may broadly describe the nature of these questions in the following terms: A certain quantity $\mathbf{\Phi}$ is characterized by a stationary distribution $W(\mathbf{\Phi})$. We first make an observation of $\mathbf{\Phi}$ at a certain instant of time $t=0$ (say) and again repeat our

[1] For historical remarks on this problem of random flights see the Bibliographical Notes at the end of the article.

[2] This is the translation of the German word "Wahrscheinlichkeitsnachwirkung" coined by M. von Smoluchowski.

observation at a later time t. We ask: What can we say about the possible values of $\mathbf{\Phi}$ which we may expect to observe at time t when we already know that $\mathbf{\Phi}$ had a particular value at $t=0$? It is clear that if the second observation were made after a sufficiently long interval of time, we should not, in general, expect any correlation with the fact that $\mathbf{\Phi}$ had a particular value at a very much earlier epoch. On the other hand as $t \to 0$ the values which we would expect to observe on the second occasion will be strongly dependent on what we observed on the earlier occasion.

An example considered by Smoluchowski in colloid statistics illustrates the nature of the problem presented in theories of probability after-effects: Suppose we observe by means of an ultramicroscope a small well-defined element of volume of a colloidal solution and count the number of particles in the element at definite intervals of time τ, 2τ, 3τ, etc., and record them consecutively. We shall further suppose that the interval τ between successive observations is not large. Then the number which is observed on any particular occasion will be correlated in a definite manner with what was observed on the immediately preceding occasion. This correlation will depend on a variety of physical factors including the viscosity of the medium: thus it is clear from general considerations that the more viscous the surrounding medium the greater will be the correlation in the numbers counted on successive occasions. We shall discuss this problem following Smoluchowski in some detail in Chapter III but pass on now to the consideration of another example typical of this theory.

We have already indicated that a fundamental problem in stellar dynamics is the specification of the distribution function $W(\boldsymbol{F})$ governing the probability of occurrence of a force $\boldsymbol{F}$ per unit mass acting on a star. Suppose that $\boldsymbol{F}$ has a definite value at a given instant of time. We can ask: How long a time should elapse on the average before the force acting on the star can be expected to have no appreciable correlation with the fact of its having had a particular value at the earlier epoch? In other words, what is the *mean life* of the state of fluctuation characterized by $\boldsymbol{F}$? In a general way it is clear that this mean life will depend on the state of stellar motions in the neighborhood of the star under consideration in contrast to the probability distribution $W(\boldsymbol{F})$ which depends only on the average number of stars per unit volume. The two examples we have cited are typical of the problems which are properly in the province of the theory dealing with probability after-effects.

A physical problem, the complete elucidation of which requires both the types of theories outlined in the preceding paragraphs, is provided by Brownian motion. We shall accordingly consider certain phases of this theory also.

CHAPTER I

THE PROBLEM OF RANDOM FLIGHTS

The problem of random flights which in its most general form we have already formulated in the introduction provides an illustrative example in reference to which we may develop several of the principal methods of the theories we wish to describe. Accordingly, in this chapter, in addition to providing the general solution of the problem, we shall also discuss it from several different points of view.

1. The Simplest One-Dimensional Problem: The Problem of Random Walk

The principal features of the solution of the problem of random flights in its most general form are disclosed and more clearly understood by considering first the following simplest version of the problem in one dimension:

A particle suffers displacements along a straight line in the form of a series of *steps* of equal length, each step being taken, either in the forward, or in backward direction with equal probability $\frac{1}{2}$. After taking N such steps the particle *could* be at any of the points[3]

$$-N,\ -N+1,\ \cdots,\ -1,\ 0,\ +1,\ \cdots,\ N-1 \text{ and } N.$$

We ask: What is the probability $W(m, N)$ that the particle arrives at the point m after suffering N displacements?

We first remark that in accordance with the conditions of the problem each individual step is equally likely to be taken either in the back-

[3] These can be regarded as the coordinates along a straight line if the unit of length be chosen to be equal to the length of a single step.

ward or in the forward direction quite independently of the direction of all the preceding ones. Hence, all possible sequences of steps each taken in a definite direction have the same probability. In other words, the probability of any given sequence of N steps is $(\frac{1}{2})^N$. The required probability $W(m, N)$ is therefore $(\frac{1}{2})^N$ times the number of distinct sequences of steps which will lead to the point m after N steps. But in order to arrive at m among the N steps, *some* $(N+m)/2$ steps should have been taken in the positive direction and the remaining $(N-m)/2$ steps in the negative direction. (Notice that m can be even or odd only according as N is even or odd.) The number of such distinct sequences is clearly

$$N!/[\tfrac{1}{2}(N+m)]![\tfrac{1}{2}(N-m)]!. \tag{2}$$

Hence

$$W(m, N)=\frac{N!}{[\frac{1}{2}(N+m)]![\frac{1}{2}(N-m)]!}\left(\frac{1}{2}\right)^N. \tag{3}$$

In terms of the binomial coefficients C_r^n's we can rewrite Eq. (3) in the form

$$W(m, N)=C^N_{(N+m)/2}\left(\frac{1}{2}\right)^N, \tag{4}$$

in other words we have a *Bernoullian distribution*. Accordingly, the expectation and the mean square deviation of $(N+m)/2$ are (see Appendix I)

$$\left.\begin{aligned} \tfrac{1}{2}\langle N+m\rangle_{\text{Av}}&=\tfrac{1}{2}N,\\ \langle[\tfrac{1}{2}(N+m)-\tfrac{1}{2}N]^2\rangle_{\text{Av}}&=\tfrac{1}{4}N. \end{aligned}\right\} \tag{5}$$

Hence,

$$\langle m\rangle_{\text{Av}}=0;\quad \langle m^2\rangle_{\text{Av}}=N. \tag{6}$$

The root mean square displacement is therefore $\sqrt{N}$.

We return to formula (3): The case of greatest interest arises when N is large and $m\ll N$. We can then simplify our formula for $W(m, N)$ by using Stirling's formula

$$\log n!=(n+\tfrac{1}{2})\log n-n +\tfrac{1}{2}\log 2\pi+O(n^{-1})\,(n\to\infty). \tag{7}$$

Accordingly when $N\to\infty$ and $m\ll N$ we have

$$\begin{aligned}\log W(m, N)\simeq{}&(N+\tfrac{1}{2})\log N\\ &-\tfrac{1}{2}(N+m+1)\log\left[\frac{N}{2}\left(1+\frac{m}{N}\right)\right]\\ &-\tfrac{1}{2}(N-m+1)\log\left[\frac{N}{2}\left(1-\frac{m}{N}\right)\right]\\ &-\tfrac{1}{2}\log 2\pi-N\log 2.\end{aligned} \tag{8}$$

But since $m\ll N$ we can use the series expansion

$$\log\left(1\pm\frac{m}{N}\right)=\pm\frac{m}{N}-\frac{m^2}{2N^2}+O(m^3/N^3). \tag{9}$$

Equation (8) now becomes

$$\begin{aligned}\log W(m, N)\simeq{}&(N+\tfrac{1}{2})\log N-\tfrac{1}{2}\log 2\pi-N\log 2\\ &-\tfrac{1}{2}(N+m+1)\left(\log N-\log 2+\frac{m}{N}-\frac{m^2}{2N^2}\right)\\ &-\tfrac{1}{2}(N-m+1)\left(\log N-\log 2-\frac{m}{N}-\frac{m^2}{2N^2}\right).\end{aligned} \tag{10}$$

Simplifying the right-hand side of this equation we obtain

$$\log W(m, N)\simeq-\tfrac{1}{2}\log N+\log 2 -\tfrac{1}{2}\log 2\pi-m^2/2N. \tag{11}$$

In other words, for large N we have the asymptotic formula

$$W(m, N)=(2/\pi N)^{\frac{1}{2}}\exp(-m^2/2N). \tag{12}$$

A numerical comparison of the two formulae (3) and (12) is made in Table I for $N=10$. We see that even for $N=10$ the asymptotic formula gives sufficient accuracy.

Now, when N is large it is convenient to introduce instead of m the net displacement x from the starting point as the variable:

$$x=ml \tag{13}$$

where l is the length of a step. Further, if we consider intervals Δx along the straight line which are large compared with the length of a

TABLE I. The problem of random walk: the distribution $W(m, N)$ for $N=10$.

m	From (3)	From (12)
0	0.24609	0.252
2	0.20508	0.207
4	0.11715	0.113
6	0.04374	0.042
8	0.00977	0.010
10	0.00098	0.002

step we can ask the probability $W(x)\Delta x$ that the particle is likely to be in the interval $x, x+\Delta x$ after N displacements. We clearly have

$$W(x, N)\Delta x = W(m, N)(\Delta x/2l), \qquad (14)$$

since m can take only even or odd values depending on whether N is even or odd. Combining Eqs. (12), (13), and (14) we obtain

$$W(x, N) = \frac{1}{(2\pi Nl^2)^{\frac{1}{2}}} \exp(-x^2/2Nl^2). \qquad (15)$$

Suppose now that the particle suffers n displacements per unit time. Then the probability $W(x, t)\Delta x$ that the particle will find itself between x and $x+\Delta x$ after a time t is given by

$$W(x, t)\Delta x = \frac{1}{2(\pi Dt)^{\frac{1}{2}}} \exp(-x^2/4Dt)\Delta x, \qquad (16)$$

where we have written

$$D = \tfrac{1}{2}nl^2. \qquad (17)$$

We shall see in §4 that the solution to the general problem of random flights has precisely this form.

2. Random Walk with Reflecting and Absorbing Barriers

In this section we shall continue the discussion of the problem of random walk in one dimension but with certain restrictions on the motion of the particle introduced by the presence of reflecting or absorbing walls. We shall first consider the influence of a reflecting barrier.

(a) A Reflecting Barrier at $m = m_1$

Without loss of generality we can suppose that $m_1 > 0$. Then, the interposition of the reflecting barrier at m_1 has simply the effect that whenever the particle arrives at m_1 it has a probability unity of retracing its step to $m_1 - 1$ when it takes the next step. We now ask the probability $W(m, N; m_1)$ that the particle will arrive at $m(\leqslant m_1)$ after N steps.

For the discussion of this problem it is convenient to trace the course of the particle in an (m, N)-plane as in Fig. 1. In this diagram, the displacement of a particle by a step means that the representative point moves upward by

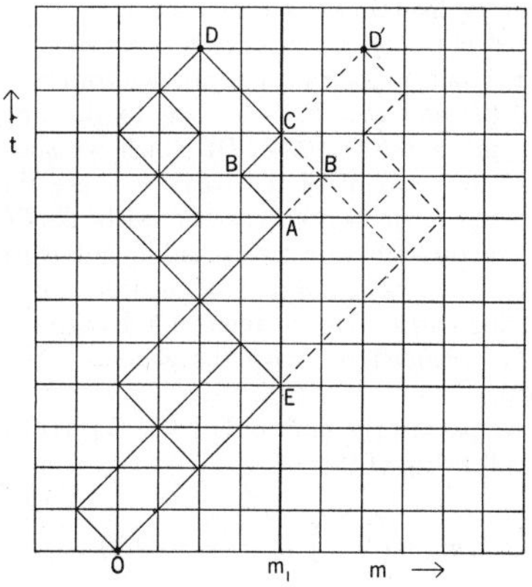

FIG. 1.

one unit while at the same time it suffers a lateral displacement also by one unit either in the positive or in the negative direction.

In the absence of a reflecting wall at $m = m_1$ the probability that the particle arrives at m after N steps is of course given by Eq. (3). But the presence of the reflecting wall requires $W(m, N)$ according to (3) to be modified to take account of the fact that a path reaching m after n reflections must be counted 2^n times since at each reflection it has a probability unity of retracing its step. It is now seen that we can take account of the relevant factors by adding to $W(m, N)$ the probability $W(2m_1 - m, N)$ of arriving at the "*image*" point $(2m_1 - m)$ after N steps (also in the absence of the reflecting wall), i.e.,

$$W(m, N; m_1) = W(m, N) + W(2m_1 - m, N). \qquad (18)$$

We can verify the truth of this assertion in the following manner: Consider first a path like OED which has suffered just one reflection at m_1. By reflecting this path about the vertical line through m_1 we obtain a trajectory leading to the image point $(2m_1 - m)$ and conversely, for every trajectory leading to the image point, having crossed the line through m_1 once, there is exactly one which leads to m after a single reflection. Thus, instead of counting twice each trajectory reflected once, we can add a uniquely defined trajectory leading to $(2m_1 - m)$. Consider next a trajectory like $OABCD$ which leads to m after two reflections. A trajectory like this should be counted four times. But there are two trajectories ($OAB'CD$ and $OABCD'$) leading to the image point and a third ($OAB'CD$) which we should exclude on account of the barrier. These three additional trajectories together with $OABCD$ give exactly four trajectories leading either to m or its image $2m_1-m$ in the absence of the reflecting barrier. In this manner the arguments can be extended to prove the general validity of (18).

If we pass to the limit of large N Eq. (18) becomes [cf. Eq. (12)]

$$W(m, N; m_1)=\left(\frac{2}{\pi N}\right)^{\frac{1}{2}}\{\exp(-m^2/2N) + \exp[-(2m_1-m)^2/2N]\}. \quad (19)$$

Again, if as in §1 we use the net displacement $x=ml$ as the variable and consider the probability $W(x, t; x_1)\Delta x$ that the particle is between x and $x+\Delta x$, $(\Delta x \gg l)$ after a time t (during which time it has taken nt steps) in the presence of a reflecting barrier at $x_1=m_1 l$, we have

$$W(x, t; x_1)=\frac{1}{2(\pi D t)^{\frac{1}{2}}}\{\exp(-x^2/4Dt) + \exp[-(2x_1-x)^2/4Dt]\}. \quad (20)$$

We may note here for future reference that according to Eq. (20)

$$(\partial W/\partial x)_{x=x_1}\equiv 0. \quad (21)$$

(*b*) *Absorbing Wall at* $m=m_1$

We shall now consider the case when there is a perfectly absorbing barrier at $m=m_1$. The interposition of the perfect absorber at m_1 means that whenever the particle arrives at m_1 it at once becomes incapable of suffering further displacements.[4] There are two questions which we should like to answer under these circumstances. The first is the analog of the problems we have considered so far, namely the probability that the particle arrives at $m(\leqslant m_1)$ after taking N steps. The second question which is characteristic of the present problem concerns the average rate at which the particle will deposit itself on the absorbing screen.

Considering first the probability $W(m, N; m_1)$, it is clear that in counting the number of distinct sequences of steps which lead to m we should be careful to exclude all sequences which include even a single arrival to m_1. In other words, if we first count *all* possible sequences which lead to m in the absence of the absorbing screen we should then exclude a certain number of "*forbidden*" sequences. It is evident, on the other hand, that every such forbidden sequence uniquely defines another sequence leading to the image $(2m_1-m)$ of m on the line $m=m_1$ in the (m, N)-plane (see Fig. 1) and conversely. For, by reflecting about the line $m=m_1$ the part of a forbidden trajectory *above* its last point of contact with the line $m=m_1$ before arriving at m we are led to a trajectory leading to the image point, and conversely for every trajectory leading to $2m_1-m$ we necessarily obtain by reflection a forbidden trajectory leading to m (since any trajectory leading to $2m_1-m$ must necessarily cross the line $m=m_1$). Hence,

$$W(m, N; m_1)=W(m, N)-W(2m_1-m, N). \quad (22)$$

For large N we have

$$W(m, N; m_1)=(2/\pi N)^{\frac{1}{2}}\{\exp(-m^2/2N) - \exp[-(2m_1-m)^2/2N]\}. \quad (23)$$

Similarly, analogous to Eq. (21) we now have

$$W(x, t; x_1)=\frac{1}{2(\pi D t)^{\frac{1}{2}}}\{\exp(-x^2/4Dt) - \exp[-(2x_1-x)^2/4Dt]\}. \quad (24)$$

We may further note that according to this equation

$$W(x_1, t; x_1)\equiv 0. \quad (25)$$

Turning next to our second question concerning the probable rate at which the particle deposits itself on the absorbing screen, we may first formulate the problem more specifically. What we wish to know is simply the probability $a(m_1, N)$ that after taking N steps the particle will arrive at m_1 *without ever* having touched or crossed the line $m=m_1$ at any earlier step.

First of all it is clear that N should have to be even or odd depending on whether m_1 is even

[4] This problem has important applications to other physical problems.

or odd. We shall suppose that this is the case. Suppose now that there is no absorbing screen. Then the arrival of the particle at m_1 after N steps implies that its position after $(N-1)$ steps must have been either (m_1-1) or (m_1+1). (See Fig. 2.) But every trajectory which arrives at (m_1, N) from $(m_1+1, N-1)$ is a forbidden one in the presence of the absorbing screen since such a trajectory must necessarily have crossed the line $m=m_1$. It does *not* however follow that *all* trajectories arriving at (m_1, N) from $(m_1-1, N-1)$ are permitted ones: For, a certain number of these trajectories will have touched or crossed the line $m=m_1$ earlier than its last step. The number of such trajectories arriving at $(m_1-1, N-1)$ but having an earlier contact with, or a crossing of, the line $m=m_1$ is equal to those arriving at $(m_1+1, N-1)$. The argument is simply that by reflection about the line $m=m_1$ we can uniquely derive from a trajectory leading to $(m_1+1, N-1)$ another leading to $(m_1-1, N-1)$ which has a forbidden character, and conversely. Thus, the number of permitted ways of arriving at m_1 for the first time after N steps is equal to *all* the possible ways of arriving at m_1 after N steps in the absence of the absorbing wall *minus* twice the number of ways of arriving at $(m_1+1, N-1)$ again in the absence of the absorbing screen: i.e.,

$$\frac{N!}{[\frac{1}{2}(N-m_1)]![\frac{1}{2}(N+m_1)]!} - 2\frac{(N-1)!}{[\frac{1}{2}(N+m_1)]![\frac{1}{2}(N-m_1-2)]!}$$
$$= \frac{N!}{[\frac{1}{2}(N-m_1)]![\frac{1}{2}(N+m_1)]!}\left(1-\frac{N-m_1}{N}\right), \quad (26)$$
$$= \frac{m_1}{N}\frac{N!}{[\frac{1}{2}(N-m_1)]![\frac{1}{2}(N+m_1)]!}.$$

The required probability $a(m_1, N)$ is therefore given by

$$a(m_1, N) = \frac{m_1}{N}W(m_1, N). \quad (27)$$

For the limiting case of large N we have

$$a(m_1, N) = \frac{m_1}{N}\left(\frac{2}{\pi N}\right)^{\frac{1}{2}} \exp(-m_1^2/2N). \quad (28)$$

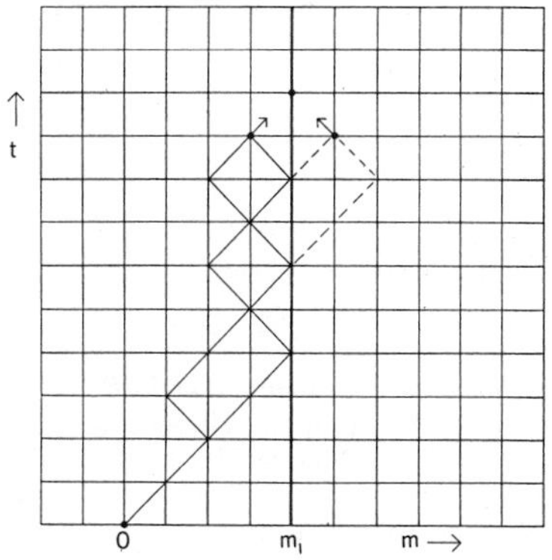

FIG. 2.

If we further write

$$x_1 = m_1 l; \quad N = nt; \quad D = \tfrac{1}{2}nl^2, \quad (29)$$

where l is the length of each step and n the number of displacements (assumed constant) which the particle suffers in unit time, then

$$a(x_1, t) = \frac{x_1}{nt}\frac{1}{(\pi Dt)^{\frac{1}{2}}} \exp(-x_1^2/4Dt). \quad (30)$$

Finally, if we ask the probability $q(x_1, t)\Delta t$ that the particle arrives at x_1 during t and $t+\Delta t$ for the first time, then

$$q(x_1, t)\Delta t = \tfrac{1}{2}a(x_1, t)n\Delta t, \quad (31)$$

since (30) is the number which arrive at x_1 in the time taken to traverse two steps. Thus,

$$q(x_1, t) = \frac{x_1}{t}\frac{1}{2(\pi Dt)^{\frac{1}{2}}} \exp(-x_1^2/4Dt). \quad (32)$$

We can interpret Eq. (32) as giving the fraction of a large number of particles initially at $x=0$ and which are deposited on the absorbing screen per unit time, at time t.

We readily verify that $q(x_1, t)$ as defined by Eq. (32) satisfies the relation

$$q(x_1, t) = -D(\partial W/\partial x)_{x=x_1}, \quad (33)$$

with W defined as in Eq. (24). This equation has an important physical interpretation to which we shall draw attention in §5.

3. The General Problem of Random Flights: Markoff's Method

In the general problem of random flights, the position $\boldsymbol{R}$ of the particle after N displacements is given by

$$\boldsymbol{R}=\sum_{i=1}^{N} \boldsymbol{r}_i, \tag{34}$$

where the $\boldsymbol{r}_i$'s $(i=1, \cdots, N)$ denote the different displacements. Further, the probability that the ith displacement lies between $\boldsymbol{r}_i$ and $\boldsymbol{r}_i+d\boldsymbol{r}_i$ is given by

$$\tau_i(x_i, y_i, z_i)dx_i dy_i dz_i=\tau_i d\boldsymbol{r}_i \quad (i=1, \cdots, N). \tag{35}$$

We require the probability $W_N(\boldsymbol{R})d\boldsymbol{R}$ that the position of the particle after N displacements lies in the interval $\boldsymbol{R}$, $\boldsymbol{R}+d\boldsymbol{R}$. In this general form the problem can be solved by using a method originally devised by A. A. Markoff. Now, Markoff's method is of such extreme generality that it actually enables us to solve the first of the two fundamental problems outlined in the introductory section. We shall accordingly describe Markoff's method in a form in which it can readily be applied to other problems besides that of random flights.

Let

$$\boldsymbol{\phi}_j=(\phi_j^1, \phi_j^2, \cdots, \phi_j^n) \quad (j=1, \cdots, N) \tag{36}$$

be N, n-dimensional vectors, the components of each of these vectors being functions of s coordinates:

$$\phi_j^k=\phi_j^k(q_j^1, \cdots, q_j^s) \quad (k=1, \cdots, n;\ j=1, \cdots, N). \tag{37}$$

The probability that the q_j^i's occur in the range

$$q_j^1, q_j^1+dq_j^1;\ q_j^2, q_j^2+dq_j^2;\ \cdots;\ q_j^s, q_j^s+dq_j^s, \quad (j=1, \cdots, N) \tag{38}$$

is given by

$$\tau_j(q_j^1, \cdots, q_j^s)dq_j^1\cdots dq_j^s=\tau_j(\boldsymbol{q}_j)d\boldsymbol{q}_j. \tag{39}$$

Further, let

$$(\Phi^1, \Phi^2, \cdots, \Phi^n)=\boldsymbol{\Phi}=\sum_{j=1}^{N} \boldsymbol{\phi}_j. \tag{40}$$

The problem is: What is the probability that

$$\boldsymbol{\Phi}_0-\tfrac{1}{2}d\boldsymbol{\Phi}_0 \leqslant \boldsymbol{\Phi} \leqslant \boldsymbol{\Phi}_0+\tfrac{1}{2}d\boldsymbol{\Phi}_0 \tag{41}$$

where $\boldsymbol{\Phi}_0$ is some preassigned value for $\boldsymbol{\Phi}$.

If we denote the required probability by

$$W_N(\boldsymbol{\Phi}_0)d\Phi_0^1\cdots d\Phi_0^n=W(\boldsymbol{\Phi}_0)d\boldsymbol{\Phi}_0, \tag{42}$$

we clearly have

$$W_N(\boldsymbol{\Phi}_0)d\boldsymbol{\Phi}_0=\int\cdots\int \prod_{j=1}^{N} \{\tau_j(\boldsymbol{q}_j)d\boldsymbol{q}_j\}, \tag{43}$$

where the integration is effected over only those parts of the Ns-dimensional configuration space $(q_1^1, \cdots, q_N^s)$ in which the inequalities (41) are satisfied.

We shall now introduce a factor $\Delta(\boldsymbol{q}_1, \cdots, \boldsymbol{q}_N)$ having the following properties:

$$\left.\begin{aligned} \Delta(\boldsymbol{q}_1, \cdots, \boldsymbol{q}_N)&=1 \quad \text{whenever} \quad \boldsymbol{\Phi}_0-\tfrac{1}{2}d\boldsymbol{\Phi}_0 \leqslant \boldsymbol{\Phi} \leqslant \boldsymbol{\Phi}_0+\tfrac{1}{2}d\boldsymbol{\Phi}_0,\\ &=0 \quad \text{otherwise.} \end{aligned}\right\} \tag{44}$$

Then,

$$W_N(\boldsymbol{\Phi}_0)d\boldsymbol{\Phi}_0=\int\cdots\int \Delta(\boldsymbol{q}_1, \cdots, \boldsymbol{q}_N) \prod_{j=1}^{N} \{\tau_j(\boldsymbol{q}_j)d\boldsymbol{q}_j\} \tag{45}$$

where, in contrast to (43), the integration is now extended over *all* the accessible regions of the configuration space. The introduction of the factor Δ under the integral sign in Eq. (45) in this manner appears at first sight as a very formal device to extend the range of integration over the entire configuration space. But the essence of Markoff's method is that an explicit expression for this factor can be given.

Consider the integrals

$$\delta_k = \frac{1}{\pi}\int_{-\infty}^{+\infty} \frac{\sin \alpha_k\rho_k}{\rho_k} \exp\,(i\rho_k\gamma_k)d\rho_k \quad (k=1,\cdots,n). \tag{46}$$

The integral defining δ_k is the well-known discontinuous integral of Dirichlet and has the property

$$\left.\begin{aligned} \delta_k &= 1 \quad \text{whenever} \quad -\alpha_k<\gamma_k<\alpha_k, \\ &= 0 \quad \text{otherwise.} \end{aligned}\right\} \tag{47}$$

Now, let

$$\alpha_k = \tfrac{1}{2}d\Phi_0^k; \quad \gamma_k = \sum_{j=1}^{N}\phi_j^k - \Phi_0^k \quad (k=1,\cdots,n). \tag{48}$$

According to Eq. (47)

$$\left.\begin{aligned} \delta_k &= 1 \quad \text{whenever} \quad \Phi_0^k - \tfrac{1}{2}d\Phi_0^k < \sum_{j=1}^{N}\phi_j^k < \Phi_0^k + \tfrac{1}{2}d\Phi_0^k, \\ &= 0 \quad \text{otherwise.} \end{aligned}\right\} \tag{49}$$

Consequently

$$\Delta = \prod_{k=1}^{n}\delta_k \tag{50}$$

has the required properties (44).

Substituting for Δ from Eqs. (46) and (50) in Eq. (45), we obtain

$$\left.\begin{aligned} W_N(\boldsymbol{\Phi}_0)d\boldsymbol{\Phi}_0 &= \frac{1}{\pi^n}\int\cdots\int_{(\boldsymbol{\varrho})}\int\cdots\int_{(\boldsymbol{q})}\left\{\prod_{j=1}^{N}\tau_j(\boldsymbol{q}_j)d\boldsymbol{q}_j\right\}\left\{\prod_{k=1}^{n}\frac{\sin\,(\frac{1}{2}d\Phi_0^k\rho_k)}{\rho_k}\right\} \\ &\qquad \times \exp\left\{i\left[\sum_{k=1}^{n}\sum_{j=1}^{N}\phi_j^k\rho_k - \sum_{k=1}^{n}\Phi_0^k\rho_k\right]\right\}d\rho_1\cdots d\rho_n \\ &= \frac{d\boldsymbol{\Phi}_0}{2^n\pi^n}\int\cdots\int \exp\,(-i\boldsymbol{\varrho}\cdot\boldsymbol{\Phi}_0)A_N(\boldsymbol{\varrho})d\boldsymbol{\varrho} \end{aligned}\right\} \tag{51}$$

where we have written

$$A_N(\boldsymbol{\varrho}) = \prod_{j=1}^{N}\int\cdots\int dq_j^1\cdots dq_j^s \exp\,(i\boldsymbol{\varrho}\cdot\boldsymbol{\phi}_j)\tau_j\,(q_j^1,\cdots,q_j^s). \tag{52}$$

The case of greatest interest is when all the functions τ_j (of the respective q_j's) are identical. Equation (52) then becomes

$$A_N(\boldsymbol{\varrho}) = \left[\int \exp\,(i\boldsymbol{\varrho}\cdot\boldsymbol{\phi})\tau(\boldsymbol{q})d\boldsymbol{q}\right]^N. \tag{53}$$

According to Eq. (51), $A_N(\boldsymbol{\varrho})$ is the n-dimensional Fourier-transform of the probability function $W(\boldsymbol{\Phi}_0)$. And Markoff's procedure illustrates a very general principle that it is the Fourier transform of the probability function, rather than the function itself, that has a more direct relation to the physical situations.

For $N\to\infty$, $A_N(\varrho)$ generally tends to the form [see §4 Eq. (91)]

$$\operatorname*{Limit}_{N\to\infty} A_N(\varrho)=\exp\left[-C(\varrho)\right]. \tag{54}$$

4. The Solution to the General Problem of Random Flights

We shall now apply Markoff's method to the problem of random flights. According to Eqs. (34), (51), and (52), the probability $W_N(\boldsymbol{R})d\boldsymbol{R}$ that the position $\boldsymbol{R}$ of the particle will be found in the interval $(\boldsymbol{R}, \boldsymbol{R}+d\boldsymbol{R})$ after N displacements is given by

$$W_N(\boldsymbol{R})=\frac{1}{8\pi^3}\int_{-\infty}^{+\infty}\exp\left(-i\varrho\cdot\boldsymbol{R}\right)A_N(\varrho)d\varrho, \tag{55}$$

where

$$A_N(\varrho)=\prod_{j=1}^{N}\int_{-\infty}^{+\infty}\tau_j(\boldsymbol{r}_j)\exp\left(i\varrho\cdot\boldsymbol{r}_j\right)d\boldsymbol{r}_j. \tag{56}$$

In Eq. (55), $\tau_j(\boldsymbol{r}_j)$ governs the probability of occurrence of a displacement $\boldsymbol{r}_j$ on the jth occasion. The explicit form which $W_N(\boldsymbol{R})$ takes will naturally depend on the assumptions made concerning the $\tau_j(\boldsymbol{r}_j)$'s. We shall now consider several cases of interest.

(a) *A Gaussian Distribution of the Displacements* $\boldsymbol{r}_j$

A case of special interest arises when

$$\tau_j=\frac{1}{(2\pi l_j^2/3)^{\frac{3}{2}}}\exp\left(-3|\boldsymbol{r}_j|^2/2l_j^2\right), \tag{57}$$

where l_j^2 denotes the mean square displacement to be expected on the jth occasion. While l_j^2 may differ from one displacement to another we assume that *all* the displacements occur in random directions.

For τ_j of the form (57), our expression for $A_N(\varrho)$ becomes

$$\begin{aligned} A_N(\varrho)&=\prod_{j=1}^{N}\frac{1}{(2\pi l_j^2/3)^{\frac{3}{2}}}\iiint_{-\infty}^{+\infty}\exp\left[i(\rho_1x_j+\rho_2y_j+\rho_3z_j)-3(x_j^2+y_j^2+z_j^2)/2l_j^2\right]dx_jdy_jdz_j \\ &=\prod_{j=1}^{N}\exp\left[-(\rho_1^2+\rho_2^2+\rho_3^2)l_j^2/6\right]=\exp\left[-(|\varrho|^2\sum_{j=1}^{N}l_j^2)/6\right]. \end{aligned} \tag{58}$$

Let $\langle l^2\rangle_{\mathsf{Av}}$ stand for

$$\langle l^2\rangle_{\mathsf{Av}}=\frac{1}{N}\sum_{j=1}^{N}l_j^2. \tag{59}$$

Equation (58) becomes

$$A_N(\varrho)=\exp\left[-N\langle l^2\rangle_{\mathsf{Av}}|\varrho|^2/6\right]. \tag{60}$$

Substituting this expression for $A_N(\varrho)$ in Eq. (55), we obtain

$$W_N(\boldsymbol{R})=\frac{1}{8\pi^3}\iiint_{-\infty}^{+\infty}\exp\left[-i(\rho_1X+\rho_2Y+\rho_3Z)-N\langle l^2\rangle_{\mathsf{Av}}(\rho_1^2+\rho_2^2+\rho_3^2)/6\right]d\rho_1d\rho_2d\rho_3. \tag{61}$$

The integrations in (61) are readily performed and we find

$$W_N(\boldsymbol{R})=\frac{1}{(2\pi N\langle l^2\rangle_{\mathsf{Av}}/3)^{\frac{3}{2}}}\exp\left[-3|\boldsymbol{R}|^2/2N\langle l^2\rangle_{\mathsf{Av}}\right]. \tag{62}$$

This is an *exact* solution valid for any value of N. That an exact solution can be found for a Gaussian distribution of the different displacements is simply a consequence of the "*addition theorem*" which these functions satisfy.

(*b*) *Each Displacement of a Constant Length But in Random Directions*

Let the displacement on the jth occasion be of length l_j in a random direction. Under these circumstances, we can define the distribution functions τ_j by

$$\tau_j = \frac{1}{4\pi l_j^3}\delta(|\boldsymbol{r}_j|^2 - l_j^2), \quad (j=1, \cdots, N) \tag{63}$$

where δ stands for Dirac's δ function.

Accordingly, our expression for $A_N(\boldsymbol{\varrho})$ becomes

$$A_N(\boldsymbol{\varrho}) = \prod_{j=1}^{N} \frac{1}{4\pi l_j^3} \int_{-\infty}^{+\infty} \exp\,(i\boldsymbol{\varrho}\cdot\boldsymbol{r}_j)\,\delta(r_j^2 - l_j^2)\,d\boldsymbol{r}_j, \tag{64}$$

or, using polar coordinates with the z axis in the direction of $\boldsymbol{\varrho}$

$$A_N(\boldsymbol{\varrho}) = \prod_{j=1}^{N} \frac{1}{4\pi l_j^3} \int_0^\infty \int_0^\pi \int_0^{2\pi} \exp\,[i|\boldsymbol{\varrho}|r_j \cos\vartheta]\,\delta(r_j^2 - l_j^2)\,r_j^2 \sin\vartheta\, dr_j d\vartheta d\omega. \tag{65}$$

The integrations over the polar and the azimuthal angles ϑ and ω are readily effected:

$$\left.\begin{aligned} A_N(\boldsymbol{\varrho}) &= \prod_{j=1}^{N} \frac{1}{2l_j^3} \int_0^\infty \int_0^\pi \exp\,(i|\boldsymbol{\varrho}|r_j\cos\vartheta)\,r_j^2\,\delta(r_j^2 - l_j^2)\sin\vartheta\, d\vartheta dr_j \\ &= \prod_{j=1}^{N} \frac{1}{l_j^3|\boldsymbol{\varrho}|} \int_0^\infty \sin\,(|\boldsymbol{\varrho}|r_j)\,r_j\,\delta(r_j^2 - l_j^2)\,dr_j \\ &= \prod_{j=1}^{N} \frac{\sin\,(|\boldsymbol{\varrho}|l_j)}{|\boldsymbol{\varrho}|l_j}. \end{aligned}\right\} \tag{66}$$

Thus,

$$W_N(\boldsymbol{R}) = \frac{1}{8\pi^3} \int_{-\infty}^{+\infty} \exp\,(-i\boldsymbol{\varrho}\cdot\boldsymbol{R}) \prod_{j=1}^{N} \frac{\sin\,(|\boldsymbol{\varrho}|l_j)}{|\boldsymbol{\varrho}|l_j}\, d\boldsymbol{\varrho}. \tag{67}$$

Again, choosing polar coordinates but with the z axis pointing this time in the direction of $\boldsymbol{R}$, we have

$$W_N(\boldsymbol{R}) = \frac{1}{8\pi^3} \int_0^\infty \int_{-1}^{+1} \int_0^{2\pi} \exp\,(-i|\boldsymbol{\varrho}|\,|\boldsymbol{R}|t) \left\{ \prod_{j=1}^{N} \frac{\sin\,(|\boldsymbol{\varrho}|l_j)}{|\boldsymbol{\varrho}|l_j} \right\} |\boldsymbol{\varrho}|^2 d\omega dt d|\boldsymbol{\varrho}|. \tag{68}$$

The integrations over ω and t are readily performed and we obtain

$$W_N(\boldsymbol{R}) = \frac{1}{2\pi^2|\boldsymbol{R}|} \int_0^\infty \sin\,(|\boldsymbol{\varrho}|\,|\boldsymbol{R}|) \left\{ \prod_{j=1}^{N} \frac{\sin\,(|\boldsymbol{\varrho}|l_j)}{|\boldsymbol{\varrho}|l_j} \right\} |\boldsymbol{\varrho}|\,d|\boldsymbol{\varrho}| \tag{69}$$

which represents the formal solution to the problem. In this form, the solution for the problem of random flights is due to Rayleigh.[5]

---

[5] Lord Rayleigh, *Collected Papers*, Vol. 6, p. 604. We may, however, draw attention to the fact that our formulation of the general problem of random flights is wider in its scope than Rayleigh's. Rayleigh's formulation of the problem corresponds to our special case (63).

The case of greatest interest arises when all the l_j's are equal. We shall assume that this is the case in the rest of our discussion:

$$l_j = l = \text{constant} \quad (j=1, \cdots, N). \tag{70}$$

Equation (69) becomes

$$W_N(\boldsymbol{R}) = \frac{1}{2\pi^2|\boldsymbol{R}|}\int_0^\infty \sin(|\varrho||\boldsymbol{R}|)\left\{\frac{\sin(|\varrho|l)}{|\varrho|l}\right\}^N |\varrho| d|\varrho|. \tag{71}$$

(*i*) *N finite.*—We shall illustrate (following Rayleigh) the method of evaluating the integral on the right-hand side of Eq. (71) for finite values of N by considering the cases $N=3$ and 4.

When $N=3$, Eq. (71) becomes

$$W_3(\boldsymbol{R}) = \frac{1}{2\pi^2|\boldsymbol{R}|l^3}\int_0^\infty \sin(|\varrho||\boldsymbol{R}|)\sin^3(|\varrho|l)\frac{d|\varrho|}{|\varrho|^2}. \tag{72}$$

But

$$\sin(|\varrho||\boldsymbol{R}|)\sin^3(|\varrho|l) = \tfrac{1}{8}\{3\cos[(|\boldsymbol{R}|-l)|\varrho|] - 3\cos[(|\boldsymbol{R}|+l)|\varrho|] - \cos[(|\boldsymbol{R}|-3l)|\varrho|] + \cos[(|\boldsymbol{R}|+3l)|\varrho|]\}. \tag{73}$$

Further

$$\left.\begin{aligned}\int_0^\infty \{\cos[(|\boldsymbol{R}|-l)|\varrho|] - \cos[(|\boldsymbol{R}|+l)|\varrho|]\}\frac{d|\varrho|}{|\varrho|^2} \\ = 2\int_0^\infty\left\{\sin^2\frac{(|\boldsymbol{R}|+l)|\varrho|}{2} - \sin^2\frac{(|\boldsymbol{R}|-l)|\varrho|}{2}\right\}\frac{d|\varrho|}{|\varrho|^2} \\ = \tfrac{1}{2}\pi(|\boldsymbol{R}|+l-||\boldsymbol{R}|-l|).\end{aligned}\right\} \tag{74}$$

We have a similar formula for the integral involving the other pair of cosines in Eq. (73). Combining these results we obtain

$$W_3(\boldsymbol{R}) = \frac{1}{32\pi|\boldsymbol{R}|l^3}\{2|\boldsymbol{R}| - 3||\boldsymbol{R}|-l| + ||\boldsymbol{R}|-3l|\}, \tag{75}$$

or, more explicitly

$$\left.\begin{aligned} W_3(\boldsymbol{R}) &= \frac{1}{8\pi l^3} && (0<|\boldsymbol{R}|<l), \\ &= \frac{1}{16\pi l^3|\boldsymbol{R}|}(3l-|\boldsymbol{R}|) && (l<|\boldsymbol{R}|<3l), \\ &= 0 && (3l<|\boldsymbol{R}|<\infty).\end{aligned}\right\} \tag{76}$$

We shall consider next the case $N=4$. According to Eq. (71) we have

$$W_4(\boldsymbol{R}) = \frac{1}{2\pi^2|\boldsymbol{R}|l^4}\int_0^\infty \frac{d|\varrho|}{|\varrho|^3}\sin(|\varrho||\boldsymbol{R}|)\sin^4(|\varrho|l). \tag{77}$$

From this equation we derive

$$\left.\begin{aligned}-\frac{d^2}{d|\boldsymbol{R}|^2}[|\boldsymbol{R}|W_4(\boldsymbol{R})]&=\frac{1}{2\pi^2l^4}\int_0^\infty\frac{d|\boldsymbol{\varrho}|}{|\boldsymbol{\varrho}|}\sin(|\boldsymbol{\varrho}||\boldsymbol{R}|)\sin^4(|\boldsymbol{\varrho}|l)\\&=\frac{1}{32\pi^2l^4}\int_0^\infty\frac{d|\boldsymbol{\varrho}|}{|\boldsymbol{\varrho}|}\{\sin[(|\boldsymbol{R}|+4l)|\boldsymbol{\varrho}|]+\sin[(|\boldsymbol{R}|-4l)|\boldsymbol{\varrho}|]\\&\qquad-4\sin[(|\boldsymbol{R}|+2l)|\boldsymbol{\varrho}|]-4\sin[(|\boldsymbol{R}|-2l)|\boldsymbol{\varrho}|]+6\sin(|\boldsymbol{R}||\boldsymbol{\varrho}|)\}\\&=\frac{1}{64\pi l^4}(1\pm1-4\mp4+6)=\frac{1}{64\pi l^4}(3\pm1\mp4),\end{aligned}\right\}\quad(78)$$

where the two alternatives in the last two steps of Eq. (78) depend, respectively, on the signs of $(|\boldsymbol{R}|-4l)$ and $(|\boldsymbol{R}|-2l)$. Thus

$$\left.\begin{aligned}64\pi l^4\frac{d^2}{d|\boldsymbol{R}|^2}[|\boldsymbol{R}|W_4(\boldsymbol{R})]&=-6\quad(0<|\boldsymbol{R}|<2l),\\&=+2\quad(2l<|\boldsymbol{R}|<4l),\\&=0\quad(4l<|\boldsymbol{R}|<\infty).\end{aligned}\right\}\quad(79)$$

We can integrate the foregoing equation working backwards from large values of $|\boldsymbol{R}|$ where all derivatives must vanish. We find

$$\left.\begin{aligned}64\pi l^4\frac{d}{d|\boldsymbol{R}|}[|\boldsymbol{R}|W_4(\boldsymbol{R})]&=2(|\boldsymbol{R}|-4l)\quad(2l<|\boldsymbol{R}|<4l),\\&=-6|\boldsymbol{R}|+8l\quad(0<|\boldsymbol{R}|<2l),\end{aligned}\right\}(80)$$

where we have used the continuity of the quantity on the left-hand side of this equation at $|\boldsymbol{R}|=2l$. Integrating Eq. (80) once again we similarly obtain

$$\left.\begin{aligned}64\pi l^4|\boldsymbol{R}|W_4(\boldsymbol{R})&=|\boldsymbol{R}|^2-8l|\boldsymbol{R}|+16l^2\\&=(4l-|\boldsymbol{R}|)^2\end{aligned}\right\}(2l<|\boldsymbol{R}|<4l),\quad(81)$$

and

$$64\pi l^4|\boldsymbol{R}|W_4(\boldsymbol{R})=-3|\boldsymbol{R}|^2+8l|\boldsymbol{R}|\,(2l>|\boldsymbol{R}|>0).\quad(82)$$

Thus, finally

$$\left.\begin{aligned}W_4(\boldsymbol{R})&=\frac{1}{64\pi l^4|\boldsymbol{R}|}(8l|\boldsymbol{R}|-3|\boldsymbol{R}|^2)\quad(0<|\boldsymbol{R}|<2l),\\&=\frac{1}{64\pi l^4|\boldsymbol{R}|}(4l-|\boldsymbol{R}|)^2\quad(2l<|\boldsymbol{R}|<4l),\\&=0\quad(4l<|\boldsymbol{R}|<\infty).\end{aligned}\right\}\quad(83)$$

In like manner it is possible, in principle, to evaluate the integral for $W_N(\boldsymbol{R})$ for any finite value of N. But the calculations become very tedious. We may however note the following solution obtained by Rayleigh for the case $N=6$.

$$
\left.\begin{aligned}
W_6(R) &= \frac{1}{2^8\pi|R|l^6}(16l^3|R|-4l|R|^3+(5/6)|R|^4) && (0<R<2l)\\
&= \frac{1}{2^8\pi|R|l^6}(-20l^4+56l^3|R|-30l^2|R|^2+6l|R|^3-(5/12)|R|^4) && (2l<|R|<4l)\\
&= \frac{1}{2^8\pi|R|l^6}(108l^4-72l^3|R|+18l^2|R|^2-2l|R|^3+(1/12)|R|^4) && (4l<|R|<6l)\\
&= 0 && (6l<|R|<\infty).
\end{aligned}\right\} \tag{84}
$$

(*ii*) $N \ll 1$.—By far the most interesting case is when N is very large. Under these circumstances

$$
\left.\begin{aligned}
\operatorname*{Limit}_{N\to\infty}\left(\frac{\sin(|\varrho|l)}{|\varrho|l}\right)^N &= \operatorname*{Limit}_{N\to\infty}(1-\tfrac{1}{6}|\varrho|^2l^2+\cdots)^N,\\
&= \exp(-N|\varrho|^2l^2/6).
\end{aligned}\right\} \tag{85}
$$

Accordingly, from Eq. (69) we conclude that for large values of N

$$
W(R)=\frac{1}{2\pi^2|R|}\int_0^\infty \exp(-Nl^2|\varrho|^2/6)|\varrho|\sin(|R||\varrho|)d|\varrho|, \tag{86}
$$

where we have written $W(R)$ for $W_N(R)$, $N\to\infty$. Evaluating the integral on the right-hand side of Eq. (86), we find

$$
W(R)=\frac{1}{(2\pi Nl^2/3)^{\frac{3}{2}}}\exp(-3|R|^2/2Nl^2). \tag{87}
$$

We notice the formal similarity of Eqs. (62) and (87). However, on our present assumptions, Eq. (87) is valid only for large values of N.

(*c*) *A Spherical Distribution of the Displacements. $N \gg 1$*

We shall assume that

$$
\tau_j(r_j)=\tau(|r_j|^2) \quad (j=1,\cdots,N). \tag{88}
$$

Then

$$
A_N(\varrho)=\left[\int_{-\infty}^{+\infty}\exp(i\varrho\cdot r)\tau(r^2)dr\right]^N. \tag{89}
$$

By using polar coordinates, the integral inside the square brackets in Eq. (89) becomes

$$
\int_{-\infty}^{+\infty}\exp(i\varrho\cdot r)\tau(r^2)dr=\int_0^\infty\int_{-1}^{+1}\int_0^{2\pi}\exp(i|\varrho|rt)r^2\tau(r^2)d\omega dt dr=4\pi\int_0^\infty\frac{\sin(|\varrho|r)}{|\varrho|r}r^2\tau(r^2)dr. \tag{90}
$$

Hence

$$
\left.\begin{aligned}
\operatorname*{Limit}_{N\to\infty}A_N(\varrho) &= \operatorname*{Limit}_{N\to\infty}\left[4\pi\int_0^\infty\frac{\sin(|\varrho|r)}{|\varrho|r}r^2\tau(r^2)dr\right]^N,\\
&= \operatorname*{Limit}_{N\to\infty}\left[4\pi\int_0^\infty(1-\tfrac{1}{6}|\varrho|^2r^2+\cdots)r^2\tau(r^2)dr\right]^N,\\
&= \exp(-N|\varrho|^2\langle r^2\rangle_{Av}/6)
\end{aligned}\right\} \tag{91}
$$

where $\langle r^2\rangle_{Av}$ is the mean square displacement to be expected on any occasion. Substituting the foregoing result in Eq. (55) we obtain

$$W(\boldsymbol{R})=\frac{1}{8\pi^3}\int_{-\infty}^{+\infty}\exp\left(-i\boldsymbol{\varrho}\cdot\boldsymbol{R}-N|\boldsymbol{\varrho}|^2\langle r^2\rangle_{Av}/6\right)d\boldsymbol{\varrho}, \tag{92}$$

or, [cf. Eq. (62)]

$$W(\boldsymbol{R})=\frac{1}{(2\pi N\langle r^2\rangle_{Av}/3)^{\frac{3}{2}}}\exp\left(-3|\boldsymbol{R}|^2/2N\langle r^2\rangle_{Av}\right). \tag{93}$$

It is seen that Eq. (93) includes the result obtained earlier in Section (b) under case (ii) [Eq. (87)] as a special case.

(d) *The Solution to the General Problem of Random Flights for $N\gg 1$*

We shall now obtain the general expression for $W_N(\boldsymbol{R})$ for large values of N with no special assumptions concerning the distribution of the different displacements except that all the τ_j's represent the same function. Accordingly, we have to examine quite generally the behavior for $N\to\infty$ of $A_N(\boldsymbol{\varrho})$ defined by [cf. Eq. (53)]

$$A_N(\boldsymbol{\varrho})=\left[\int_{-\infty}^{+\infty}\exp\left(i\boldsymbol{\varrho}\cdot\boldsymbol{r}\right)\tau(\boldsymbol{r})d\boldsymbol{r}\right]^N. \tag{94}$$

Let ρ_1, ρ_2, ρ_3 denote the components of $\boldsymbol{\varrho}$ in some fixed system of coordinates. Then

$$\left.\begin{aligned}A_N(\boldsymbol{\varrho})&=\left[\int_{-\infty}^{+\infty}\int_{-\infty}^{+\infty}\int_{-\infty}^{+\infty}\exp\left[i(\rho_1x+\rho_2y+\rho_3z)\right]\tau(x,y,z)dxdydz\right]^N,\\&=\left[\int_{-\infty}^{+\infty}\int_{-\infty}^{+\infty}\int_{-\infty}^{+\infty}\{1+i(\rho_1x+\rho_2y+\rho_3z)-\tfrac{1}{2}(\rho_1^2x^2+\rho_2^2y^2+\rho_3^2z^2+2\rho_1\rho_2xy\right.\\&\qquad\left.+2\rho_2\rho_3yz+2\rho_3\rho_1zx)+\cdots\}\tau(x,y,z)dxdydz\right]^N,\\&=[1+i(\rho_1\langle x\rangle+\rho_2\langle y\rangle+\rho_3\langle z\rangle)-\tfrac{1}{2}(\rho_1^2\langle x^2\rangle+\rho_2^2\langle y^2\rangle+\rho_3^2\langle z^2\rangle+2\rho_1\rho_2\langle xy\rangle\\&\qquad+2\rho_2\rho_3\langle yz\rangle+2\rho_3\rho_1\langle zx\rangle)+\cdots]^N\end{aligned}\right\} \tag{95}$$

where $\langle x\rangle,\cdots,\langle zx\rangle$ denote the various first and second moments of the function $\tau(x, y, z)$. Hence for $N\to\infty$ we have

$$A_N(\boldsymbol{\varrho})=\exp\left[iN(\rho_1\langle x\rangle+\rho_2\langle y\rangle+\rho_3\langle z\rangle)-\tfrac{1}{2}NQ(\boldsymbol{\varrho})\right] \tag{96}$$

where $Q(\boldsymbol{\varrho})$ stands for the homogeneous quadratic form

$$Q(\boldsymbol{\varrho})=\langle x^2\rangle\rho_1^2+\langle y^2\rangle\rho_2^2+\langle z^2\rangle\rho_3^2+2\langle xy\rangle\rho_1\rho_2+2\langle yz\rangle\rho_2\rho_3+2\langle zx\rangle\rho_3\rho_1. \tag{97}$$

Substituting for $A_N(\boldsymbol{\varrho})$ from Eq. (96) in Eq. (55) we obtain for the probability distribution for large values of N the expression:

$$W(\boldsymbol{R})=\frac{1}{8\pi^3}\int_{-\infty}^{+\infty}\int_{-\infty}^{+\infty}\int_{-\infty}^{+\infty}\exp\left[-\tfrac{1}{2}NQ(\boldsymbol{\varrho})-i\{\rho_1(X-N\langle x\rangle)+\rho_2(Y-N\langle y\rangle)+\rho_3(Z-N\langle z\rangle)\}\right]d\rho_1d\rho_2d\rho_3. \tag{98}$$

To evaluate this integral we first rotate our coordinate system to bring the quadratic form $Q(\boldsymbol{\varrho})$ to its diagonal form.

$$Q(\boldsymbol{\varrho})=\langle\xi^2\rangle\rho_\xi^2+\langle\eta^2\rangle\rho_\eta^2+\langle\zeta^2\rangle\rho_\zeta^2. \tag{99}$$

In Eq. (99) $\langle\xi^2\rangle$, $\langle\eta^2\rangle$ and $\langle\zeta^2\rangle$ are the eigenvalues of the symmetric matrix formed by the second moments:

$$\begin{vmatrix} \langle x^2\rangle & \langle xy\rangle & \langle xz\rangle \\ \langle yx\rangle & \langle y^2\rangle & \langle yz\rangle \\ \langle zx\rangle & \langle zy\rangle & \langle z^2\rangle \end{vmatrix} \tag{100}$$

Further, the three eigenvectors of the matrix (100) form an orthogonal system which we have denoted by (ξ, η, ζ). Let

$$\boldsymbol{R}=(\Xi,\ \mathrm{H},\ \mathrm{Z}) \tag{101}$$

in this system of coordinates. Equation (98) now reduces to

$$W(\boldsymbol{R})=\frac{1}{8\pi^3}\int_{-\infty}^{+\infty}\int_{-\infty}^{+\infty}\int_{-\infty}^{+\infty}\exp\left[-\tfrac{1}{2}N(\langle\xi^2\rangle\rho_\xi^2+\langle\eta^2\rangle\rho_\eta^2+\langle\zeta^2\rangle\rho_\zeta^2)\right.$$
$$\left.-i\{\rho_\xi(\Xi-N\langle\xi\rangle)+\rho_\eta(\mathrm{H}-N\langle\eta\rangle)+\rho_\zeta(\mathrm{Z}-N\langle\zeta\rangle)\}\right]d\rho_\xi d\rho_\eta d\rho_\zeta. \tag{102}$$

The integrations over ρ_ξ, ρ_η and ρ_ζ are now readily performed, and we find

$$W(\boldsymbol{R})=\frac{1}{(8\pi^3N^3\langle\xi^2\rangle\langle\eta^2\rangle\langle\zeta^2\rangle)^{\frac{1}{2}}}\exp\left[-\frac{(\Xi-N\langle\xi\rangle)^2}{2N\langle\xi^2\rangle}-\frac{(\mathrm{H}-N\langle\eta\rangle)^2}{2N\langle\eta^2\rangle}-\frac{(\mathrm{Z}-N\langle\zeta\rangle)^2}{2N\langle\zeta^2\rangle}\right]. \tag{103}$$

According to Eq. (103), the probability distribution $W(\boldsymbol{R})$ of the position $\boldsymbol{R}$ of the particle after suffering a large number of displacements (governed by a basic distribution function $\tau[x, y, z]$) is an *ellipsoidal distribution* centered at $(N\langle\xi\rangle, N\langle\eta\rangle, N\langle\zeta\rangle)$—in other words the particle suffers an average systematic net displacement of amount $(N\langle\xi\rangle, N\langle\eta\rangle, N\langle\zeta\rangle)$ and superposed on this a general random distribution.

The principal axes of this ellipsoidal distribution are along the principal directions of the moment-ellipsoid defined by (100) and the mean square net displacements about $(N\langle\xi\rangle, N\langle\eta\rangle, N\langle\zeta\rangle)$ along the three principal directions are

$$\langle(\Xi-N\langle\xi\rangle)^2\rangle_{\mathrm{Av}}=N\langle\xi^2\rangle;\quad \langle(\mathrm{H}-N\langle\eta\rangle)^2\rangle_{\mathrm{Av}}=N\langle\eta^2\rangle;\quad \langle(\mathrm{Z}-N\langle\zeta\rangle)^2\rangle_{\mathrm{Av}}=N\langle\zeta^2\rangle. \tag{104}$$

5. The Passage to a Differential Equation: The Reduction of the Problem of Random Flights for Large N to a Boundary Value Problem

In the preceding sections we have obtained the solution to the problem of random flights under various conditions. Though in each case the problem was first formulated and solved for a finite number of displacements, the greatest interest is attached to the limiting form of the solutions for large values of N. And, for large values of N the solutions invariably take very simple forms. Thus, according to Eq. (93) a particle starting from the origin and suffering n displacements per unit time, each displacement $\boldsymbol{r}$ being governed by a probability distribution $\tau(|\boldsymbol{r}|^2)$, will find itself in the element of volume defined by $\boldsymbol{R}$ and $\boldsymbol{R}+d\boldsymbol{R}$ after a time t with the probability

$$W(\boldsymbol{R})d\boldsymbol{R}=\frac{1}{(2\pi n\langle r^2\rangle_{\mathrm{Av}}t/3)^{\frac{3}{2}}}\exp\left(-3|\boldsymbol{R}|^2/2n\langle r^2\rangle_{\mathrm{Av}}t\right)d\boldsymbol{R}. \tag{105}$$

In the foregoing equation $\langle r^2\rangle_{\mathrm{Av}}$ denotes the mean square displacement that is to be expected on any given occasion. If we put

$$D=n\langle r^2\rangle_{\mathrm{Av}}/6 \tag{106}$$

Eq. (105) takes the form [cf. Eq. (16)]

$$W(\boldsymbol{R})d\boldsymbol{R}=\frac{1}{(4\pi Dt)^{\frac{3}{2}}}\exp\left(-|\boldsymbol{R}|^2/4Dt\right)d\boldsymbol{R}. \tag{107}$$

In view of the simplicity of this and the other solutions, the question now arises whether we cannot obtain the asymptotic distributions directly, without passing to the limit of large N, in each case, individually. This problem is of particular importance when restrictions on the motion of the particle in the form of reflecting and absorbing barriers are introduced. Our discussion in §2 of the simple problem of random walk in one dimension with such restrictions already indicates how very complicated the method of enumeration must become under even somewhat more general conditions than those contemplated in §2. The fact, however, that for the solutions obtained in §2, W vanishes on an absorbing wall [Eq. (25)] while grad W vanishes on a reflecting wall [Eq. (21)] suggests that the solutions perhaps correspond to solving a partial differential equation with appropriate boundary conditions. We shall now show how this passage to a differential equation and a boundary value problem is to be achieved.

First, we shall introduce a somewhat different language from that we have used so far in discussing the problem of random flights. Up to the present we have spoken of a *single* particle suffering displacements according to a given probability law, and asking for the probability of finding this particle in some given element of volume at a later time. It is clear that we can instead imagine a very large number of particles starting under the same initial conditions and undergoing the displacements without any mutual interference, and ask the *fraction* of the original number which will be found in a given element of volume at a later time. On this picture, the interpretation of the quantity on the right-hand side of Eq. (106) is that it represents the fraction of a large number of particles which will be found between $\boldsymbol{R}$ and $\boldsymbol{R}+d\boldsymbol{R}$ at time t if all the particles started from $\boldsymbol{R}=0$ at $t=0$. However, the two methods of interpretation are fully equivalent and we shall adopt the language of whichever of the two happens to be more convenient.

We pass on to considerations which lead to a differential equation for $W(\boldsymbol{R}, t)$:

Let Δt denote an interval of time long enough for a particle to suffer a large number of displacements but *still* short enough for the net mean square increment $\langle|\Delta\boldsymbol{R}|^2\rangle_{\text{Av}}$ in $\boldsymbol{R}$ to be small. Under these circumstances, the probability that a particle suffers a net displacement $\Delta\boldsymbol{R}$ in time Δt is given by

$$\psi(\Delta\boldsymbol{R};\,\Delta t)=\frac{1}{(4\pi D\Delta t)^{3/2}}\exp\left(-|\Delta\boldsymbol{R}|^2/4D\Delta t\right) \tag{108}$$

and is independent of $\boldsymbol{R}$. With Δt chosen in this manner, we seek to derive the probability distribution $W(\boldsymbol{R}, t+\Delta t)$ at time $t+\Delta t$ from the distribution $W(\boldsymbol{R}, t)$ at the earlier time t. In view of (108) and its independence of $\boldsymbol{R}$ we have the integral equation

$$W(\boldsymbol{R}, t+\Delta t)=\int_{-\infty}^{+\infty} W(\boldsymbol{R}-\Delta\boldsymbol{R}, t)\psi(\Delta\boldsymbol{R};\,\Delta t)d(\Delta\boldsymbol{R}). \tag{109}$$

Since $\langle|\Delta\boldsymbol{R}|^2\rangle_{\text{Av}}$ is assumed to be small we can expand $W(\boldsymbol{R}-\Delta\boldsymbol{R}, t)$ under the integral sign in (109) in a Taylor series and integrate term by term. We find

$$\left.\begin{aligned}
W(\boldsymbol{R}, t+\Delta t)=&\frac{1}{(4\pi D\Delta t)^{3/2}}\int_{-\infty}^{+\infty}\int_{-\infty}^{+\infty}\int_{-\infty}^{+\infty}\exp\left(-|\Delta\boldsymbol{R}|^2/4D\Delta t\right)\bigg\{W(\boldsymbol{R}, t)-\Delta X\frac{\partial W}{\partial X}-\Delta Y\frac{\partial W}{\partial Y}\\
&-\Delta Z\frac{\partial W}{\partial Z}+\frac{1}{2}\bigg[(\Delta X)^2\frac{\partial^2 W}{\partial X^2}+(\Delta Y)^2\frac{\partial^2 W}{\partial Y^2}+(\Delta Z)^2\frac{\partial^2 W}{\partial Z^2}+2\Delta X\Delta Y\frac{\partial^2 W}{\partial X\partial Y}\\
&\qquad+2\Delta Y\Delta Z\frac{\partial^2 W}{\partial Y\partial Z}+2\Delta Z\Delta X\frac{\partial^2 W}{\partial Z\partial X}\bigg]+\cdots\bigg\}d(\Delta X)d(\Delta Y)d(\Delta Z)\\
=&\,W(\boldsymbol{R}, t)+D\Delta t\left(\frac{\partial^2 W}{\partial X^2}+\frac{\partial^2 W}{\partial Y^2}+\frac{\partial^2 W}{\partial Z^2}\right)+O([\Delta t]^2).
\end{aligned}\right\} \tag{110}$$

Accordingly,

$$\frac{\partial W}{\partial t}\Delta t+O([\Delta t]^2)=D\left(\frac{\partial^2 W}{\partial X^2}+\frac{\partial^2 W}{\partial Y^2}+\frac{\partial^2 W}{\partial Z^2}\right)\Delta t+O([\Delta t]^2). \tag{111}$$

Passing now to the limit of $\Delta t=0$ we obtain

$$\frac{\partial W}{\partial t}=D\left(\frac{\partial^2 W}{\partial X^2}+\frac{\partial^2 W}{\partial Y^2}+\frac{\partial^2 W}{\partial Z^2}\right) \tag{112}$$

which is the required differential equation. And, it is seen that $W(\boldsymbol{R}, t)$ defined according to Eq. (107) is indeed the fundamental solution of this differential equation.

Equation (112) is the standard form of the *equation of diffusion* or of heat conduction. This analogy that exists between our differential Eq. (112) to the equation of diffusion provides a new interpretation of the problem of random flights in terms of a *diffusion coefficient D*.

It is well known that in the *macroscopic* theory of diffusion if $W(\boldsymbol{R}, t)$ denotes the concentration of the diffusing substance at $\boldsymbol{R}$ and at time t, then the amount crossing an area $\Delta\sigma$ in time Δt is given by

$$-D(\mathbf{1}_{\Delta\sigma}\cdot\text{grad } W)\Delta\sigma\Delta t, \tag{113}$$

where $\mathbf{1}_{\Delta\sigma}$ is a unit vector normal to the element of area $\Delta\sigma$. The diffusion equation is an elementary consequence of this fact. Consequently, we may describe the motion of a large number of particles describing random flights without mutual interference as a process of diffusion with the diffusion coefficient

$$D=n\langle r^2\rangle_{\text{Av}}/6. \tag{114}$$

With this visualization of the problem, the boundary conditions

$$W=0 \text{ on an element of surface which is a perfect absorber} \tag{115}$$

and

$$\text{grad } W=0 \text{ normal to an element surface which is a perfect reflector} \tag{116}$$

become intelligible. Further, according to Eq. (113), the rate at which particles appear on an absorbing screen per unit area, and per unit time, is given by

$$-D(\mathbf{1}\cdot\text{grad } W)_{W=0} \tag{117}$$

where $\mathbf{1}$ is a unit vector normal to the absorbing surface. This is in agreement with Eq. (33).

We shall now derive the differential equation for the problem of random flights in its general form considered in §4, subsection (d). This problem differs from the one we have just considered in that the probability distribution $\tau(\boldsymbol{r})$ governing the individual displacements $\boldsymbol{r}$ is now a function with no special symmetry properties. Accordingly, the first moments of τ cannot be assumed to vanish; further, the second moments define a general symmetric tensor of the second rank. Under these circumstances, the probability of finding the particle between $\boldsymbol{R}$ and $\boldsymbol{R}+d\boldsymbol{R}$ after it has suffered a large number of displacements is given by [cf. Eq. (103)]

$$W(\boldsymbol{R})d\boldsymbol{R}=\frac{1}{(8\pi^3N^3\langle x^2\rangle\langle y^2\rangle\langle z^2\rangle)^{\frac{1}{2}}}\exp\left[-\frac{(X-N\langle x\rangle)^2}{2N\langle x^2\rangle}-\frac{(Y-N\langle y\rangle)^2}{2N\langle y^2\rangle}-\frac{(Z-N\langle z\rangle)^2}{2N\langle z^2\rangle}\right]d\boldsymbol{R}. \tag{118}$$

In writing the probability distribution $W(\boldsymbol{R})$ in this form we have supposed that the coordinate system has been so chosen that the X, Y, and Z directions are along the principal axes of the moment ellipsoid.

Assuming that, on the average, the particle suffers n displacements per unit time we can rewrite our expression for $W(\boldsymbol{R})$ more conveniently in the form

$$W(\boldsymbol{R})=\frac{1}{8(\pi t)^{\frac{3}{2}}(D_1D_2D_3)^{\frac{1}{2}}}\exp\left[-\frac{(X+\beta_1 t)^2}{4D_1 t}-\frac{(Y+\beta_2 t)^2}{4D_2 t}-\frac{(Z+\beta_3 t)^2}{4D_3 t}\right] \tag{119}$$

where we have written

$$\left.\begin{aligned} D_1=\tfrac{1}{2}n\langle x^2\rangle;\quad D_2=\tfrac{1}{2}n\langle y^2\rangle;\quad D_3=\tfrac{1}{2}n\langle z^2\rangle,\\ \beta_1=-n\langle x\rangle;\quad \beta_2=-n\langle y\rangle;\quad \beta_3=-n\langle z\rangle.\end{aligned}\right\} \tag{120}$$

To make the passage to a differential equation, we consider, as before, an interval Δt which is long enough for the particle to suffer a large number of individual displacements but short enough for the mean square increment $\langle|\Delta \boldsymbol{R}|^2\rangle_{\text{Av}}$ to be small. The probability that the particle suffers an increment $\Delta \boldsymbol{R}$ in the interval Δt is therefore governed by the distribution function

$$\psi(\Delta\boldsymbol{R};\Delta t)=\frac{1}{8(\pi\Delta t)^{\frac{3}{2}}(D_1D_2D_3)^{\frac{1}{2}}}\exp\left[-\frac{(\Delta X+\beta_1\Delta t)^2}{4D_1\Delta t}-\frac{(\Delta Y+\beta_2\Delta t)^2}{4D_2\Delta t}-\frac{(\Delta Z+\beta_3\Delta t)^2}{4D_3\Delta t}\right]. \tag{121}$$

Hence, analogous to Eqs. (109) and (110) we now have

$$\left.\begin{aligned} W(\boldsymbol{R},t+\Delta t)&=W(\boldsymbol{R},t)+\frac{\partial W}{\partial t}\Delta t+O([\Delta t]^2)=\int_{-\infty}^{+\infty}W(\boldsymbol{R}-\Delta\boldsymbol{R},t)\psi(\Delta\boldsymbol{R};\Delta t)d(\Delta\boldsymbol{R})\\ &=\frac{1}{8(\pi\Delta t)^{\frac{3}{2}}(D_1D_2D_3)^{\frac{1}{2}}}\int_{-\infty}^{+\infty}\int_{-\infty}^{+\infty}\int_{-\infty}^{+\infty}\exp\left[-\frac{(\Delta X+\beta_1\Delta t)^2}{4D_1\Delta t}-\frac{(\Delta Y+\beta_2\Delta t)^2}{4D_2\Delta t}\right.\\ &\quad\left.-\frac{(\Delta Z+\beta_3\Delta t)^2}{4D_3\Delta t}\right]\left\{W(\boldsymbol{R},t)-\left(\Delta X\frac{\partial W}{\partial X}+\Delta Y\frac{\partial W}{\partial Y}+\Delta Z\frac{\partial W}{\partial Z}\right)\right.\\ &\quad+\frac{1}{2}\left(\Delta X^2\frac{\partial^2 W}{\partial X^2}+\Delta Y^2\frac{\partial^2 W}{\partial Y^2}+\Delta Z^2\frac{\partial^2 W}{\partial Z^2}+2\Delta X\Delta Y\frac{\partial^2 W}{\partial X\partial Y}+2\Delta Y\Delta Z\frac{\partial^2 W}{\partial Y\partial Z}\right.\\ &\quad\left.\left.+2\Delta Z\Delta X\frac{\partial^2 W}{\partial Z\partial X}\right)-\cdots\right\}d(\Delta X)d(\Delta Y)d(\Delta Z).\end{aligned}\right\} \tag{122}$$

Since for the distribution function (121)

$$\langle\Delta X\rangle_{\text{Av}}=-\beta_1\Delta t;\quad \langle\Delta Y\rangle_{\text{Av}}=-\beta_2\Delta t;\quad \langle\Delta Z\rangle_{\text{Av}}=-\beta_3\Delta t, \tag{123}$$

and

$$\left.\begin{aligned} \langle\Delta X^2\rangle_{\text{Av}}=2D_1\Delta t+\beta_1^2\Delta t^2;\quad \langle\Delta Y\Delta Z\rangle_{\text{Av}}=\beta_2\beta_3\Delta t^2,\\ \langle\Delta Y^2\rangle_{\text{Av}}=2D_2\Delta t+\beta_2^2\Delta t^2;\quad \langle\Delta Z\Delta X\rangle_{\text{Av}}=\beta_3\beta_1\Delta t^2,\\ \langle\Delta Z^2\rangle_{\text{Av}}=2D_3\Delta t+\beta_3^2\Delta t^2;\quad \langle\Delta X\Delta Y\rangle_{\text{Av}}=\beta_1\beta_2\Delta t^2,\end{aligned}\right\} \tag{124}$$

we conclude from Eq. (122) that

$$\frac{\partial W}{\partial t}\Delta t+O([\Delta t]^2)=\left(\beta_1\frac{\partial W}{\partial X}+\beta_2\frac{\partial W}{\partial Y}+\beta_3\frac{\partial W}{\partial Z}\right)\Delta t+\left(D_1\frac{\partial^2 W}{\partial X^2}+D_2\frac{\partial^2 W}{\partial Y^2}+D_3\frac{\partial^2 W}{\partial Z^2}\right)\Delta t+O([\Delta t]^2). \tag{125}$$

Passing now to the limit $\Delta t=0$ we obtain

$$\frac{\partial W}{\partial t}=\beta_1\frac{\partial W}{\partial X}+\beta_2\frac{\partial W}{\partial Y}+\beta_3\frac{\partial W}{\partial Z}+D_1\frac{\partial^2 W}{\partial X^2}+D_2\frac{\partial^2 W}{\partial Y^2}+D_3\frac{\partial^2 W}{\partial Z^2}, \tag{126}$$

which is the required differential equation. According to this equation we can describe the phenomenon under discussion as a general process of diffusion in which the number of particles crossing elements of area normal to the X, Y, and Z direction per unit area and per unit time are given, respectively, by

$$-\beta_1 W - D_1\frac{\partial W}{\partial X}; \quad -\beta_2 W - D_2\frac{\partial W}{\partial Y}; \quad -\beta_3 W - D_3\frac{\partial W}{\partial Z}. \tag{127}$$

For the purposes of solving the differential Eq. (126) it is convenient to introduce a change in the independent variable. Let

$$W = U\exp\left[-\frac{\beta_1}{2D_1}(X-X_0)-\frac{\beta_2}{2D_2}(Y-Y_0)-\frac{\beta_3}{2D_3}(Z-Z_0)-\frac{\beta_1^2}{4D_1}t-\frac{\beta_2^2}{4D_2}t-\frac{\beta_3^2}{4D_3}t\right]. \tag{128}$$

We verify that Eq. (126) now reduces to

$$\frac{\partial U}{\partial t} = D_1\frac{\partial^2 U}{\partial X^2}+D_2\frac{\partial^2 U}{\partial Y^2}+D_3\frac{\partial^2 U}{\partial Z^2}. \tag{129}$$

The fundamental solution of this differential equation is

$$U = \frac{\text{Constant}}{(D_1D_2D_3t^3)^{\frac{1}{2}}}\exp\left[-\frac{(X-X_0)^2}{4D_1t}-\frac{(Y-Y_0)^2}{4D_2t}-\frac{(Z-Z_0)^2}{4D_3t}\right]. \tag{130}$$

Returning to the variable W, we have

$$W = \frac{\text{Constant}}{(D_1D_2D_3t^3)^{\frac{1}{2}}}\exp\left[-\frac{(X-X_0+\beta_1 t)^2}{4D_1t}-\frac{(Y-Y_0+\beta_2 t)^2}{4D_2t}-\frac{(Z-Z_0+\beta_3 t)^2}{4D_3t}\right]. \tag{131}$$

In other words, the distribution (119) does indeed represent the fundamental solution of the differential Eq. (126).

CHAPTER II

THE THEORY OF THE BROWNIAN MOTION

1. Introductory Remarks. Langevin's Equation

In the studies on Brownian motion we are principally concerned with the perpetual irregular motions exhibited by small grains or particles of colloidal size immersed in a fluid. As is now well known, we witness in Brownian movement the phenomenon of molecular agitation on a reduced scale by particles very large on the molecular scale—so large in fact as to be readily visible in an ultramicroscope. The perpetual motions of the Brownian particles are maintained by fluctuations in the collisions with the molecules of the surrounding fluid. Under normal conditions, in a liquid, a Brownian particle will suffer about 10^{21} collisions per second and this is so frequent that we cannot really speak of separate collisions. Also, since each collision can be thought of as producing a kink in the path of the particle, it follows that we cannot hope to follow the path in any detail—indeed, to our senses the details of the path are impossibly fine.

The modern theory of the Brownian motion of a *free particle* (i.e., in the absence of an external field of force) generally starts with Langevin's equation

$$d\mathbf{u}/dt = -\beta\mathbf{u} + \mathbf{A}(t), \tag{132}$$

where $\mathbf{u}$ denotes the velocity of the particle. According to this equation, the influence of the surrounding medium on the motion of the particle can be split up into two parts: first, a systematic part $-\beta\mathbf{u}$ representing a *dynamical friction* experienced by the particle and second, a fluctuating part $\mathbf{A}(t)$ which is characteristic of the Brownian motion.

Regarding the frictional term $-\beta \boldsymbol{u}$ it is assumed that this is governed by Stokes' law which states that the frictional force decelerating a spherical particle of radius a and mass m is given by $6\pi a\eta \boldsymbol{u}/m$ where η denotes the coefficient of viscosity of the surrounding fluid. Hence

$$\beta = 6\pi a\eta/m. \tag{133}$$

As for the fluctuating part $\boldsymbol{A}(t)$ the following principal assumptions are made:

(i) $\boldsymbol{A}(t)$ is independent of $\boldsymbol{u}$.
(ii) $\boldsymbol{A}(t)$ varies extremely rapidly compared to the variations of $\boldsymbol{u}$.

The second assumption implies that time intervals of duration Δt exist such that during Δt the variations in $\boldsymbol{u}$ that are to be expected are very small indeed while during the same interval $\boldsymbol{A}(t)$ may undergo several fluctuations. Alternatively, we may say that though $\boldsymbol{u}(t)$ and $\boldsymbol{u}(t+\Delta t)$ are expected to differ by a negligible amount, no correlation between $\boldsymbol{A}(t)$ and $\boldsymbol{A}(t+\Delta t)$ exists. (The assumptions which are made here are quite analogous to those made in Chapter I, §5 in the passage to the differential equation for the problem of random flights; also see §§2 and 4 in this chapter.)

We shall show in the following sections how with the assumptions made in the foregoing paragraphs, we can derive from Langevin's equation all the physically significant relations concerning the motions of the Brownian particles. But we should draw attention even at this stage to the very drastic nature of assumptions implicit in the very writing of an equation of the form (132). For we have in reality supposed that we can divide the phenomenon into two parts, one in which the discontinuity of the events taking place is essential while in the other it is trivial and can be ignored. In view of the discontinuities in all matter and all events, this is a prima facie, an *ad-hoc* assumption. They are however made with reliance on physical intuition and the *aposteriori* justification by the success of the hypothesis. However, the correct procedure would be to treat the phenomenon in its entirety without appealing to the laws of continuous physics except insofar as they can be explicitly justified. As we shall see in Chapter IV a problem which occurs in stellar dynamics appears to provide a model in which the rigorous procedure can be explicitly followed.

2. The Theory of the Brownian Motion of a Free Particle

Our problem is to solve the stochastic differential equation (132) subject to the restrictions on $\boldsymbol{A}(t)$ stated in the preceding section. But "solving" a stochastic differential equation like (132) is not the same thing as solving any ordinary differential equation. For one thing, Eq. (132) involves the function $\boldsymbol{A}(t)$ which, as we shall presently see, has only statistically defined properties. Consequently, "solving" the Langevin Eq. (132) has to be understood rather in the sense of specifying a probability distribution $W(\boldsymbol{u}, t; \boldsymbol{u}_0)$ which governs the probability of occurrence of the velocity $\boldsymbol{u}$ at time t given that $\boldsymbol{u}=\boldsymbol{u}_0$ at $t=0$. Of this function $W(\boldsymbol{u}, t; \boldsymbol{u}_0)$ we should clearly require that, as $t\to 0$,

$$W(\boldsymbol{u}, t; \boldsymbol{u}_0)\to\delta(u_x-u_{x,0})\delta(u_y-u_{y,0})\delta(u_z-u_{z,0}) \quad (t\to 0), \tag{134}$$

where the δ's are Dirac's δ functions. Further, the physical circumstances of the problem require that we demand of $W(\boldsymbol{u}, t; \boldsymbol{u}_0)$ that it tend to a Maxwellian distribution for the temperature T of the surrounding fluid, *independently* of $\boldsymbol{u}_0$ as $t\to\infty$:

$$W(\boldsymbol{u}, t; \boldsymbol{u}_0)\to\left(\frac{m}{2\pi kT}\right)^{\frac{3}{2}} \exp\left(-m|\boldsymbol{u}|^2/2kT\right) \quad (t\to\infty). \tag{135}$$

This last demand on $W(\boldsymbol{u}, t; \boldsymbol{u}_0)$ conversely requires that $\boldsymbol{A}(t)$ satisfy certain statistical requirements. For, according to the Langevin equation we have the formal solution

$$\boldsymbol{u}-\boldsymbol{u}_0e^{-\beta t}=e^{-\beta t}\int_0^t e^{\beta\xi}\boldsymbol{A}(\xi)d\xi. \tag{136}$$

Consequently, the statistical properties of

$$u - u_0 e^{-\beta t} \tag{137}$$

must be the same as those of

$$e^{-\beta t}\int_0^t e^{\beta\xi} A(\xi)d\xi. \tag{138}$$

And, as $t\rightarrow\infty$ the quantity (137) tends to u; hence the distribution of

$$\underset{t\rightarrow\infty}{\text{Limit}}\left\{e^{-\beta t}\int_0^t e^{\beta\xi}A(\xi)d\xi\right\} \tag{139}$$

must be the Maxwellian distribution

$$(m/2\pi kT)^{\frac{3}{2}} \exp\left(-m|u|^2/2kT\right). \tag{140}$$

Now one of our principal assumptions concerning $A(t)$ is that it varies extremely rapidly compared to any of the other quantities which enter into our discussion. Further, the fluctuating acceleration experienced by the Brownian particles is statistical in character in the sense that Brownian particles having the same initial coordinates and/or velocities will suffer accelerations which will differ from particle to particle both in magnitude and in their dependence on time. However, on account of the rapidity of these fluctuations, we can always divide an interval of time which is long enough for any of the physical parameters like the position or the velocity of a Brownian particle to change appreciably, into a very large number of subintervals of duration Δt such that during each of these subintervals we can treat all functions of time except $A(t)$ which enter in our formulae as constants. Thus, the quantity on the right-hand side of Eq. (136) can be written as

$$e^{-\beta t}\sum_j e^{\beta j\Delta t}\int_{j\Delta t}^{(j+1)\Delta t} A(\xi)d\xi. \tag{141}$$

Let

$$B(\Delta t) = \int_t^{t+\Delta t} A(\xi)d\xi. \tag{142}$$

The physical meaning of $B(\Delta t)$ is that it represents the net acceleration which a Brownian particle may suffer on a given occasion during an interval of time Δt.

Equation (136) becomes

$$u - u_0 e^{-\beta t} = \sum_j e^{\beta(j\Delta t - t)} B(\Delta t), \tag{143}$$

and we require that as $t\rightarrow\infty$ the quantity on the right-hand side tends to the Maxwellian distribution (140). We now assert that this requires *the probability of occurrence of different values for* $B(\Delta t)$ *be governed by the distribution function*

$$w(B[\Delta t]) = \frac{1}{(4\pi q\Delta t)^{\frac{3}{2}}} \exp\left(-|B(\Delta t)|^2/4q\Delta t\right) \tag{144}$$

where

$$q = \beta kT/m. \tag{145}$$

To prove this assertion we have to show that the distribution function $W(u, t; u_0)$ derived on the basis of Eqs. (143) and (144) does in fact tend to the Maxwellian distribution (140) as $t\rightarrow\infty$. We shall presently show that this is the case but we may remark meantime on the formal similarity of Eq. (144) giving the probability distribution of the acceleration $B(\Delta t)$ suffered by a Brownian particle in time Δt and Eq. (108) giving the probability distribution of the increment ΔR in the position of a particle describing random flights in time Δt. It will be recalled that for the validity of Eq. (108) it is necessary that Δt be long enough for a large number of individual displacements to occur; analogously, our expression for $w(\boldsymbol{B}[\Delta t])$ is valid only for times Δt large compared to the average period of a single fluctuation of $\boldsymbol{A}(t)$. Now, the period of fluctuation of $\boldsymbol{A}(t)$ is clearly of the order of the time between successive collisions between the Brownian particle and the molecules of the surrounding fluid; in a liquid this is generally of the order of 10^{-21} sec. Accordingly, the similarity of our expression for $w(\boldsymbol{B}[\Delta t])$ with Eq. (108) in the theory of random flights, leads us to interpret the acceleration $\boldsymbol{B}(\Delta t)$ suffered by a Brownian particle (in a time Δt large compared with the frequency of collisions with the surrounding molecules) as the result of superposition of the large number of random accelerations caused by collisions with the individual molecules. This is of course eminently reasonable; but the reason why q in Eq. (144) has to be precisely that given by Eq. (145) is due to our requirement that $W(\boldsymbol{u}, t; \boldsymbol{u}_0)$ tend to the Maxwellian distribution (140) as $t \to \infty$. We shall return to these questions again in §5.

We now proceed to prove our assertion concerning Eqs. (143), (144) and (145):

We first prove the following lemma:

Lemma I. Let

$$\boldsymbol{R} = \int_0^t \psi(\xi) \boldsymbol{A}(\xi) d\xi. \tag{146}$$

Then, the probability distribution of $\boldsymbol{R}$ *is given by*

$$W(\boldsymbol{R}) = \frac{1}{\left[4\pi q \int_0^t \psi^2(\xi) d\xi\right]^{3/2}} \exp\left(-|\boldsymbol{R}|^2 \Big/ 4q \int_0^t \psi^2(\xi) d\xi\right). \tag{147}$$

In order to prove this, we first divide the interval $(0, t)$ into a large number of subintervals of duration Δt. We can then write

$$\boldsymbol{R} = \sum_j \psi(j\Delta t) \int_{j\Delta t}^{(j+1)\Delta t} \boldsymbol{A}(\xi) d\xi. \tag{148}$$

Remembering our definition of $\boldsymbol{B}(\Delta t)$ [Eq. (142)] we can express $\boldsymbol{R}$ in the form

$$\boldsymbol{R} = \sum_j \boldsymbol{r}_j, \tag{149}$$

where

$$\boldsymbol{r}_j = \psi(j\Delta t) \boldsymbol{B}(\Delta t) = \psi_j \boldsymbol{B}(\Delta t). \tag{150}$$

According to Eq. (144) the probability distribution of $\boldsymbol{r}_j$ is governed by

$$\tau(\boldsymbol{r}_j) = \frac{1}{(2\pi l_j^2/3)^{3/2}} \exp(-3|\boldsymbol{r}_j|^2/2l_j^2), \tag{151}$$

where we have written

$$l_j^2 = 6q\psi_j^2 \Delta t. \tag{152}$$

A comparison of Eqs. (149) and (151) with Eqs. (34) and (57) shows that we have reduced our present problem to the special case in the theory of random flights considered in Chapter I, §4 case (a). Hence, [cf. Eqs. (59) and (62)]

$$W(\boldsymbol{R}) = \frac{1}{(2\pi \sum l_j^2/3)^{3/2}} \exp(-3|\boldsymbol{R}|^2/2\sum l_j^2). \tag{153}$$

But

$$\left.\begin{aligned} \sum l_j^2 &= 6q \sum_j \psi_j^2 \Delta t = 6q \sum_j \psi^2(j\Delta t)\Delta t, \\ &= 6q \int_0^t \psi^2(\xi) d\xi. \end{aligned}\right\} \tag{154}$$

We therefore have

$$W(\boldsymbol{R})=\frac{1}{\left[4\pi q\int_0^t \psi^2(\xi)d\xi\right]^{\frac{3}{2}}}\exp\left(-|\boldsymbol{R}|^2\Big/4q\int_0^t \psi^2(\xi)d\xi\right), \tag{155}$$

which proves the lemma.

Returning to Eq. (136) we notice that we can express the right-hand side of this equation in the form

$$\int_0^t \psi(\xi)\boldsymbol{A}(\xi)d\xi \tag{156}$$

if we define

$$\psi(\xi)=e^{\beta(\xi-t)}. \tag{157}$$

We can therefore apply lemma I and with the foregoing definition of $\psi(\xi)$, Eq. (155) governs the probability distribution of

$$\boldsymbol{u}-\boldsymbol{u}_0 e^{-\beta t}. \tag{158}$$

Since, now,

$$\int_0^t \psi^2(\xi)d\xi=\int_0^t e^{2\beta(\xi-t)}d\xi=\frac{1}{2\beta}(1-e^{-2\beta t}), \tag{159}$$

and remembering that according to Eq. (145)

$$q/\beta=kT/m \tag{160}$$

we have proved that

$$W(\boldsymbol{u}, t; \boldsymbol{u}_0)=\left[\frac{m}{2\pi kT(1-e^{-2\beta t})}\right]^{\frac{3}{2}}\exp\left[-m|\boldsymbol{u}-\boldsymbol{u}_0e^{-\beta t}|^2/2kT(1-e^{-2\beta t})\right]. \tag{161}$$

We verify that according to this equation

$$W(\boldsymbol{u}, t; \boldsymbol{u}_0)\rightarrow\left(\frac{m}{2\pi kT}\right)^{\frac{3}{2}}\exp\left(-m|\boldsymbol{u}|^2/2kT\right) \quad (t\rightarrow\infty) \tag{162}$$

i.e., the Maxwellian distribution (140). This proves the assertion we made that with the statistical properties of $\boldsymbol{B}(\Delta t)$ implied in Eqs. (144) and (145), Eq. (143) leads to a distribution $W(\boldsymbol{u}, t; \boldsymbol{u}_0)$ which tends to be Maxwellian independent of $\boldsymbol{u}_0$ as $t\rightarrow\infty$.

We shall now show how with the assumptions already made concerning $\boldsymbol{B}(\Delta t)$ we can further derive the distribution of the displacement $\boldsymbol{r}$ of a Brownian particle at time t given that the particle is at $\boldsymbol{r}_0$ with a velocity $\boldsymbol{u}_0$ at time $t=0$:

Since

$$\boldsymbol{r}-\boldsymbol{r}_0=\int_0^t \boldsymbol{u}(t)dt, \tag{163}$$

we have according to Eq. (136)

$$\boldsymbol{r}-\boldsymbol{r}_0=\int_0^t d\eta\left\{\boldsymbol{u}_0e^{-\beta\eta}+e^{-\beta\eta}\int_0^\eta d\xi e^{\beta\xi}\boldsymbol{A}(\xi)\right\} \tag{164}$$

or

$$\boldsymbol{r}-\boldsymbol{r}_0-\beta^{-1}\boldsymbol{u}_0(1-e^{-\beta t})=\int_0^t d\eta e^{-\beta\eta}\int_0^\eta d\xi e^{\beta\xi}\boldsymbol{A}(\xi). \tag{165}$$

We can simplify the right-hand side of this equation by an integration by parts. We find

$$\boldsymbol{r}-\boldsymbol{r}_0-\beta^{-1}\boldsymbol{u}_0(1-e^{-\beta t})=-\beta^{-1}e^{-\beta t}\int_0^t e^{\beta\xi}\boldsymbol{A}(\xi)d\xi+\beta^{-1}\int_0^t \boldsymbol{A}(\xi)d\xi. \tag{166}$$

Again, we can reduce this equation to the form

$$\mathbf{r}-\mathbf{r}_0-\beta^{-1}\mathbf{u}_0(1-e^{-\beta t})=\int_0^t \psi(\xi)\mathbf{A}(\xi)d\xi, \tag{167}$$

by defining

$$\psi(\xi)=\beta^{-1}(1-e^{\beta(\xi-t)}). \tag{168}$$

Thus lemma I can be applied and with the definition of $\psi(\xi)$ according to Eq. (168), Eq. (155) governs the probability distribution of

$$\mathbf{r}-\mathbf{r}_0-\beta^{-1}\mathbf{u}_0(1-e^{-\beta t}) \tag{169}$$

i.e., of $\mathbf{r}$ at time t for given $\mathbf{r}_0$ and $\mathbf{u}_0$. Since,

$$\left.\begin{aligned}\int_0^t \psi^2(\xi)d\xi&=\frac{1}{\beta^2}\int_0^t (1-e^{\beta(\xi-t)})^2 d\xi,\\ &=\frac{1}{2\beta^3}(2\beta t-3+4e^{-\beta t}-e^{-2\beta t}),\end{aligned}\right\} \tag{170}$$

we have

$$W(\mathbf{r},t;\mathbf{r}_0,\mathbf{u}_0)=\left\{\frac{m\beta^2}{2\pi kT[2\beta t-3+4e^{-\beta t}-e^{-2\beta t}]}\right\}^{\frac{3}{2}}\exp-\left\{\frac{m\beta^2|\mathbf{r}-\mathbf{r}_0-\mathbf{u}_0(1-e^{-\beta t})/\beta|^2}{2kT[2\beta t-3+4e^{-\beta t}-e^{-2\beta t}]}\right\}. \tag{171}$$

For intervals of time long compared to β^{-1} the foregoing expression simplifies considerably. For, under these circumstances we can ignore the exponential and the constant terms as compared to $2\beta t$. Further, as we shall presently show, $\langle|\mathbf{r}-\mathbf{r}_0|^2\rangle_{\text{Av}}$ is of order t [cf. Eq. (174)]; hence we can also neglect $\mathbf{u}_0(1-e^{-\beta t})\beta^{-1}$ compared to $\mathbf{r}-\mathbf{r}_0$. Thus Eq. (171) reduces to

$$W(\mathbf{r},t;\mathbf{r}_0,\mathbf{u}_0)\simeq\frac{1}{(4\pi Dt)^{\frac{3}{2}}}\exp(-|\mathbf{r}-\mathbf{r}_0|^2/4Dt)\quad(t\gg\beta^{-1}) \tag{172}$$

where we have introduced the "diffusion coefficient" D defined by

$$D=kT/m\beta=kT/6\pi a\eta. \tag{173}$$

In Eq. (173) we have substituted for β according to Eq. (133).

From Eq. (172) we obtain for the mean square displacement along any given direction (say, the x direction) the formula

$$\langle(x-x_0)^2\rangle_{\text{Av}}=\tfrac{1}{3}\langle|\mathbf{r}-\mathbf{r}_0|^2\rangle_{\text{Av}}=2Dt=(kT/3\pi a\eta)t. \tag{174}$$

This is Einstein's result. Equation (174) has been verified by Perrin to lead to consistent and satisfactory values for the Boltzmann constant k by observation of $\langle(x-x_0^2)\rangle_{\text{Av}}/t$ over wide ranges of T, η and a.

The law of distribution of displacements (172) has been exhaustively tested by observation. Perrin gives the following sets of counts of the displacements of a grain of radius 2.1×10^{-5} cm at 30 sec. intervals. Out of a number N of such observations the number of observed values of x displacements between x_1 and x_2 should be

$$\frac{N}{\pi^{\frac{1}{2}}}\int_{x_1}^{x_2}\exp(-x^2/4Dt)\frac{dx}{(4Dt)^{\frac{1}{2}}}.$$

The agreement is satisfactory. See Table II.

Comparing Eq. (172) with the solution for the problem of random flights obtained in Eq. (107) we conclude that for times $t\gg\beta^{-1}$ we can regard the motion of a Brownian particle as one of random

TABLE II. Observations and calculations of the distribution of the displacements of a Brownian particle.

Range $x \times 10^4$ cm	1st set Obs.	1st set Calc.	2nd set Obs.	2nd set Calc.	Total Obs.	Total Calc.
0 - 3.4	82	91	86	84	168	175
3.4- 6.8	66	70	65	63	131	132
6.8–10.2	46	39	31	36	77	75
10.2–17.0	27	23	23	21	50	44

flights. And therefore, according to the ideas of *I* §5, describe the motion of Brownian particles also as one of diffusion and governed by the diffusion equation. We shall return to this connection with the diffusion equation from a more general point of view in §4.

Returning to Eq. (171) we see that, quite generally, we have

$$\langle |\mathbf{r}-\mathbf{r}_0|^2\rangle_{Av} = \frac{|\mathbf{u}_0|^2}{\beta^2}(1-e^{-\beta t})^2 + 3\frac{kT}{m\beta^2}(2\beta t - 3 + 4e^{-\beta t} - e^{-2\beta t}). \tag{175}$$

Averaging this equation over all values of $\mathbf{u}_0$ and remembering that $\langle |\mathbf{u}_0|^2\rangle_{Av} = 3kT/m$ we obtain

$$\langle\langle |\mathbf{r}-\mathbf{r}_0|^2\rangle\rangle_{Av} = 6\frac{kT}{m\beta^2}(\beta t - 1 + e^{-\beta t}). \tag{175'}$$

For $t\to\infty$, Eq. (175′) is in agreement with our result (174), while for $t\to 0$ we have instead

$$\langle\langle |\mathbf{r}-\mathbf{r}_0|^2\rangle\rangle_{Av} = 3\frac{kT}{m}t^2 = \langle |\mathbf{u}_0|^2\rangle_{Av} t^2. \tag{175''}$$

So far we have only inquired into the law of distributions of $\mathbf{u}$ and $\mathbf{r}$ separately. But we can also ask for the distribution $W(\mathbf{r}, \mathbf{u}, t; \mathbf{u}_0, \mathbf{r}_0)$ governing the probability of the simultaneous occurrence of the velocity $\mathbf{u}$ and the position $\mathbf{r}$ at time t, given that $\mathbf{u}=\mathbf{u}_0$ and $\mathbf{r}=\mathbf{r}_0$ at $t=0$. The solution to this problem can be obtained from the following lemma:

Lemma II. *Let*

$$\mathbf{R} = \int_0^t \psi(\xi)\mathbf{A}(\xi)d\xi, \tag{176}$$

and

$$\mathbf{S} = \int_0^t \phi(\xi)\mathbf{A}(\xi)d\xi. \tag{177}$$

Then, the bivariate probability distribution of $\mathbf{R}$ *and* $\mathbf{S}$ *is given by*

$$W(\mathbf{R}, \mathbf{S}) = \frac{1}{8\pi^3(FG-H^2)^{\frac{3}{2}}} \exp\left[-(G|\mathbf{R}|^2 - 2H\mathbf{R}\cdot\mathbf{S} + F|\mathbf{S}|^2)/2(FG-H^2)\right] \tag{178}$$

where

$$F = 2q\int_0^t \psi^2(\xi)d\xi;\quad G = 2q\int_0^t \phi^2(\xi)d\xi;\quad H = 2q\int_0^t \phi(\xi)\psi(\xi)d\xi. \tag{179}$$

The lemma is proved by writing $\mathbf{R}$ and $\mathbf{S}$ in the forms [cf. Eqs. (149) and (150)]

$$\mathbf{R} = \sum_j \psi(j\Delta t)\mathbf{B}(\Delta t);\quad \mathbf{S} = \sum_j \phi(j\Delta t)\mathbf{B}(\Delta t) \tag{180}$$

and remembering that the distribution of $\mathbf{B}$ is Gaussian according to Eq. (144). The problem then reduces to the one considered in Appendix II and the solution stated readily follows.

To obtain the distribution $W(\mathbf{r}, \mathbf{u}, t; \mathbf{u}_0, \mathbf{r}_0)$ we have only to set [cf. Eqs. (157), (158), (167) and (168)]

$$\left.\begin{aligned} \mathbf{R}&=\mathbf{r}-\mathbf{r}_0-\beta^{-1}\mathbf{u}_0(1-e^{-\beta t}); \quad &\psi(\xi)=\beta^{-1}(1-e^{\beta(\xi-t)}),\\ \mathbf{S}&=\mathbf{u}-\mathbf{u}_0e^{-\beta t}; &\phi(\xi)=e^{\beta(\xi-t)}, \end{aligned}\right\} \tag{181}$$

and [cf. Eqs. (159) and (170)]

$$F=q\beta^{-3}(2\beta t-3+4e^{-\beta t}-e^{-2\beta t}); \quad G=q\beta^{-1}(1-e^{-2\beta t}), \tag{182}$$

and finally

$$H=2q\beta^{-1}\int_0^t e^{\beta(\xi-t)}(1-e^{\beta(\xi-t)})dt=q\beta^{-2}(1-e^{-\beta t})^2. \tag{183}$$

3. The Theory of the Brownian Motion of a Particle in a Field of Force. The Harmonically Bound Particle

In the presence of an external field of force, the Langevin Eq. (132) is generalized to

$$d\mathbf{u}/dt=-\beta\mathbf{u}+\mathbf{A}(t)+\mathbf{K}(\mathbf{r}, t) \tag{184}$$

where $\mathbf{K}(\mathbf{r}, t)$ is the acceleration produced by the field. In writing this equation we are making the same general assumptions as are involved in writing the original Langevin equation (cf. the remarks at the end of §1).

In solving the stochastic equation (184) we attribute to $\mathbf{A}(t)$ or more particularly for

$$\mathbf{B}(\Delta t)=\int_t^{t+\Delta t}\mathbf{A}(\xi)d\xi \tag{185}$$

the statistical properties already assigned in the preceding section [Eq. (144)]. The method of solution is illustrated sufficiently by a one-dimensional harmonic oscillator describing Brownian motion. The appropriate stochastic equation is

$$du/dt=-\beta u+A(t)-\omega^2 x, \tag{186}$$

where ω denotes the circular frequency of the oscillator. We can write Eq. (184) alternatively in the form

$$d^2x/dt^2+\beta dx/dt+\omega^2 x=A(t). \tag{187}$$

What we seek from this equation are, of course, the probability distributions $W(x, t; x_0, u_0)$, $W(u, t; x_0, u_0)$ and $W(x, u, t; x_0, u_0)$. To obtain these distributions we first write down the formal solution of Eq. (187) regarded as an ordinary differential equation. The method of solution most appropriate for our present purposes is that of the variation of the parameters. In this method, as applied to Eq. (187), we express the solution in terms of that of the homogeneous equation:

$$x=a_1\exp(\mu_1 t)+a_2\exp(\mu_2 t) \tag{188}$$

where μ_1 and μ_2 are the roots of

$$\mu^2+\beta\mu+\omega^2=0; \tag{189}$$

i.e.,

$$\mu_1=-\tfrac{1}{2}\beta+(\tfrac{1}{4}\beta^2-\omega^2)^{\frac{1}{2}}; \quad \mu_2=-\tfrac{1}{2}\beta-(\tfrac{1}{4}\beta^2-\omega^2)^{\frac{1}{2}}. \tag{190}$$

We assume that the solution of Eq. (187) is of the form (188) where a_1 and a_2 are functions of time restricted however to satisfy the equation

$$\exp(\mu_1 t)(da_1/dt)+\exp(\mu_2 t)(da_2/dt)=0. \tag{191}$$

From Eq. (187) we derive the further relation

$$\mu_1 \exp(\mu_1 t)(da_1/dt) + \mu_2 \exp(\mu_2 t)(da_2/dt) = A(t). \tag{192}$$

Solving Eqs. (191) and (192) we readily obtain the integrals

$$\left.\begin{aligned} a_1 &= +\frac{1}{\mu_1-\mu_2}\int_0^t \exp(-\mu_1\xi)A(\xi)d\xi + a_{10}, \\ a_2 &= -\frac{1}{\mu_1-\mu_2}\int_0^t \exp(-\mu_2\xi)A(\xi)d\xi + a_{20}, \end{aligned}\right\} \tag{193}$$

where a_{10} and a_{20} are constants. Accordingly, we have the solution

$$x = \frac{1}{\mu_1-\mu_2}\left\{\exp(\mu_1 t)\int_0^t \exp(-\mu_1\xi)A(\xi)d\xi - \exp(\mu_2 t)\int_0^t \exp(-\mu_2\xi)A(\xi)d\xi\right\} + a_{10}\exp(\mu_1 t) + a_{20}\exp(\mu_2 t). \tag{194}$$

From the foregoing equation we obtain for the velocity u the formula

$$u = \frac{1}{\mu_1-\mu_2}\left\{\mu_1\exp(\mu_1 t)\int_0^t \exp(-\mu_1\xi)A(\xi)d\xi - \mu_2\exp(\mu_2 t)\int_0^t \exp(-\mu_2\xi)A(\xi)d\xi\right\} + \mu_1 a_{10}\exp(\mu_1 t) + \mu_2 a_{20}\exp(\mu_2 t). \tag{195}$$

The constants a_{10} and a_{20} can now be determined from the conditions that $x = x_0$ and $u = u_0$ at $t = 0$. We find

$$a_{10} = -\frac{x_0\mu_2 - u_0}{\mu_1-\mu_2}; \quad a_{20} = +\frac{x_0\mu_1 - u_0}{\mu_1-\mu_2}. \tag{196}$$

Thus, we have the solutions

$$x + \frac{1}{\mu_1-\mu_2}\left[(x_0\mu_2-u_0)\exp(\mu_1 t) - (x_0\mu_1-u_0)\exp(\mu_2 t)\right] = \int_0^t A(\xi)\psi(\xi)d\xi, \tag{197}$$

and

$$u + \frac{1}{\mu_1-\mu_2}\left[\mu_1(x_0\mu_2-u_0)\exp(\mu_1 t) - \mu_2(x_0\mu_1-u_0)\exp(\mu_2 t)\right] = \int_0^t A(\xi)\phi(\xi)d\xi, \tag{198}$$

where we have written

$$\left.\begin{aligned} \psi(\xi) &= \frac{1}{\mu_1-\mu_2}\left[\exp[\mu_1(t-\xi)] - \exp[\mu_2(t-\xi)]\right], \\ \phi(\xi) &= \frac{1}{\mu_1-\mu_2}\left[\mu_1\exp[\mu_1(t-\xi)] - \mu_2\exp[\mu_2(t-\xi)]\right]. \end{aligned}\right\} \tag{199}$$

It is now seen that the quantities on the right-hand sides of Eqs. (197) and (198) are of the forms considered in lemmas I and II in §2. Accordingly, we can at once write down the distribution functions $W(x, t; x_0, u_0)$, $W(u, t; x_0, u_0)$ and $W(x, u, t; x_0, u_0)$ in terms of the integrals

$$\int_0^t \psi^2(\xi)d\xi; \quad \int_0^t \phi^2(\xi)d\xi \quad \text{and} \quad \int_0^t \psi(\xi)\phi(\xi)d\xi. \tag{200}$$

With $\psi(\xi)$ and $\phi(\xi)$ defined as in Eqs. (199) we readily verify that

$$\int_0^t \psi^2(\xi)d\xi=\frac{1}{(\mu_1-\mu_2)^2}\left[\frac{1}{2\mu_1\mu_2}(\mu_2\exp(2\mu_1t)+\mu_1\exp(2\mu_2t))-\frac{2}{\mu_1+\mu_2}(\exp[(\mu_1+\mu_2)t]-1)-\frac{\mu_1+\mu_2}{2\mu_1\mu_2}\right], \quad (201)$$

$$\int_0^t \phi^2(\xi)d\xi=\frac{1}{(\mu_1-\mu_2)^2}\left[\tfrac{1}{2}(\mu_1\exp(2\mu_1t)+\mu_2\exp(2\mu_2t))-\frac{2\mu_1\mu_2}{\mu_1+\mu_2}(\exp[(\mu_1+\mu_2)t]-1)-\tfrac{1}{2}(\mu_1+\mu_2)\right], \quad (202)$$

and

$$\int_0^t \psi(\xi)\phi(\xi)d\xi=\frac{1}{2(\mu_1-\mu_2)^2}(\exp(\mu_1t)-\exp(\mu_2t))^2. \quad (203)$$

At this point it is convenient to introduce in the foregoing expressions the values of μ_1 and μ_2 explicitly according to Eq. (190): We find that the quantities on the left-hand sides of Eqs. (197) and (198) become, respectively,

$$x-x_0e^{-\beta t/2}\cosh\tfrac{1}{2}\beta_1t-\frac{x_0\beta+2u_0}{\beta_1}e^{-\beta t/2}\sinh\tfrac{1}{2}\beta_1t, \quad (204)$$

and

$$u-u_0e^{-\beta t/2}\cosh\tfrac{1}{2}\beta_1t+\frac{2x_0\omega^2+\beta u_0}{\beta_1}e^{-\beta t/2}\sinh\tfrac{1}{2}\beta_1t, \quad (205)$$

where we have introduced the quantity β_1 defined by

$$\beta_1=(\beta^2-4\omega^2)^{\frac{1}{2}}. \quad (206)$$

Similarly, we find

$$\int_0^t \psi^2(\xi)d\xi=\frac{1}{2\omega^2\beta}-\frac{e^{-\beta t}}{2\omega^2\beta_1^2\beta}(2\beta^2\sinh^2\tfrac{1}{2}\beta_1t+\beta\beta_1\sinh\beta_1t+\beta_1^2), \quad (207)$$

$$\int_0^t \phi^2(\xi)d\xi=\frac{1}{2\beta}-\frac{e^{-\beta t}}{2\beta_1^2\beta}(2\beta^2\sinh^2\tfrac{1}{2}\beta_1t-\beta\beta_1\sinh\beta_1t+\beta_1^2), \quad (208)$$

and

$$\int_0^t \psi(\xi)\phi(\xi)d\xi=2\beta_1^{-2}e^{-\beta t}\sinh^2\tfrac{1}{2}\beta_1t. \quad (209)$$

It is seen that all the foregoing expressions remain finite and real even when β_1 is zero or imaginary. Thus, while all the expressions remain valid as they stand in the "overdamped" case (β_1 real) the formulae appropriate for the periodic (β_1 imaginary) and the aperiodic (β_1 zero) cases can be readily written down by replacing

$$\cosh\tfrac{1}{2}\beta_1t,\ \beta_1^{-1}\sinh\tfrac{1}{2}\beta_1t \quad\text{and}\quad \beta_1^{-1}\sinh\beta_1t, \quad (210)$$

respectively, by

$$\cos\omega_1t,\ \frac{1}{2\omega_1}\sin\omega_1t \quad\text{and}\quad \frac{1}{2\omega_1}\sin 2\omega_1t \text{ where } \omega_1=(\omega^2-\tfrac{1}{4}\beta^2)^{\frac{1}{2}} \quad (211)$$

in the periodic case, and by

$$1,\ \tfrac{1}{2}t \quad\text{and}\quad t \quad (212)$$

in the aperiodic case.

As we have already remarked, we can immediately write down the distribution functions for the quantities on the left-hand sides of the Eqs. (197) and (198) [i.e., the quantities (204) and (205)] according to lemmas I and II of §2 in terms of the integrals (207)–(209). Thus,

$$W(x,\, t;\, x_0,\, u_0)=\left[\frac{m}{4\pi\beta kT\int_0^t \psi^2(\xi)d\xi}\right]^{\frac{1}{2}}\exp-\frac{\left(x-x_0e^{-\beta t/2}\left[\cosh\tfrac{1}{2}\beta_1t+\dfrac{\beta}{\beta_1}\sinh\tfrac{1}{2}\beta_1t\right]-\dfrac{2u_0}{\beta_1}e^{-\beta t/2}\sinh\tfrac{1}{2}\beta_1t\right)^2}{\dfrac{2kT}{m\omega^2}\left\{1-e^{-\beta t}\left(\dfrac{2\beta^2}{\beta_1^2}\sinh^2\tfrac{1}{2}\beta_1t+\dfrac{\beta}{\beta_1}\sinh\beta_1t+1\right)\right\}}. \tag{213}$$

We have similar expressions for $W(u,\, t;\, x_0,\, u_0)$ and $W(x,\, u,\, t;\, x_0,\, u_0)$.

The quantities of greatest interest are the moments $\langle x\rangle_{Av}$, $\langle u\rangle_{Av}$, $\langle x^2\rangle_{Av}$, $\langle u^2\rangle_{Av}$ and $\langle xu\rangle_{Av}$. We find

$$\left.\begin{aligned}
\langle x\rangle_{Av}&=x_0e^{-\beta t/2}\left(\cosh\tfrac{1}{2}\beta_1t+\frac{\beta}{\beta_1}\sinh\tfrac{1}{2}\beta_1t\right)+\frac{2u_0}{\beta_1}e^{-\beta t/2}\sinh\tfrac{1}{2}\beta_1t,\\
\langle u\rangle_{Av}&=u_0e^{-\beta t/2}\left(\cosh\tfrac{1}{2}\beta_1t-\frac{\beta}{\beta_1}\sinh\tfrac{1}{2}\beta_1t\right)-\frac{2x_0\omega^2}{\beta_1}e^{-\beta t/2}\sinh\tfrac{1}{2}\beta_1t,\\
\langle x^2\rangle_{Av}&=\langle x\rangle_{Av}^2+\frac{kT}{m\omega^2}\left\{1-e^{-\beta t}\left(2\frac{\beta^2}{\beta_1^2}\sinh^2\tfrac{1}{2}\beta_1t+\frac{\beta}{\beta_1}\sinh\beta_1t+1\right)\right\},\\
\langle u^2\rangle_{Av}&=\langle u\rangle_{Av}^2+\frac{kT}{m}\left\{1-e^{-\beta t}\left(2\frac{\beta^2}{\beta_1^2}\sinh^2\tfrac{1}{2}\beta_1t-\frac{\beta}{\beta_1}\sinh\beta_1t+1\right)\right\},\\
\langle xu\rangle_{Av}&=\langle x\rangle_{Av}\langle u\rangle_{Av}+\frac{4\beta kT}{\beta_1^2m}e^{-\beta t}\sinh^2\tfrac{1}{2}\beta_1t.
\end{aligned}\right\} \tag{214}$$

The foregoing expressions are the average values of the various quantities at time t for assigned values of x and u (namely, x_0 and u_0) at time $t=0$. We see that

$$\left.\begin{aligned}
&\langle x\rangle_{Av}\to 0;\quad \langle u\rangle_{Av}\to 0;\quad \langle xu\rangle_{Av}\to 0,\\
&\langle x^2\rangle_{Av}\to kT/m\omega^2;\quad \langle u^2\rangle_{Av}=kT/m,
\end{aligned}\right\}\quad t\to\infty. \tag{215}$$

By averaging the various moments over all values of u_0 and remembering that

$$\langle u_0\rangle_{Av}=0;\quad \langle u_0^2\rangle_{Av}=kT/m. \tag{216}$$

we obtain from Eqs. (214) that

$$\left.\begin{aligned}
\langle\langle x\rangle\rangle_{Av}&=x_0e^{-\beta t/2}\left(\cosh\tfrac{1}{2}\beta_1t+\frac{\beta}{\beta_1}\sinh\tfrac{1}{2}\beta_1t\right),\\
\langle\langle u\rangle\rangle_{Av}&=-\frac{2x_0\omega^2}{\beta_1}e^{-\beta t/2}\sinh\tfrac{1}{2}\beta_1t,\\
\langle\langle x^2\rangle\rangle_{Av}&=\frac{kT}{m\omega^2}+\left(x_0^2-\frac{kT}{m\omega^2}\right)e^{-\beta t}\left(\cosh\tfrac{1}{2}\beta_1t+\frac{\beta}{\beta_1}\sinh\tfrac{1}{2}\beta_1t\right)^2,\\
\langle\langle u^2\rangle\rangle_{Av}&=\frac{kT}{m}+\frac{4\omega}{\beta_1^2}\left(x_0^2-\frac{kT}{m\omega^2}\right)e^{-\beta t}\sinh^2\tfrac{1}{2}\beta_1t,\\
\langle\langle xu\rangle\rangle_{Av}&=\frac{2\omega^2}{\beta_1}\left(\frac{kT}{m\omega^2}-x_0^2\right)e^{-\beta t}\sinh\tfrac{1}{2}\beta_1t\left(\cosh\tfrac{1}{2}\beta_1t+\frac{\beta}{\beta_1}\sinh\tfrac{1}{2}\beta_1t\right).
\end{aligned}\right\} \tag{217}$$

Equations (214) and (217) show how the equipartition values (215) are reached as $t\to\infty$.

4. The Fokker-Planck Equation. The Generalization of Liouville's Theorem

As we have already remarked on several occasions, in an analysis of the Brownian movement we regard as impracticable a detailed description of the motions of the individual particles. Instead, we emphasize the essential stochastic nature of the phenomenon and seek a description in terms of the probability distributions of position and/or velocity at a later time starting from given initial distributions. Thus, in our discussion of the Brownian movement of a free particle in §2 we obtain explicitly the distribution functions $W(\boldsymbol{u}, t; \boldsymbol{u}_0)$, $W(\boldsymbol{r}, t; \boldsymbol{u}_0, \boldsymbol{r}_0)$ and $W(\boldsymbol{r}, \boldsymbol{u}, t; \boldsymbol{r}_0, \boldsymbol{u}_0)$ for given initial values of $\boldsymbol{r}_0$ and $\boldsymbol{u}_0$; similarly, in §3 we determined the distributions $W(u, t; x_0, u_0)$, $W(x, t; x_0, u_0)$ and $W(x, u, t; x_0, u_0)$ for a harmonically bound particle describing Brownian motion. In deriving these distributions in §§2 and 3 we started with the Langevin equation [Eq. (132) in the field-free case, and Eq. (184) when an external field is present] and solved it in a manner appropriate to the problem. We shall now consider the question whether we cannot reduce the determination of these distribution functions to appropriate boundary value problems of suitably chosen partial differential equations. We have in mind a reduction similar to that achieved in Chapter I, §5 where we showed how, under certain circumstances, the solution to the problem of random flights can be obtained as solutions of boundary-value problems long familiar in the theory of diffusion or conduction of heat. That a similar reduction should be possible under our present circumstances is apparent when we recall that the interpretation of the problem of random flights as one in diffusion (or heat conduction) is possible only if there exist time intervals Δt long enough for the particle to suffer a large number of individual displacements but still short enough for the *net* mean square displacement $\langle |\Delta \boldsymbol{R}|^2 \rangle_{\text{Av}}$ to be small and of $O(\Delta t)$. And, it is in the essence of Brownian motion that there exist time intervals Δt during which the physical parameters (like position and velocity of the Brownian particle) change by "infinitesimal" amounts while there occur a very large number of fluctuations characteristic of the motion and arising from the collisions with the molecules of the surrounding fluid.

It is clear that for the solutions of the most general problem we shall require the density function $W(\boldsymbol{r}, \boldsymbol{u}, t)$; in other words, we should really consider the problem in the six-dimensional phase space. Accordingly, we may state our principal objective by the remark that what we are seeking is essentially a generalization of Liouville's theorem of classical dynamics to include Brownian motion. But before we proceed to establish such a general theorem it will be instructive to consider the simplest problem of the Brownian motion of a free particle in the velocity space and obtain a differential equation for $W(\boldsymbol{u}, t)$; this leads us to the discussion of the Fokker-Planck equation in its most familiar form.

(i) The Fokker-Planck Equation in Velocity Space to Describe the Brownian Motion of a Free Particle

Let Δt denote an interval of time long compared to the periods of fluctuations of the acceleration $\boldsymbol{A}(t)$ occurring in the Langevin equation but short compared to intervals during which the velocity of a Brownian particle changes by appreciable amounts. Under these circumstances we should expect to derive the distribution function $W(\boldsymbol{u}, t+\Delta t)$ governing the probability of occurrence of $\boldsymbol{u}$ at time $t+\Delta t$ from the distribution $W(\boldsymbol{u}, t)$ at time t and a knowledge of the *transition probability* $\psi(\boldsymbol{u}; \Delta\boldsymbol{u})$ that $\boldsymbol{u}$ suffers an increment $\Delta\boldsymbol{u}$ in time Δt. More particularly, we expect the relation

$$W(\boldsymbol{u}, t+\Delta t) = \int W(\boldsymbol{u}-\Delta\boldsymbol{u}, t)\psi(\boldsymbol{u}-\Delta\boldsymbol{u}; \Delta\boldsymbol{u})d(\Delta\boldsymbol{u}), \tag{218}$$

to be valid. We may parenthetically remark that in expecting this integral equation between $W(\boldsymbol{u}, t+\Delta t)$ and $W(\boldsymbol{u}, t)$ to be true we are actually supposing that the course which a Brownian particle will take depends only on the instantaneous values of its physical parameters and is entirely independent of its whole previous history. In general probability theory, a stochastic process which has this characteristic, namely, that what happens at a given instant of time t depends *only* on the state of the system at time t is said to be a *Markoff* process. We may describe a Markoff process picturesquely by the statement that it represents "the gradual unfolding of a transition probability" in exactly the same sense as the development of a conservative dynamical system can be described as "the gradual unfolding of a contact transformation." That we should be able to idealize Brownian motion as a Markoff process appears very reasonable. But we should be careful not to conclude too hastily that every stochastic process is necessarily of the Markoff type. For, it can happen that the future course of a system is conditioned by its past history: i.e., what happens at a given instant of time t may depend on what has already happened during all time preceding t.

Returning to Eq. (218), for the case under discussion we have

$$\psi(\boldsymbol{u};\Delta\boldsymbol{u})=\frac{1}{(4\pi q\Delta t)^{\frac{3}{2}}}\exp\left(-|\Delta\boldsymbol{u}+\beta\boldsymbol{u}\Delta t|^2/4q\Delta t\right)\quad(q=\beta kT/m). \tag{219}$$

For, according to the Langevin equation [cf. Eq. (142)]

$$\Delta\boldsymbol{u}=-\beta\boldsymbol{u}\Delta t+\boldsymbol{B}(\Delta t) \tag{220}$$

where $\boldsymbol{B}(\Delta t)$ denotes the net acceleration arising from fluctuations which a Brownian particle suffers in time Δt; and, since the distribution of $\boldsymbol{B}(\Delta t)$ is given by Eq. (144), the transition probability (218) follows at once.

Expanding $W(\boldsymbol{u}, t+\Delta t)$, $W(\boldsymbol{u}-\Delta\boldsymbol{u}, t)$ and $\psi(\boldsymbol{u}-\Delta\boldsymbol{u};\Delta\boldsymbol{u})$ in Eq. (218) in the form of Taylor series, we obtain

$$\begin{aligned}W(\boldsymbol{u},t)&+\frac{\partial W}{\partial t}\Delta t+O(\Delta t^2)\\&=\int_{-\infty}^{+\infty}\int_{-\infty}^{+\infty}\int_{-\infty}^{+\infty}\left\{W(\boldsymbol{u},t)-\sum_i\frac{\partial W}{\partial u_i}\Delta u_i+\frac{1}{2}\sum_i\frac{\partial^2 W}{\partial u_i^2}\Delta u_i^2+\sum_{i<j}\frac{\partial^2 W}{\partial u_i\partial u_j}\Delta u_i\Delta u_j+\cdots\right\}\\&\quad\times\left\{\psi(\boldsymbol{u};\Delta\boldsymbol{u})-\sum_i\frac{\partial\psi}{\partial u_i}\Delta u_i+\frac{1}{2}\sum_i\frac{\partial^2\psi}{\partial u_i^2}\Delta u_i^2+\sum_{i<j}\frac{\partial^2\psi}{\partial u_i\partial u_j}\Delta u_i\Delta u_j+\cdots\right\}d(\Delta u_1)d(\Delta u_2)d(\Delta u_3)\end{aligned} \tag{221}$$

or, writing

$$\left.\begin{aligned}\langle\Delta u_i\rangle_{Av}&=\int_{-\infty}^{+\infty}\Delta u_i\psi(\boldsymbol{u};\Delta\boldsymbol{u})d(\Delta\boldsymbol{u}),\\\langle\Delta u_i^2\rangle_{Av}&=\int_{-\infty}^{+\infty}\Delta u_i^2\psi(\boldsymbol{u};\Delta\boldsymbol{u})d(\Delta\boldsymbol{u}),\\\langle\Delta u_i\Delta u_j\rangle_{Av}&=\int_{-\infty}^{+\infty}\Delta u_i\Delta u_j\psi(\boldsymbol{u};\Delta\boldsymbol{u})d(\Delta\boldsymbol{u}),\end{aligned}\right\} \tag{222}$$

we have

$$\begin{aligned}\frac{\partial W}{\partial t}\Delta t+O(\Delta t^2)=&-\sum_i\frac{\partial W}{\partial u_i}\langle\Delta u_i\rangle_{Av}+\frac{1}{2}\sum_i\frac{\partial^2 W}{\partial u_i^2}\langle\Delta u_i^2\rangle_{Av}+\sum_{i<j}\frac{\partial^2 W}{\partial u_i\partial u_j}\langle\Delta u_i\Delta u_j\rangle_{Av}-\sum_i W\frac{\partial}{\partial u_i}\langle\Delta u_i\rangle_{Av}\\&+\sum_i\frac{\partial}{\partial u_i}\langle\Delta u_i^2\rangle_{Av}\frac{\partial W}{\partial u_i}+\sum_{i\neq j}\frac{\partial W}{\partial u_i}\frac{\partial}{\partial u_j}\langle\Delta u_i\Delta u_j\rangle_{Av}+\frac{1}{2}\sum_i\frac{\partial^2}{\partial u_i^2}\langle\Delta u_i^2\rangle_{Av}W\\&+\sum_{i<j}W\frac{\partial^2}{\partial u_i\partial u_j}\langle\Delta u_i\Delta u_j\rangle_{Av}+O(\langle\Delta u_i\Delta u_j\Delta u_k\rangle_{Av}),\end{aligned} \tag{223}$$

where the remainder term involves the averages of the quantities

$$\Delta u_i^3, \quad \Delta u_i^2 \Delta u_j \quad \text{and} \quad \Delta u_i \Delta u_j \Delta u_k, \quad (i, j, k=1, 2, 3).$$

Equation (223) can be written more conveniently as

$$\frac{\partial W}{\partial t}\Delta t+O(\Delta t^2)=-\sum_i \frac{\partial}{\partial u_i}(W\langle\Delta u_i\rangle_{\text{Av}})+\frac{1}{2}\sum_i \frac{\partial^2}{\partial u_i^2}(W\langle\Delta u_i^2\rangle_{\text{Av}})$$
$$+\sum_{i<j}\frac{\partial^2}{\partial u_i \partial u_j}(W\langle\Delta u_i \Delta u_j\rangle_{\text{Av}})+O(\langle\Delta u_i \Delta u_j \Delta u_k\rangle_{\text{Av}}), \quad (224)$$

which is the *Fokker-Planck equation* in its most general form.

For the transition probability (219),

$$\langle\Delta u_i\rangle_{\text{Av}}=-\beta u_i \Delta t; \quad \langle\Delta u_i \Delta u_j\rangle_{\text{Av}}=O(\Delta t^2); \quad \langle\Delta u_i^2\rangle_{\text{Av}}=2q\Delta t+O(\Delta t^2). \quad (225)$$

Hence, Eq. (224) reduces in our case to

$$\frac{\partial W}{\partial t}\Delta t+O(\Delta t^2)=\{\beta \operatorname{div}_{\boldsymbol{u}}(W\boldsymbol{u})+q\nabla_{\boldsymbol{u}}^2 W\}\Delta t+O(\Delta t^2), \quad (226)$$

and passing now to the limit $\Delta t=0$ we have

$$\partial W/\partial t=\beta \operatorname{div}_{\boldsymbol{u}}(W\boldsymbol{u})+q\nabla_{\boldsymbol{u}}^2 W. \quad (227)$$

We shall now show that the distribution function $W(\boldsymbol{u}, t; \boldsymbol{u}_0)$ obtained in §2, Eq. (161) is the fundamental solution of the Fokker-Planck Eq. (227) in the sense that this is the solution which tends to the δ function

$$\delta(u_1-u_{1,0})\delta(u_2-u_{2,0})\delta(u_3-u_{3,0}) \quad (228)$$

as $t\to 0$. To prove this, we first note that but for the Laplacian term, Eq. (227) is a linear partial differential equation of the first order. Hence, it is natural to expect that the general solution of Eq. (227) will be intimately connected with that of the associated first-order equation

$$(\partial W/\partial t)-\beta \operatorname{div}_{\boldsymbol{u}}(W\boldsymbol{u})=0. \quad (229)$$

The general solution of this first-order equation involves the three first integrals of the Lagrangian subsidiary system

$$d\boldsymbol{u}/dt=-\beta\boldsymbol{u}. \quad (230)$$

The required first integrals are therefore

$$\boldsymbol{u}e^{\beta t}=\boldsymbol{u}_0=\text{constant}. \quad (231)$$

Accordingly, for solving Eq. (227) we introduce a new vector $\boldsymbol{\varrho}$ defined by

$$\boldsymbol{\varrho}=(\xi, \eta, \zeta)=\boldsymbol{u}e^{\beta t}. \quad (232)$$

Equation (227) now becomes

$$\partial W/\partial t=3\beta W+qe^{2\beta t}\nabla_{\varrho}^2 W. \quad (233)$$

This equation can be further simplified by introducing the variable

$$\chi=We^{-3\beta t}. \quad (234)$$

We have

$$\frac{\partial \chi}{\partial t}=qe^{2\beta t}\left(\frac{\partial^2\chi}{\partial\xi^2}+\frac{\partial^2\chi}{\partial\eta^2}+\frac{\partial^2\chi}{\partial\zeta^2}\right). \quad (235)$$

The solution of this equation can be readily written down by using the following lemma:

Lemma I. If $\phi(t)$ is an arbitrary function of time, the solution of the partial differential equation

$$\partial\chi/\partial t=\phi^2(t)\nabla_{\varrho}^2\chi \tag{236}$$

which has a source at $\varrho=\varrho_0$ at time $t=0$ is

$$\chi=\frac{1}{\left[4\pi\int_0^t\phi^2(t)dt\right]^{\frac{3}{2}}}\exp\left(-|\varrho-\varrho_0|^2\Big/4\int_0^t\phi^2(t)dt\right). \tag{237}$$

We shall omit the proof of this lemma as it is very elementary.

Applying this lemma to Eq. (235) we have the fundamental solution

$$\chi=\frac{1}{\left[4\pi q\int_0^t e^{2\beta t}dt\right]^{\frac{3}{2}}}\exp\left(-|\boldsymbol{u}e^{\beta t}-\boldsymbol{u}_0|^2\Big/4q\int_0^t e^{2\beta t}dt\right), \tag{238}$$

or, returning to the variable W according to Eq. (234) we have

$$W(\boldsymbol{u},t;\boldsymbol{u}_0)=\frac{1}{[2\pi q(1-e^{-2\beta t})/\beta]^{\frac{3}{2}}}\exp\left[-\beta|\boldsymbol{u}-\boldsymbol{u}_0e^{-\beta t}|^2/2q(1-e^{-2\beta t})\right] \tag{239}$$

which agrees with our earlier result in §2, Eq. (161).

(ii) The Generalization of Liouville's Theorem to Include Brownian Motion

We shall now consider the general problem of a particle describing Brownian motion and under the influence of an external field of force.

Let Δt again denote an interval of time which is long compared to the periods of fluctuations of the acceleration $\boldsymbol{A}(t)$ occurring in the Langevin Eq. (184) but short compared to the intervals in which any of the physical parameters change appreciably. Then, the increments $\Delta\boldsymbol{r}$ and $\Delta\boldsymbol{u}$ in position and velocity which the particle suffers during Δt are

$$\Delta\boldsymbol{r}=\boldsymbol{u}\Delta t;\quad \Delta\boldsymbol{u}=-(\beta\boldsymbol{u}-\boldsymbol{K})\Delta t+\boldsymbol{B}(\Delta t), \tag{240}$$

where $\boldsymbol{K}$ denotes the acceleration per unit mass caused by the external field of force and $\boldsymbol{B}(\Delta t)$ the net acceleration arising from fluctuations which the particle suffers in time Δt. The distribution of $\boldsymbol{B}(\Delta t)$ is again given by Eq. (144).

Assuming as before that the Brownian movement can be idealized as a Markoff process the probability distribution $W(\boldsymbol{r},\boldsymbol{u},t+\Delta t)$ in *phase space* at time $t+\Delta t$ can be derived from the distribution $W(\boldsymbol{r},\boldsymbol{u},t)$ at the earlier time t by means of the integral equation

$$W(\boldsymbol{r},\boldsymbol{u},t+\Delta t)=\int\int W(\boldsymbol{r}-\Delta\boldsymbol{r},\boldsymbol{u}-\Delta\boldsymbol{u},t)\Psi(\boldsymbol{r}-\Delta\boldsymbol{r},\boldsymbol{u}-\Delta\boldsymbol{u};\Delta\boldsymbol{r},\Delta\boldsymbol{u})d(\Delta\boldsymbol{r})d(\Delta\boldsymbol{u}). \tag{241}$$

According to the Eqs. (240) we can write

$$\Psi(\boldsymbol{r},\boldsymbol{u};\Delta\boldsymbol{r},\Delta\boldsymbol{u})=\psi(\boldsymbol{r},\boldsymbol{u};\Delta\boldsymbol{u})\delta(\Delta x-u_1\Delta t)\delta(\Delta y-u_2\Delta t)\delta(\Delta z-u_3\Delta t), \tag{242}$$

where the δ's denote Dirac's δ functions and $\psi(\boldsymbol{r},\boldsymbol{u};\Delta\boldsymbol{u})$ the transition probability in the velocity space. With this form for the transition probability in the phase space the integration over $\Delta\boldsymbol{r}$ in

Eq. (241) is immediately performed and we get

$$W(\boldsymbol{r}, \boldsymbol{u}, t+\Delta t) = \int W(\boldsymbol{r}-\boldsymbol{u}\Delta t, \boldsymbol{u}-\Delta\boldsymbol{u}, t)\psi(\boldsymbol{r}-\boldsymbol{u}\Delta t, \boldsymbol{u}-\Delta\boldsymbol{u}; \Delta\boldsymbol{u})d(\Delta\boldsymbol{u}). \tag{243}$$

Alternatively, we can write

$$W(\boldsymbol{r}+\boldsymbol{u}\Delta t, \boldsymbol{u}, t+\Delta t) = \int W(\boldsymbol{r}, \boldsymbol{u}-\Delta\boldsymbol{u}, \Delta t)\psi(\boldsymbol{r}, \boldsymbol{u}-\Delta\boldsymbol{u}; \Delta\boldsymbol{u})d(\Delta\boldsymbol{u}). \tag{244}$$

Expanding the various functions in the foregoing equation in the form of Taylor series and proceeding as in our derivation of the Fokker-Planck equation, we obtain [cf. Eq. (221)]

$$\left(\frac{\partial W}{\partial t}+\boldsymbol{u}\cdot\text{grad}_r\, W\right)\Delta t+O(\Delta t^2) = -\sum_i \frac{\partial}{\partial u_i}(W\langle\Delta u_i\rangle_{\text{Av}})+\frac{1}{2}\sum_i \frac{\partial^2}{\partial u_i^2}(W\langle\Delta u_i^2\rangle_{\text{Av}})$$
$$+\sum_{i<j}\frac{\partial^2}{\partial u_i\partial u_j}(W\langle\Delta u_i\Delta u_j\rangle_{\text{Av}})+O(\langle\Delta u_i\Delta u_j\Delta u_k\rangle_{\text{Av}}). \tag{245}$$

This is the complete analog in the phase space of the Fokker-Planck equation in the velocity space.

For the case (240), the transition probability $\psi(\boldsymbol{u}; \Delta\boldsymbol{u})$ is given by [cf. Eq. (144)]

$$\psi(\boldsymbol{u}; \Delta\boldsymbol{u}) = \frac{1}{(4\pi q\Delta t)^{\frac{3}{2}}}\exp\left(-|\Delta\boldsymbol{u}+(\beta\boldsymbol{u}-\boldsymbol{K})\Delta t|^2/4q\Delta t\right). \tag{246}$$

And with this expression for the transition probability we clearly have

$$\langle\Delta u_i\rangle_{\text{Av}} = -(\beta u_i - K_i)\Delta t;\quad \langle\Delta u_i^2\rangle_{\text{Av}} = 2q\Delta t+O(\Delta t^2);\quad \langle\Delta u_i\Delta u_j\rangle_{\text{Av}} = O(\Delta t^2). \tag{247}$$

Accordingly Eq. (245) simplifies to

$$\left\{\frac{\partial W}{\partial t}+\boldsymbol{u}\cdot\text{grad}_r\, W\right\}\Delta t+O(\Delta t^2) = \left\{\sum_i \frac{\partial}{\partial u_i}[(\beta u_i - K_i)W]+q\sum_i\frac{\partial^2 W}{\partial u_i^2}\right\}\Delta t+O(\Delta t^2), \tag{248}$$

and now passing to the limit $\Delta t=0$ and after some minor rearranging of the terms we finally obtain

$$\partial W/\partial t+\boldsymbol{u}\cdot\text{grad}_r\, W+\boldsymbol{K}\cdot\text{grad}_u\, W = \beta\, \text{div}_u\,(W\boldsymbol{u})+q\nabla_u^2 W. \tag{249}$$

The foregoing equation represents the complete generalization of the Fokker-Planck Eq. (227) to the phase space. At the same time Eq. (249) represents also the generalization of Liouville's theorem of classical dynamics to include Brownian motion; more particularly, on the right-hand side of Eq. (249) we have the terms arising from Brownian motion while on the left-hand side we have the usual Stokes operator D/Dt acting on W.

(iii) The Solution of Equation (249) *for the Field Free Case*

When no external field is present Eq. (249) becomes

$$\partial W/\partial t+\boldsymbol{u}\cdot\text{grad}_r\, W = 3\beta W+\beta\boldsymbol{u}\cdot\text{grad}_u\, W+q\nabla_u^2 W. \tag{250}$$

To solve this equation we again note that the equation

$$\partial W/\partial t+\boldsymbol{u}\cdot\text{grad}_r\, W = 3\beta W+\beta\boldsymbol{u}\cdot\text{grad}_u\, W \tag{251}$$

derived from (250) by ignoring the Laplacian term $q\nabla_u^2 W$ is a linear homogeneous first-order partial differential equation for $We^{-3\beta t}$. Accordingly, the general solution of Eq. (251) can be expressed in terms of any six independent integrals of the Lagrangian subsidiary system

$$d\boldsymbol{u}/dt = -\beta\boldsymbol{u}; \quad d\boldsymbol{r}/dt = \boldsymbol{u}. \tag{252}$$

Two vector integrals of this system are

$$\boldsymbol{u}e^{\beta t} = \boldsymbol{I}_1; \quad \boldsymbol{r} + \boldsymbol{u}/\beta = \boldsymbol{I}_2. \tag{253}$$

Accordingly, to solve Eq. (119) we introduce the new variables

$$\boldsymbol{\varrho} = (\xi, \eta, \zeta) = \boldsymbol{u}e^{\beta t}; \quad \boldsymbol{P} = (X, Y, Z) = \boldsymbol{r} + \boldsymbol{u}/\beta. \tag{254}$$

For this transformation of the variables we have

$$\left.\begin{aligned} \frac{\partial W}{\partial t} &= \frac{\partial}{\partial t} W(\boldsymbol{\varrho}, \boldsymbol{P}, t) + \beta\boldsymbol{\varrho} \cdot \text{grad}_{\boldsymbol{\varrho}}\, W, \\ \text{grad}_{\boldsymbol{r}}\, W &= \text{grad}_{\boldsymbol{P}}\, W, \\ \text{grad}_{\boldsymbol{u}}\, W &= e^{\beta t}\, \text{grad}_{\boldsymbol{\varrho}}\, W + (1/\beta)\, \text{grad}_{\boldsymbol{P}}\, W, \end{aligned}\right\} \tag{255}$$

and finally

$$\nabla_{\boldsymbol{u}}^2 W = e^{2\beta t}\nabla_{\boldsymbol{\varrho}}^2 W + (2/\beta)e^{\beta t}\nabla_{\boldsymbol{\varrho}} \cdot \nabla_{\boldsymbol{P}} W + (1/\beta^2)\nabla_{\boldsymbol{P}}^2 W. \tag{256}$$

Substituting the foregoing equations in Eq. (250) we obtain

$$\partial W/\partial t = 3\beta W + q\{e^{2\beta t}\nabla_{\boldsymbol{\varrho}}^2 W + (2/\beta)e^{\beta t}\nabla_{\boldsymbol{\varrho}} \cdot \nabla_{\boldsymbol{P}} W + (1/\beta^2)\nabla_{\boldsymbol{P}}^2 W\}. \tag{257}$$

Again, we introduce the variable

$$\chi = W e^{-3\beta t}. \tag{258}$$

Equation (257) reduces to

$$\partial\chi/\partial t = q\{e^{2\beta t}\nabla_{\boldsymbol{\varrho}}^2\chi + (2/\beta)e^{\beta t}\nabla_{\boldsymbol{\varrho}} \cdot \nabla_{\boldsymbol{P}}\chi + (1/\beta^2)\nabla_{\boldsymbol{P}}^2\chi\}, \tag{259}$$

or, written out explicitly

$$\frac{\partial\chi}{\partial t} = q\left\{e^{2\beta t}\left(\frac{\partial^2\chi}{\partial\xi^2} + \frac{\partial^2\chi}{\partial\eta^2} + \frac{\partial^2\chi}{\partial\zeta^2}\right) + \frac{2}{\beta}e^{\beta t}\left(\frac{\partial^2\chi}{\partial\xi\partial X} + \frac{\partial^2\chi}{\partial\eta\partial Y} + \frac{\partial^2\chi}{\partial\zeta\partial Z}\right) + \frac{1}{\beta^2}\left(\frac{\partial^2\chi}{\partial X^2} + \frac{\partial^2\chi}{\partial Y^2} + \frac{\partial^2\chi}{\partial Z^2}\right)\right\}. \tag{260}$$

To solve this equation we first prove the following lemma:

Lemma II. Let $\phi(t)$ and $\psi(t)$ be two arbitrary functions of time. The solution of the differential equation

$$\frac{\partial\chi}{\partial t} = \phi^2(t)\frac{\partial^2\chi}{\partial\xi^2} + 2\phi(t)\psi(t)\frac{\partial^2\chi}{\partial\xi\partial X} + \psi^2(t)\frac{\partial^2\chi}{\partial X^2} \tag{261}$$

which has a source at $\xi = X = 0$ at $t = 0$ is

$$\chi = \frac{1}{2\pi\Delta^{\frac{1}{2}}}\exp\left[-(a\xi^2 + 2h\xi X + bX^2)/2\Delta\right] \tag{262}$$

where

$$a = 2\int_0^t \psi^2(t)dt; \quad h = -2\int_0^t \phi(t)\psi(t)dt; \quad b = 2\int_0^t \phi^2(t)dt, \tag{263}$$

and

$$\Delta = ab - h^2. \tag{264}$$

To prove this lemma we substitute for χ according to Eq. (262) in the differential Eq. (261). After some minor reductions we find that we are left with

$$\frac{1}{\Delta}\frac{d\Delta}{dt}+\xi^2\frac{da_1}{dt}+2\xi X\frac{dh_1}{dt}+X^2\frac{db_1}{dt}+2\phi^2(a_1^2\xi^2+2a_1h_1\xi X+h_1^2X^2-a_1)$$
$$+4\phi\psi(a_1h_1\xi^2+h_1b_1X^2+\xi X[h_1^2+a_1b_1]-h_1)+2\psi^2(h_1^2\xi^2+2h_1b_1\xi X+b_1^2X^2-b_1)=0, \quad (265)$$

where we have written

$$a_1=a/\Delta;\quad h_1=h/\Delta;\quad b_1=b/\Delta. \quad (266)$$

Equating the coefficients of ξ^2, ξX and X^2 in (265) we obtain the set of equations

$$\left.\begin{aligned} da_1/dt&=-2(a_1\phi+h_1\psi)^2,\\ db_1/dt&=-2(h_1\phi+b_1\psi)^2,\\ dh_1/dt&=-2(a_1\phi+h_1\psi)(h_1\phi+b_1\psi),\end{aligned}\right\} \quad (267)$$

and

$$d\Delta/dt=2\Delta(a_1\phi^2+2h_1\phi\psi+b_1\psi^2). \quad (268)$$

It is readily verified that Eq. (268) is consistent with the Eq. (267) [see Eqs. (271) and (272) below].

Since [cf. Eqs. (266)]

$$da/dt=\Delta(da_1/dt)+a_1(d\Delta/dt), \quad (269)$$

we have according to Eqs. (267) and (268)

$$da/dt=-2\Delta(a_1\phi+h_1\psi)^2+2\Delta(a_1^2\phi^2+2a_1h_1\phi\psi+a_1b_1\psi^2)=2\Delta(a_1b_1-h_1^2)\psi^2, \quad (270)$$

or

$$da/dt=2\psi^2. \quad (271)$$

Similarly we prove that

$$db/dt=2\phi^2;\quad dh/dt=-2\phi\psi. \quad (272)$$

Hence,

$$a=2\int^t\psi^2dt;\quad h=-2\int^t\phi\psi dt;\quad b=2\int^t\phi^2dt. \quad (273)$$

The lemma now follows as an immediate consequence of the boundary conditions at $t=0$ stated.

In order to apply the foregoing lemma to Eq. (260) we first notice that the equation is separable in the pairs of variables (ξ, X), (η, Y) and (ζ, Z). Expressing therefore the solution in the form

$$\chi=\chi_1(\xi, X)\chi_2(\eta, Y)\chi_3(\zeta, Z), \quad (274)$$

we see that each of the functions χ_1, χ_2 and χ_3 satisfies an equation of the form (261) with

$$\phi(t)=q^{\frac{1}{2}}e^{\beta t};\quad \psi(t)=q^{\frac{1}{2}}/\beta. \quad (275)$$

Hence, the solution of Eq. (260) with the boundary condition

$$\boldsymbol{\varrho}=\boldsymbol{\varrho}_0,\quad \boldsymbol{P}=\boldsymbol{P}_0\quad\text{at}\quad t=0 \quad (276)$$

is

$$\chi=\frac{1}{8\pi^3\Delta^{\frac{3}{2}}}\exp\{-[a|\boldsymbol{\varrho}-\boldsymbol{\varrho}_0|^2+2h(\boldsymbol{\varrho}-\boldsymbol{\varrho}_0)\cdot(\boldsymbol{P}-\boldsymbol{P}_0)+b|\boldsymbol{P}-\boldsymbol{P}_0|^2]/2\Delta\} \quad (277)$$

where

$$\left.\begin{aligned} a&=2q\beta^{-2}\int_0^t dt=2q\beta^{-2}t,\\ b&=2q\int_0^t e^{2\beta t}dt=q\beta^{-1}(e^{2\beta t}-1),\\ h&=-2q\beta^{-1}\int_0^t e^{\beta t}dt=-2q\beta^{-2}(e^{\beta t}-1),\end{aligned}\right\} \quad (278)$$

and

$$\varrho-\varrho_0=e^{\beta t}\boldsymbol{u}-\boldsymbol{u}_0;\quad \boldsymbol{P}-\boldsymbol{P}_0=\boldsymbol{r}+\boldsymbol{u}/\beta-\boldsymbol{r}_0-\boldsymbol{u}_0/\beta. \tag{279}$$

In Eq. (279) $\boldsymbol{r}_0$ and $\boldsymbol{u}_0$ denote the position and velocity of the Brownian particle at time $t=0$. Finally,

$$W=\frac{e^{3\beta t}}{8\pi^3\Delta^{\frac{3}{2}}}\exp\{-[a|\varrho-\varrho_0|^2+2h(\varrho-\varrho_0)\cdot(\boldsymbol{P}-\boldsymbol{P}_0)+b|\boldsymbol{P}-\boldsymbol{P}_0|^2]/2\Delta\}. \tag{280}$$

We shall now verify that the foregoing solution for W obtained as the fundamental solution of Eq. (250) agrees with what we obtained in §2 through a discussion of the Langevin equation: With $\boldsymbol{R}$ and $\boldsymbol{S}$ as defined in Eqs. (181) we have

$$\varrho-\varrho_0=e^{\beta t}\boldsymbol{S};\quad \boldsymbol{P}-\boldsymbol{P}_0=\boldsymbol{R}+(1/\beta)\boldsymbol{S}. \tag{281}$$

Accordingly,

$$\left.\begin{aligned} a|\varrho-\varrho_0|^2+2h(\varrho-\varrho_0)\cdot(\boldsymbol{P}-\boldsymbol{P}_0)+b|\boldsymbol{P}-\boldsymbol{P}_0|^2&=ae^{2\beta t}|\boldsymbol{S}|^2+2he^{\beta t}(\boldsymbol{R}\cdot\boldsymbol{S}+(1/\beta)|\boldsymbol{S}|^2)+b|\boldsymbol{R}+(1/\beta)\boldsymbol{S}|^2,\\ &=e^{2\beta t}(F|\boldsymbol{S}|^2-2H\boldsymbol{R}\cdot\boldsymbol{S}+G|\boldsymbol{R}|^2),\end{aligned}\right\} \tag{282}$$

where

$$F=a+2h\beta^{-1}e^{-\beta t}+b\beta^{-2}e^{-2\beta t};\quad G=be^{-2\beta t};\quad H=-(he^{-\beta t}+b\beta^{-1}e^{-2\beta t}). \tag{283}$$

With a, b and h as given by Eqs. (278) we find that

$$F=q\beta^{-3}(2\beta t-3+4e^{-\beta t}-e^{-2\beta t});\quad G=q\beta^{-1}(1-e^{-2\beta t});\quad H=q\beta^{-2}(1-e^{-\beta t})^2. \tag{284}$$

Further,

$$FG-H^2=(ab-h^2)e^{-2\beta t}=\Delta e^{-2\beta t}. \tag{285}$$

Thus the solution (280) can be expressed alternatively in the form

$$W=\frac{1}{8\pi^3(FG-H^2)^{\frac{3}{2}}}\exp[-(F|\boldsymbol{S}|^2-2H\boldsymbol{R}\cdot\boldsymbol{S}+G|\boldsymbol{R}|^2)/2(FG-H^2)]. \tag{286}$$

Comparing Eqs. (284) and (286) with Eqs. (178), (182) and (183) we see that the verification is complete.

(iv) *The Solution of Equation* (249) *for the Case of a Harmonically Bound Particle*

The method of solution is sufficiently illustrated by considering the case of a one-dimensional oscillator describing Brownian motion. Equation (249) then reduces to

$$\frac{\partial W}{\partial t}+u\frac{\partial W}{\partial x}-\omega^2x\frac{\partial W}{\partial u}=\beta u\frac{\partial W}{\partial u}+\beta W+q\frac{\partial^2 W}{\partial u^2}. \tag{287}$$

As in our discussion in the two preceding sections we introduce as variables two first integrals of the associated subsidiary system:

$$dx/dt=u;\quad du/dt=-\beta u-\omega^2 x. \tag{288}$$

Two independent first integrals of Eqs. (288) are readily seen to be

$$(x\mu_1-u)\exp(-\mu_2 t)\quad\text{and}\quad(x\mu_2-u)\exp(-\mu_1 t) \tag{289}$$

where μ_1 and μ_2 have the same meanings as in §3 [cf. Eqs. (189) and (190)]. Accordingly we set

$$\xi=(x\mu_1-u)\exp(-\mu_2 t);\quad \eta=(x\mu_2-u)\exp(-\mu_1 t). \tag{290}$$

In these variables Eq (287) becomes

$$\frac{\partial W}{\partial t}=\beta W+q\left(\exp(-2\mu_2 t)\frac{\partial^2 W}{\partial \xi^2}+2\exp(-(\mu_1+\mu_2)t)\frac{\partial^2 W}{\partial\xi\partial\eta}+\exp(-2\mu_1 t)\frac{\partial^2 W}{\partial\eta^2}\right). \tag{291}$$

Introducing the further transformation

$$W=\chi e^{\beta t}, \tag{292}$$

we finally obtain

$$\frac{\partial \chi}{\partial t}=q\left(\exp(-2\mu_2 t)\frac{\partial^2 \chi}{\partial \xi^2}+2\exp[-(\mu_1+\mu_2)t]\frac{\partial^2 \chi}{\partial\xi\partial\eta}+\exp(-2\mu_1 t)\frac{\partial^2 \chi}{\partial\eta^2}\right). \tag{293}$$

This equation is of the same form as Eq. (261) in lemma II. Hence the solution of this equation which tends to $\delta(\xi-\xi_0)\delta(\eta-\eta_0)$ as $t\rightarrow 0$ is

$$\chi=\frac{1}{2\pi\Delta^{\frac{1}{2}}}\exp\{-[a(\xi-\xi_0)^2+2h(\xi-\xi_0)(\eta-\eta_0)+b(\eta-\eta_0)^2]/2\Delta\}, \tag{294}$$

where

$$\left.\begin{aligned}
a&=2q\int_0^t \exp(-2\mu_1 t)dt=\frac{q}{\mu_1}[1-\exp(-2\mu_1 t)],\\
b&=2q\int_0^t \exp(-2\mu_2 t)dt=\frac{q}{\mu_2}[1-\exp(-2\mu_2 t)],\\
h&=-2q\int_0^t \exp[-(\mu_1+\mu_2)t]dt=-\frac{2q}{\mu_1+\mu_2}\{1-\exp[-(\mu_1+\mu_2)t]\}.
\end{aligned}\right\} \tag{295}$$

Further,

$$\xi_0=x_0\mu_1-u_0;\quad \eta_0=x_0\mu_2-u_0, \tag{296}$$

where x_0 and u_0 denote the position and velocity of the particle at time $t=0$. It is again verified that the solution

$$W=\frac{e^{\beta t}}{2\pi\Delta^{\frac{1}{2}}}\exp\{-[a(\xi-\xi_0)^2+2h(\xi-\xi_0)(\eta-\eta_0)+b(\eta-\eta_0)^2]/2\Delta\}, \tag{297}$$

obtained as the fundamental solution of Eq. (287) is in agreement with the distributions obtained in §3 through a discussion of the Langevin equation.

(v) *The General Case*

Our discussion in the two preceding sections suggests that in dealing with Eq. (249) quite generally we may introduce as new variables six independent first integrals of the equations of motion

$$d\mathbf{r}/dt=\mathbf{u};\quad d\mathbf{u}/dt=-\beta\mathbf{u}+\mathbf{K}. \tag{298}$$

These are the Lagrangian subsidiary equations of the linear first-order equation derived from (249) after ignoring the Laplacian term $q\nabla_u^2 W$. If $I_1, \cdots, I_6$ are six such integrals, we introduce

$$I_1(\mathbf{r},\mathbf{u},t),\cdots,I_6(\mathbf{r},\mathbf{u},t) \tag{299}$$

as the new independent variables. If we further set

$$W=\chi e^{3\beta t}, \tag{300}$$

Eq. (249) will transform to

$$\partial\chi/\partial t=q[\nabla_u^2\chi]_{I_1,\cdots I_6}, \tag{301}$$

where the Laplacian of χ on the right-hand side has to be expressed in terms of the new variables $I_1, \cdots, I_6$.

We shall thus be left with a general linear partial differential equation of the second order for χ; and we seek a solution of this equation of the form

$$\chi = \frac{1}{8\pi^3 \Delta^{\frac{1}{2}}} e^{-Q/2\Delta}, \tag{302}$$

where Q stands for a general homogeneous quadratic form in the six variables $I_1, \cdots, I_6$ with coefficients which are functions of time only. Further, in Eq. (302) Δ is the determinant of the matrix formed by the coefficients of the quadratic form. In this manner we can expect to solve the general problem.

(vi) The Differential Equation for the Displacement ($t \gg \beta^{-1}$). The Smoluchowski Equation

We have seen that all the physically significant questions concerning the motion of a free Brownian particle can be answered by solving Eq. (250) with appropriate boundary conditions. However, if we are interested only in time intervals very large compared to the "*time of relaxation*" β^{-1} we can apply the method of the Fokker-Planck equation to configuration space ($\mathbf{r}$) independently of the velocity space. For, according to Eq. (172), we may say that for a free Brownian particle, the transition probability that $\mathbf{r}$ suffers an increment $\Delta\mathbf{r}$ in time $\Delta t \gg \beta^{-1}$ is given by

$$\psi(\Delta\mathbf{r}) = \frac{1}{(4\pi D\Delta t)^{\frac{3}{2}}} \exp\left(-|\Delta\mathbf{r}|^2/4D\Delta t\right), \tag{303}$$

where

$$D = q/\beta^2 = kT/m\beta. \tag{304}$$

Thus, again with the understanding that $\Delta t \gg \beta^{-1}$ we can write [cf. Eq. (218) and the remarks following it]

$$w(\mathbf{r}, t+\Delta t) = \int w(\mathbf{r}-\Delta\mathbf{r}, t)\psi(\Delta\mathbf{r})d(\Delta\mathbf{r}). \tag{305}$$

Applying now to this equation the procedure that was followed in the derivation of the Fokker-Planck equation in the velocity space we readily obtain the "diffusion equation"

$$\partial w/\partial t = D\nabla_r^2 w. \tag{306}$$

That we should be led to the diffusion equation is not surprising since Eq. (303) implies that for time intervals $\Delta t \gg \beta^{-1}$ the motion of the particle reduces to the elementary case of the problem of random flights (Chapter I, §4 case [c]) and the analysis of *I* §5 leading to Eq. (112) applies.

Equation (306) is valid for a free Brownian particle. To extend this result for the case when an external field is acting we start from Eq. (249) which is quite generally true in phase space. We first rewrite this equation in the form

$$\frac{\partial W}{\partial t} = \beta\left(\operatorname{div}_{\mathbf{u}} - \frac{1}{\beta}\operatorname{div}_{\mathbf{r}}\right)\left(W\mathbf{u} + \frac{q}{\beta}\operatorname{grad}_{\mathbf{u}} W - \frac{\mathbf{K}}{\beta}W + \frac{q}{\beta^2}\operatorname{grad}_{\mathbf{r}} W\right) + \operatorname{div}_{\mathbf{r}}\left(\frac{q}{\beta^2}\operatorname{grad}_{\mathbf{r}} W - \frac{\mathbf{K}}{\beta}W\right). \tag{307}$$

We now integrate this equation along the straight line

$$\mathbf{r} + \mathbf{u}/\beta = \text{constant} = \mathbf{r}_0, \tag{308}$$

from $\mathbf{u} = -\infty$ to $+\infty$. We obtain

$$\frac{\partial}{\partial t}\int_{\mathbf{r}+\mathbf{u}\beta^{-1}=\mathbf{r}_0} W d\mathbf{u} = \int_{\mathbf{r}+\mathbf{u}\beta^{-1}=\mathbf{r}_0} \operatorname{div}_{\mathbf{r}}\left(\frac{q}{\beta^2}\operatorname{grad}_{\mathbf{r}} W - \frac{\mathbf{K}}{\beta}W\right)d\mathbf{u}. \tag{309}$$

We shall now suppose that $\boldsymbol{K}(\boldsymbol{r})$ does not change appreciably over distances of the order of $(q/\beta^3)^{\frac{1}{2}}$. Then, starting from an arbitrary initial distribution $W(\boldsymbol{r}, \boldsymbol{u}, 0)$ at time $t=0$ we should expect that a Maxwellian distribution of the velocities will be established at all points after time intervals $\Delta t \gg \beta^{-1}$. Consequently, if we are not interested in time intervals of the order of β^{-1} we can write

$$W(\boldsymbol{r}, \boldsymbol{u}, t) \simeq \left(\frac{m}{2\pi kT}\right)^{\frac{3}{2}} \exp\left(-m|\boldsymbol{u}|^2/2kT\right) w(\boldsymbol{r}, t). \tag{310}$$

With these assumptions Eq. (309) becomes

$$\frac{\partial w}{\partial t} \simeq \operatorname{div}_{\boldsymbol{r}_0} \left\{ \frac{q}{\beta^2} \operatorname{grad}_{\boldsymbol{r}_0} w(\boldsymbol{r}_0) - \frac{\boldsymbol{K}(\boldsymbol{r}_0)}{\beta} w(\boldsymbol{r}_0) \right\} \tag{311}$$

The passage from Eqs. (309) to (311) is the result of our supposition that in the domain of $\boldsymbol{u}$ from which the dominant contribution to the integral on the right-hand side of Eq. (309) arises (namely, $|\boldsymbol{u}| \lesssim (kT/m)^{\frac{1}{2}} = (q/\beta)^{\frac{1}{2}}$) the variation of $\boldsymbol{r}$ (which is of the order $|\boldsymbol{u}|/\beta \simeq (q/\beta^3)^{\frac{1}{2}}$) is small compared to the distances in the configuration space in which $\boldsymbol{K}$ and w change appreciably. The required generalization of Eq. (306) is therefore

$$\frac{\partial w}{\partial t} = \operatorname{div}_{\boldsymbol{r}} \left(\frac{q}{\beta^2} \operatorname{grad}_{\boldsymbol{r}} w - \frac{\boldsymbol{K}}{\beta} w \right). \tag{312}$$

Equation (312) is sometimes called Smoluchowski's equation.

An immediate consequence of Eq. (312) may be noted. According to this equation a *stationary diffusion current* $\boldsymbol{j}$ obeys the law

$$\boldsymbol{j} = \beta^{-1} \boldsymbol{K} w - q\beta^{-2} \operatorname{grad} w = \text{constant}. \tag{313}$$

If $\boldsymbol{K}$ can be derived from a potential $\mathfrak{V}$ so that

$$\boldsymbol{K} = -\operatorname{grad} \mathfrak{V} \tag{314}$$

Eq. (313) can be rewritten in the form

$$\boldsymbol{j} = -q\beta^{-2} \exp(-\beta\mathfrak{V}/q) \operatorname{grad}(w \exp(\beta\mathfrak{V}/q)), \tag{315}$$

where it may be noted $q/\beta = kT/m$. Integrating Eq. (315) between any two points A and B we obtain

$$\boldsymbol{j} \cdot \int_A^B \beta \exp(\beta\mathfrak{V}/q) d\boldsymbol{s} = \frac{kT}{m} w \exp(\beta\mathfrak{V}/q) \Big|_B^A, \tag{316}$$

an important equation, first derived by Kramers.

We may finally again draw attention to the fact that Eqs. (306) and (312) are valid only if we ignore effects which happen in time intervals of the order of β^{-1} and space intervals of the order of $(q/\beta^3)^{\frac{1}{2}}$; when such effects are of interest we should go back to Eqs. (249) or (250) which are rigorously valid in phase space.

(vii) *General Remarks*

So far we have only shown that the discussion based on Eq. (249) and its various special forms leads to results in agreement with those already derived on the basis of the Langevin equation. However, the special importance of the partial differential equations arises when further restrictions on the problem are imposed. For, these additional restrictions can also be expressed in the form of boundary conditions which the solutions will have to satisfy and the consequent reduction to a boundary value problem in partial differential equations provides a very direct method for obtaining the necessary solutions. The alternative analysis based on the Langevin equation would in general be too involved.

Further examples of the use of the partial differential equations obtained in this section will be found in Chapters III and IV.

5. General Remarks

A general characteristic of the stochastic processes of the type considered in the preceding sections is that the increment in the velocity, $\Delta \boldsymbol{u}$ which a particle suffers in a time Δt long compared to the periods of the elementary fluctuations can be expressed as the sum of two distinct terms: a term $\boldsymbol{K}\Delta t$ which represents the action of the external field of force, and a term $\delta \boldsymbol{u}(\Delta t)$ which denotes a fluctuating quantity with a definite law of distribution. Thus

$$\Delta \boldsymbol{u} = \boldsymbol{K}\Delta t + \delta \boldsymbol{u}(\Delta t); \tag{317}$$

the corresponding increment in the position, $\Delta \boldsymbol{r}$ is given by

$$\Delta \boldsymbol{r} = \boldsymbol{u}\Delta t, \tag{318}$$

where $\boldsymbol{u}$ is the instantaneous velocity of the particle.

When dealing with stochastic processes of the *strictly* Brownian motion type we further suppose that the term $\delta \boldsymbol{u}(\Delta t)$ in Eq. (317) can in turn be decomposed into two parts: a part $-\beta \boldsymbol{u}\Delta t$ representing the deceleration caused by the dynamical friction $-\beta \boldsymbol{u}$ and a fluctuating part $\boldsymbol{B}(\Delta t)$ which is really the vector sum of a very large number of very "minute" accelerations arising from collisions with individual molecules of the surrounding fluid:

$$\delta \boldsymbol{u}(\Delta t) = -\beta \boldsymbol{u}\Delta t + \boldsymbol{B}(\Delta t). \tag{319}$$

It is *this* particular decomposition of $\delta \boldsymbol{u}(\Delta t)$ that is peculiarly characteristic of stochastic processes of the Brownian type.

Concerning $\boldsymbol{B}(\Delta t)$ in Eq. (319) we have supposed in §§2, 3, and 4 that it is governed by the distribution function [cf. Eqs. (144) and (145)]

$$w(\boldsymbol{B}[\Delta t]) = \frac{1}{(4\pi q \Delta t)^{\frac{3}{2}}} \exp\left(-|\boldsymbol{B}(\Delta t)|^2/4q\Delta t\right), \tag{320}$$

where

$$q = \beta k T/m. \tag{321}$$

In this choice of the distribution function for $\boldsymbol{B}(\Delta t)$ we were guided by two considerations: *First*, that starting from any arbitrarily assigned distribution of the velocities we shall always be led to the Maxwellian distribution as $t\rightarrow\infty$ (or, alternatively that the Maxwellian distribution of the velocities is invariant to stochastic processes of the type considered); and *second* that during a time Δt in which the position and the velocity of the particle will change by an "infinitesimal" amount of order Δt the particle will in reality suffer an *exceedingly* large number of individual accelerations by collisions with the molecules of the surrounding fluid. This second consideration would suggest, from analogy with the simple case of the problem of random flights [Eq. (108)], a formula of the *form* (320). The particular value of q (321) then follows from the first requirement.

Combining Eqs. (319) and (320) we obtain for the *transition probability* $\psi(\boldsymbol{u}; \delta \boldsymbol{u})$ for $\boldsymbol{u}$ to suffer an increment $\delta \boldsymbol{u}$ due to the Brownian forces only, the expression

$$\psi(\boldsymbol{u}; \delta \boldsymbol{u}) = \frac{1}{(4\pi q \Delta t)^{\frac{3}{2}}} \exp\left(-|\beta \boldsymbol{u}\Delta t + \delta \boldsymbol{u}|^2/4q\Delta t\right). \tag{322}$$

We shall now briefly re-examine the problem of continuous stochastic processes more generally from the point of view of the invariance of the *Maxwell-Boltzmann* distribution

$$W=\text{constant}\exp\{-[m|\boldsymbol{u}|^2+2m\mathfrak{B}(\boldsymbol{r})]/2kT\};\quad \boldsymbol{K}=-\text{grad}\,\mathfrak{B} \tag{323}$$

to processes governed by Eqs. (317) and (318) *only* i.e., without making the further assumptions included in Eqs. (319)–(322).

Assuming, as we have done hitherto, that the stochastic process we are considering is of the Markoff type we can write the integral equation [cf. Eq. (241)]

$$W(\boldsymbol{r},\boldsymbol{u},t+\Delta t)=\int\int W(\boldsymbol{r}-\Delta\boldsymbol{r},\boldsymbol{u}-\Delta\boldsymbol{u},t)\Psi(\boldsymbol{r}-\Delta\boldsymbol{r},\boldsymbol{u}-\Delta\boldsymbol{u};\Delta\boldsymbol{r},\Delta\boldsymbol{u})d(\Delta\boldsymbol{r})d(\Delta\boldsymbol{u}). \tag{324}$$

According to Eqs. (318) we expect that [cf. Eq. (242)]

$$\Psi(\boldsymbol{r},\boldsymbol{u};\Delta\boldsymbol{r},\Delta\boldsymbol{u})=\psi(\boldsymbol{r},\boldsymbol{u};\Delta\boldsymbol{u})\delta(\Delta x-u_1\Delta t)\delta(\Delta y-u_2\Delta t)\delta(\Delta z-u_3\Delta t). \tag{325}$$

Equation (324) becomes

$$W(\boldsymbol{r},\boldsymbol{u},t+\Delta t)=\int W(\boldsymbol{r}-\boldsymbol{u}\Delta t,\boldsymbol{u}-\boldsymbol{K}\Delta t-\delta\boldsymbol{u},t)\psi(\boldsymbol{r}-\boldsymbol{u}\Delta t,\boldsymbol{u}-\boldsymbol{K}\Delta t-\delta\boldsymbol{u};\boldsymbol{K}\Delta t+\delta\boldsymbol{u})d(\delta\boldsymbol{u}), \tag{326}$$

where we have further substituted for $\Delta\boldsymbol{u}$ according to Eq. (317). Equation (326) can be written alternatively as

$$W(\boldsymbol{r}+\boldsymbol{u}\Delta t,\boldsymbol{u}+\boldsymbol{K}\Delta t,t+\Delta t)=\int W(\boldsymbol{r},\boldsymbol{u}-\delta\boldsymbol{u},t)\psi(\boldsymbol{r},\boldsymbol{u}-\delta\boldsymbol{u};\delta\boldsymbol{u})d(\delta\boldsymbol{u}). \tag{327}$$

Applying to this equation the same procedure as was adopted in the derivation of the Fokker-Planck and the generalized Liouville equations in §4, we readily find that [cf. Eq. (245)]

$$\left\{\frac{\partial W}{\partial t}+\boldsymbol{u}\cdot\text{grad}_r\,W+\boldsymbol{K}\cdot\text{grad}_u\,W\right\}\Delta t+O(\Delta t^2)=-\sum_i\frac{\partial}{\partial u_i}(W\langle\delta u_i\rangle_{Av})+\frac{1}{2}\sum_i\frac{\partial^2}{\partial u_i^2}(W\langle\delta u_i^2\rangle_{Av})$$
$$+\sum_{i<j}\frac{\partial^2}{\partial u_i\partial u_j}(W\langle\delta u_i\delta u_j\rangle_{Av})+O(\langle\delta u_i\delta u_j\delta u_k\rangle_{Av}) \tag{328}$$

where $\langle\delta u_i\rangle_{Av}$ etc., denote the various moments of the transition probability $\psi(\boldsymbol{r},\boldsymbol{u};\delta\boldsymbol{u})$.

We shall now suppose that

$$\langle\delta u_i\rangle_{Av}=\mu_i\Delta t+O(\Delta t^2);\quad\langle\delta u_i^2\rangle_{Av}=\mu_{ii}\Delta t+O(\Delta t^2);\quad\langle\delta u_i\delta u_j\rangle_{Av}=\mu_{ij}\Delta t+O(\Delta t^2), \tag{329}$$

and that all averages of quantities like $\delta u_i\delta u_j\delta u_k$ are of order higher than one in Δt. With this understanding we shall obtain from Eq. (328), on passing to the limit $\Delta t=0$ the result

$$\frac{\partial W}{\partial t}+\boldsymbol{u}\cdot\text{grad}_r\,W+\boldsymbol{K}\cdot\text{grad}_u\,W=-\sum_i\frac{\partial}{\partial u_i}(W\mu_i)+\frac{1}{2}\sum_i\frac{\partial^2}{\partial u_i^2}(W\mu_{ii})+\sum_{i<j}\frac{\partial^2}{\partial u_i\partial u_j}(W\mu_{ij}). \tag{330}$$

We now require that the Maxwell-Boltzmann distribution (323) satisfy Eq. (330) identically. On substituting this distribution in Eq. (330) we find that the left-hand side of this equation vanishes and we are left with

$$-\sum_i\frac{\partial}{\partial u_i}[\exp(-m|\boldsymbol{u}|^2/2kT)\mu_i]+\frac{1}{2}\sum_i\frac{\partial^2}{\partial u_i^2}[\exp(-m|\boldsymbol{u}|^2/2kT)\mu_{ii}]$$
$$+\sum_{i<j}\frac{\partial^2}{\partial u_i\partial u_j}[\exp(-m|\boldsymbol{u}|^2/2kT)\mu_{ij}]=0. \tag{331}$$

Equation (331) is to be regarded as the general condition on the moments.

For the distribution (322)

$$\mu_i = -\beta u_i; \quad \mu_{ii} = 2q = 2\beta kT/m; \quad \mu_{ij} = 0. \tag{332}$$

Also, the third and higher moments of δu do not contain terms linear in Δt.

We readily verify that with the μ's given by (332) we satisfy Eq. (331). It is not, however, to be expected that (332) represents the most general solution for the μ's which will satisfy Eq. (331). It would clearly be a matter of considerable interest to investigate Eq. (331) (or the generalization of this equation to include terms involving μ_{ijk} etc.,) with a view to establishing the nature of the restrictions on the μ's implied by Eq. (331). Such an investigation might lead to the discovery of new classes of Markoff processes which will leave the Maxwell-Boltzmann distribution invariant but which will not be of the classical Brownian motion type. It is not proposed to undertake this investigation in this article. We may, however, draw special attention to the fact that according to Eqs. (331) and (332), β can very well depend on the spatial coordinates (though $q/\beta[=kT/m]$ must be a constant throughout the system). Thus, the generalized Liouville Eq. (249) and the Smoluchowski Eq. (312) are valid as they stand, also when $\beta = \beta(r)$.

CHAPTER III

PROBABILITY AFTER-EFFECTS: COLLOID STATISTICS; THE SECOND LAW OF THERMODYNAMICS. THE THEORY OF COAGULATION, SEDIMENTATION, AND THE ESCAPE OVER POTENTIAL BARRIERS

In this chapter we shall consider certain problems in the theory of Brownian motion which require the more explicit introduction than we had occasion hitherto, of the notion of *probability after-effects*. The fundamental ideas underlying this notion have already been described in the introductory section where we have also seen that colloid statistics (or, more generally, the phenomenon of density fluctuations in a medium of constant average density) provides a very direct illustration of the problem. The theory of this phenomenon which has been developed along very general lines by Smoluchowski has found beautiful confirmation in the experiments of Svedberg, Westgren, and others. This theory of Smoluchowski in addition to providing a striking application of the principles of Brownian motion has also important applications to the elucidation of the statistical nature of the second law of thermodynamics. In view, therefore, of the fundamental character of Smoluchowski's theory we shall give a somewhat detailed account of it in this chapter (§§1–3). (In the later sections of this chapter we consider further miscellaneous applications of the theory of Brownian motion which have bearings on problems considered in Chapter IV.)

1. The General Theory of Density Fluctuations for Intermittent Observations. The Mean Life and the Average Time of Recurrence of a State of Fluctuation*

Consider a geometrically well-defined small element of volume v in a solution containing Brownian particles under conditions of diffusion equilibrium. (More generally, we may also consider v as an element in a very much larger volume containing a large number of particles in equilibrium.) Suppose now that we observe the number of particles contained in v systematically at constant intervals of time τ apart. Then the frequency $W(n)$ with which different numbers of particles will be observed in v will follow the *Poisson distribution* (see Appendix III),

$$W(n) = e^{-\nu}\nu^n/n!, \tag{333}$$

where ν denotes the average number of particles that will be contained in v:

$$\langle n\rangle_{\text{Av}} = \sum_{n=0}^{\infty} nW(n) = e^{-\nu}\nu \sum_{n=1}^{\infty} \frac{\nu^{n-1}}{(n-1)!} = \nu. \tag{334}$$

In other words, the number of particles that will be observed in v is subject to *fluctuations* and the different *states of fluctuations* (which, in this case, can be labelled by n) occur with definite frequencies.

According to Eq. (333) the mean square deviation δ^2 from the average value ν is given by

$$\delta^2 = \langle(n-\nu)^2\rangle_{\text{Av}} = \langle n^2\rangle_{\text{Av}} - \nu^2, \tag{335}$$

*Professor M. S. Bartlett has pointed out (Nature 165, 727, 1950) that the treatment of the average time of recurrence in this section is not free of errors.

or, since

$$\left.\begin{aligned}\langle n^2\rangle_{\text{Av}} &= \sum_{n=0}^{\infty} n^2 \frac{e^{-\nu}\nu^n}{n!}\\ &= e^{-\nu}\nu\left\{\nu\sum_{n=2}^{\infty}\frac{\nu^{n-2}}{(n-2)!}+\sum_{n=1}^{\infty}\frac{\nu^{n-1}}{(n-1)!}\right\}\\ &= \nu^2+\nu,\end{aligned}\right\} \qquad (336)$$

we have

$$\delta^2 = \nu. \qquad (337)$$

It is seen that the frequency with which the different states of fluctuation n occur is independent of all physical parameters describing the particle (e.g., radius and density) and the surrounding fluid (e.g., viscosity). The situation is, however, completely changed when we consider the *speed* with which the different states of fluctuations follow each other in time. More specifically, consider the number of particles n and m contained in v at an interval of time τ apart. We expect that the number m observed on the second occasion will be correlated with the number n observed on the first occasion. This correlation should be such, that as $t\rightarrow 0$ the result of the second observation can be predicted with certainty as n, while as $t\rightarrow\infty$ we shall observe on the second occasion numbers which will increasingly be distributed according to the Poisson distribution (333). For finite intervals of time τ we can therefore ask for the *transition probability* $W(n; m)$ that m particles will be counted in v after a time τ from the instant when there was observed to be n particles in it.

In solving the problem stated toward the end of the preceding paragraph we shall make, following Smoluchowski, the two assumptions: (1) that the motions of the individual particles are not mutually influenced and are independent of each other and (2) that all positions in the element of volume considered have equal *a priori* probability. Under these circumstances we can expect to define *a probability P that a particle somewhere inside v will have emerged from it during the time τ*. The exact value of this *probability after-effect factor P* will depend on the precise circumstances of the problem including the geometry of the volume v. In §2 we shall obtain the explicit formula for P when the motions of the individual particles are governed by the laws of Brownian motion [Eq. (380)]; and similarly in §3 we shall obtain the formula for P for the case when the particles describe linear trajectories [Eq. (413)]. Meantime, we shall continue the discussion of the speed of fluctuations on the assumption that the factor P as defined can be unambiguously evaluated depending, however, on circumstances.

It is clear that the required transition probability $W(n; m)$ can be written down in an entirely elementary way if we know the probabilities with which particles enter and leave the element of volume. More precisely, let $A_i^{(n)}$ denote the probability that starting from an initial situation in which there are n particles inside v *some i* particles will have emerged from it during τ; this probability of emergence of a certain number of particles will clearly depend on the initial number of particles inside v. Similarly, let E_i denote the probability that i particles will have entered the element of volume v during τ. Since one of our principal assumptions is that the motions of the particles are not mutually influenced, the probability of entrance of a certain number of particles cannot depend on the number already contained in it. We shall now obtain explicit expressions for these two probabilities in terms of P.

The expression for $A_i^{(n)}$ can be written down at once when we recall that this must be equal to the product of the probability P^i that some particular group of i particles leaves v during τ, the probability $(1-P)^{n-i}$ that the remaining $(n-i)$ particles do not leave v during τ, and the number of distinct ways C_i^n of selecting i particles from the initial group of n. Accordingly,

$$A_i^{(n)} = C_i^n P^i(1-P)^{n-i} = \frac{n!}{i!(n-i)!}P^i(1-P)^{n-i}, \qquad (338)$$

which is a Bernoulli distribution.

To obtain the expression for E_i we first remark that this must equal the probability that i particles *emerge* from the element of volume v on an *arbitrary* occasion; since, under *equilibrium conditions* the *a priori* probabilities for the entrance and emergence of particles must be equal. Remembering further, that E_i is independent of the number of particles initially contained in v, we clearly have

$$E_i=\langle A_i^{(n)}\rangle_{Av}=\sum_{n=i}^{\infty} W(n)A_i^{(n)}, \qquad (339)$$

where $W(n)$ is the probability that v initially contained n particles; $W(n)$ accordingly is given by (333). Combining Eqs. (333), (338), and (339) we therefore have

$$\left.\begin{aligned} E_i&=\sum_{n=i}^{\infty}\frac{e^{-\nu}\nu^n}{n!}\frac{n!}{i!(n-i)!}P^i(1-P)^{n-i},\\ &=\frac{e^{-\nu}(\nu P)^i}{i!}\sum_{n=i}^{\infty}\frac{\nu^{n-i}(1-P)^{n-i}}{(n-i)!},\\ &=\frac{e^{-\nu}(\nu P)^i}{i!}e^{\nu(1-P)}.\end{aligned}\right\} \qquad (340)$$

Thus,

$$E_i=e^{-\nu P}(\nu P)^i/i!, \qquad (341)$$

in other words, a Poisson distribution with variance νP.

Using the formulae (338) and (341) for $A_i^{(n)}$ and E_i we can at once write down the expression for the transition-probability $W(n;n+k)$ that there is an increase in the number of particles from n to $n+k$. We clearly have

$$W(n;n+k)=\sum_{i=0}^{n} A_i^{(n)}E_{i+k}. \qquad (342)$$

Similarly, for the transition probability $W(n;n-k)$ that there is a decrease in the number of particles from n to $n-k$ we have

$$W(n;n-k)=\sum_{i=k}^{n} A_i^{(n)}E_{i-k}, \quad (k\leqslant n). \qquad (343)$$

From Eqs. (338), (341), (342), and (343) we therefore obtain

$$W(n;n+k)=e^{-\nu P}\sum_{i=0}^{n} C_i^nP^i(1-P)^{n-i} \times(\nu P)^{i+k}/(i+k)!, \qquad (344)$$

and

$$W(n;n-k)=e^{-\nu P}\sum_{i=k}^{n} C_i^nP^i(1-P)^{n-i} \times(\nu P)^{i-k}/(i-k)! \qquad (345)$$

The foregoing expressions for the transition probabilities are due to Smoluchowski.

The formulae (344) and (345) in spite of their apparent complexity have in reality very simple structures. To see this we first introduce the Bernoulli and the Poisson distributions

$$w_1^{(n)}(x)=C_x^n(1-P)^xP^{n-x} \quad (0\leqslant x\leqslant n), \qquad (346)$$

and

$$w_2(y)=e^{-\nu P}(\nu P)^y/y! \quad (0\leqslant y<\infty). \qquad (347)$$

$w_1^{(n)}(x)$ is the probability that *some* x particles *remain* in v after a time τ when initially there were n particles in it; similarly, $w_2(y)$ is the probability that y particles enter v in time τ. In terms of the distributions (346) and (347) we can rewrite Eqs. (344) and (345) as

$$W(n;n+k)=\sum_{i=0}^{n} w_1^{(n)}(n-i)w_2(i+k), \qquad (348)$$

and

$$W(n;n-k)=\sum_{i=k}^{n} w_1^{(n)}(n-i)w_2(i-k), \qquad (349)$$

or, writing m for $n+k$, respectively $n-k$, we see that both Eqs. (348) and (349) can be included in the single formula

$$W(n,m)=\sum_{x+y=m} w_1^{(n)}(x)w_2(y). \qquad (350)$$

In other words, *the distribution* $W(n,m)$ for a fixed value of n *is the "sum" of the two distributions* (346) *and* (347). And, therefore, the mean and the mean square deviation for the distribution of m according to (350) is the sum of the means and the mean square deviations of the component distributions (346) and (347) (see Appendix IV). Since [cf. Eqs. (334) and (335) and Appendix I Eqs. (621) and (624)]

$$\langle x\rangle_{Av}=n(1-P);\quad \langle(x-\langle x\rangle_{Av})^2\rangle_{Av}=nP(1-P), \qquad (351)$$

and

$$\langle y\rangle_{Av}=\nu P;\quad \langle(y-\langle y\rangle_{Av})^2\rangle_{Av}=\nu P, \qquad (352)$$

we conclude that

$$\langle m\rangle_{Av}=n(1-P)+\nu P, \qquad (353)$$

and

$$\langle(m-\langle m\rangle_{Av})^2\rangle_{Av}=nP(1-P)+\nu P. \qquad (354)$$

Let

$$\Delta_n=m-n. \qquad (355)$$

Then, according to Eqs. (354) and (355)

$$\langle\Delta_n\rangle_{Av}=\langle m\rangle_{Av}-n=(\nu-n)P, \qquad (356)$$

and

$$\left.\begin{aligned}\langle\Delta_n^2\rangle_{Av}&=\langle(m-\langle m\rangle_{Av}+\langle m\rangle_{Av}-n)^2\rangle_{Av}\\ &=\langle(m-\langle m\rangle_{Av})^2\rangle_{Av}+(\langle m\rangle_{Av}-n)^2\\ &=nP(1-P)+\nu P+(\nu-n)^2P^2,\end{aligned}\right\} \qquad (357)$$

or

$$\langle\Delta_n^2\rangle_{Av}=P^2[(\nu-n)^2-n]+(n+\nu)P. \qquad (358)$$

It is seen that according to Eq. (356) the number of particles inside v changes, on the average, in the direction of making n approach its mean value, namely ν. In other words, the density fluctuations studied here in terms of a "microscopic" analysis of the stochastic motions of the individual particles are in complete agreement with the macroscopic theory of diffusion.

The quantities $\langle\Delta_n\rangle_{Av}$ and $\langle\Delta_n^2\rangle_{Av}$ represent the mean and the mean square of the differences that are to be expected in the numbers observed on two occasions at an interval of time τ apart when on the first occasion n particles were observed. If now, we further average $\langle\Delta_n\rangle_{Av}$ and $\langle\Delta_n^2\rangle_{Av}$ over all values of n with the weight function $W(n)$ we shall obtain the mean and the mean square of the differences in the numbers of particles observed on consecutive occasions in a long sequence of observations made at constant intervals τ apart. Thus [cf. Eq. (334)]

$$\langle\Delta\rangle_{Av}=\langle\langle\Delta_n\rangle_{Av}\rangle_{Av}=\langle\nu-n\rangle_{Av}P=0, \qquad (359)$$

a result which is to be expected. On the other hand [cf. Eq. (337)]

$$\left.\begin{aligned}\langle\Delta^2\rangle_{Av}&=\langle\langle\Delta_n^2\rangle_{Av}\rangle_{Av}\\ &=P^2[\langle(\nu-n)^2\rangle_{Av}-\langle n\rangle_{Av}]+\langle n+\nu\rangle_{Av}P\\ &=P^2(\delta^2-\langle n\rangle_{Av})+(\langle n\rangle_{Av}+\nu)P,\end{aligned}\right\} \qquad (360)$$

or

$$\langle\Delta^2\rangle_{Av}=2\nu P. \qquad (361)$$

Equation (361) suggests a direct method for the experimental determination of the probability after-effect factor P from the simple evaluation of the mean square differences $\langle\Delta^2\rangle_{Av}$ from long sequences of observations of n (see §2 below). Further, according to Eq. (361)

$$\langle\Delta^2\rangle_{Av}=2\nu \quad \text{when} \quad P=1. \qquad (362)$$

This result.is in agreement with what we should expect, since, when $P=1$ there will be no correlation between the numbers that will be observed on two occasions at an interval τ apart; $\langle\Delta^2\rangle_{Av}$ then simply becomes the mean square of the differences between two numbers each of which (without correlation) is governed by the same Poisson distribution; and, therefore [cf. Eqs. (333) and (336)],

$$\begin{aligned}\langle\Delta^2\rangle_{Av}&=\langle(n-m)^2\rangle_{Av}=\langle n^2\rangle_{Av}+\langle m^2\rangle_{Av}-2\langle n\rangle_{Av}\langle m\rangle_{Av}\\ &=2(\nu^2+\nu)-2\nu^2=2\nu, \quad (P=1). \qquad (363)\end{aligned}$$

We shall now show how we can define the *mean life* and *the average time of recurrence* for a given state of fluctuation in terms of the transition probability $W(n;n)$:

$$W(n;n)=e^{-\nu P}\sum_{i=0}^{n} C_i^n P^i(1-P)^{n-i}(\nu P)^i/i!, \qquad (364)$$

which gives the probability that n will be observed on two consecutive occasions. Accordingly, the probability $\phi_n(k\tau)$ that the same number n will be observed on $(k-1)$ consecutive occasions (at constant intervals τ apart) and that on the kth occasion some number different from n will be observed is given by

$$\phi_n(k\tau)=W^{k-1}(n;n)[1-W(n;n)]. \qquad (365)$$

On the other hand, in terms of $\phi_n(k\tau)$ we can give a natural definition to the mean life to the state of fluctuation n by the equation

$$T_n=\sum_{k=1}^{\infty} k\tau\phi_n(k\tau). \qquad (366)$$

Combining Eqs. (365) and (366) we obtain

$$T_n=\tau[1-W(n;n)]\sum_{k=1}^{\infty} kW^{k-1}(n;n). \qquad (367)$$

The infinite series in Eq. (367) is readily evaluated and we find

$$T_n=\frac{\tau}{1-W(n;n)}. \qquad (368)$$

In an analogous manner we can define the time of recurrence of the state n by the equation

$$\Theta_n=\sum_{k=1}^{\infty} k\tau\psi_n(k\tau), \qquad (369)$$

where $\psi_n(k\tau)$ denotes the probability that starting from *an arbitrary state which is not n* we shall observe on $k-1$ successive occasions states which are not n and on the kth occasion observe the state n. If

$$W(Nn;Nn) \qquad (370)$$

denotes the probability that from an arbitrary state $\neq n$ we shall have a transition to a state which is also $\neq n$, then clearly

$$\psi_n(k\tau)=W^{k-1}(Nn;Nn)[1-W(Nn;Nn)]. \qquad (371)$$

Substituting the foregoing expression for $\psi_n(k\tau)$ in Eq. (369) we obtain [cf. Eqs. (365) and (368)]

$$\Theta_n=\frac{\tau}{1-W(Nn;\,Nn)}. \tag{372}$$

We shall now obtain a formula for $W(Nn;\,Nn)$. First of all it is clear that

$$1-W(Nn;\,Nn)=W(Nn;\,n), \tag{373}$$

where $W(Nn;\,n)$ is the probability that from an arbitrary state $\neq n$ we shall have a transition to the state n. Now, under equilibrium conditions, the number of transitions from states $\neq n$ to the state n must equal the number of transitions from the state n to states $\neq n$; accordingly

$$[1-W(n)]W(Nn;\,n) = W(n)[1-W(n;\,n)], \tag{374}$$

where $W(n)$ is given by Eq. (333). Hence,

$$W(Nn;\,n)=W(n)\frac{1-W(n;\,n)}{1-W(n)}. \tag{375}$$

Combining Eqs. (372), (373), and (375) we obtain

$$\Theta_n=\frac{\tau}{1-W(n;\,n)}\,\frac{1-W(n)}{W(n)}. \tag{376}$$

Finally, we may note that between T_n and Θ_n we have the relation

$$\Theta_n=T_n\frac{1-W(n)}{W(n)}. \tag{377}$$

In the next section we shall give a brief account of the experiments of Svedberg and Westgren on colloid statistics which have provided complete confirmation of Smoluchowski's theory of density fluctuations which we have developed in this section. Also, the formulae for T_n and Θ_n which we have derived have important applications to the elucidation of the second law of thermodynamics to which we shall return in §4.

2. Experimental Verification of Smoluchowski's Theory: Colloid Statistics

In the experiments of Svedberg, Westgren, and others on colloid statistics observations are

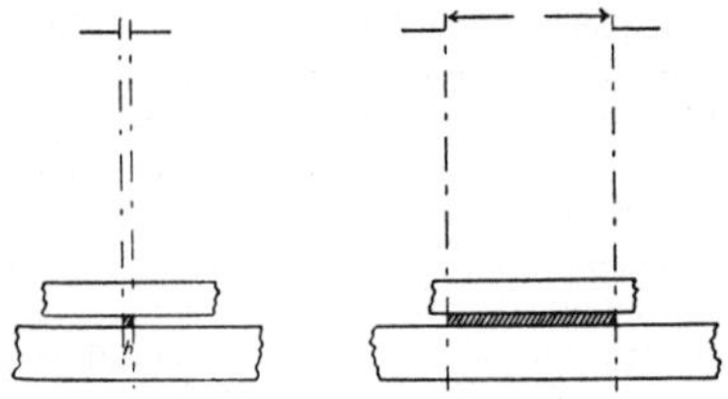

FIG. 3.

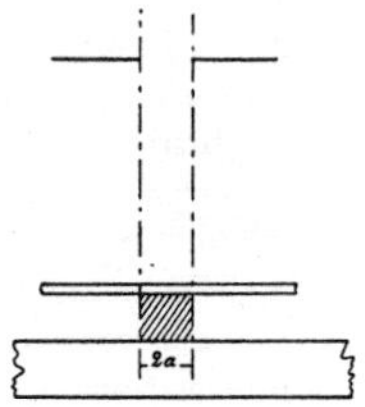

FIG. 4.

made by means of an ultramicroscope on the numbers of particles in a well-defined element of volume in a colloidal solution. These observations, made systematically at constant intervals τ apart, are secured either by the use of intermittent illumination (Svedberg) or by counting on the ticks of a metronome (Westgren). The volumes in which the counts are made are defined either optically by illuminating only plane parallel layers several microns in thickness (Svedberg) or mechanically by having the solution under observation sealed between the objective of the microscope and a glass plate and observing with the help of a cardioid condenser (Westgren). The dimensions of the element of volume at right angles to the line of sight are defined directly by limiting the field of observation (see Figs. 3 and 4).

The colloidal particles describe Brownian motion and since the intervals of time we are normally interested in are never less than a few hundredths of a second we can suppose that the motions of the particles are governed by the diffusion equation [cf. Eqs. (133), (304), and (306)]

$$\partial w/\partial t=D\nabla^2 w;\qquad D=q/\beta^2=kT/m\beta=kT/6\pi a\eta. \tag{378}$$

For, according to our discussion in Chapter II, §§2 and 4 the validity of the diffusion equation requires that we only ignore what happens in time intervals of order β^{-1} and for colloidal gold particles of radius $a=50\mu\mu$ this time of relaxation is of the order of 10^{-9}–10^{-10} second.

From Eq. (378) we conclude that the probability of occurrence of a particle at $\mathbf{r}_2$ at time t when it was at $\mathbf{r}_1$ at time $t=0$ is given by [cf. Eq. (172)]

$$\frac{1}{(4\pi Dt)^{\frac{3}{2}}}\exp\left(-|\mathbf{r}_2-\mathbf{r}_1|^2/4Dt\right). \tag{379}$$

On this basis we can readily write down a general formula for the probability after-effect factor P introduced in §1. For, by definition, P denotes the probability that a particle somewhere inside the given element of volume v (with uniform probability) at time $t=0$ will find itself outside of it at time $t=\tau$. Accordingly

$$P=\frac{1}{(4\pi D\tau)^{\frac{3}{2}}v}\int\int\exp\left(-|\mathbf{r}_1-\mathbf{r}_2|^2/4D\tau\right)d\mathbf{r}_1d\mathbf{r}_2, \tag{380}$$

where the integration over $\mathbf{r}_1$ is extended over all points in the interior of v while that over $\mathbf{r}_2$ is extended over all points exterior to v. Alternatively, we can also write

$$1-P=\frac{1}{(4\pi D\tau)^{\frac{3}{2}}v}\int_{\mathbf{r}_1\epsilon v}\int_{\mathbf{r}_2\epsilon v}\times\exp\left(-|\mathbf{r}_1-\mathbf{r}_2|^2/4D\tau\right)d\mathbf{r}_1d\mathbf{r}_2, \tag{381}$$

where, now, the integrations over *both* $\mathbf{r}_1$ and $\mathbf{r}_2$ are extended over all points inside v (indicated by the symbols $\mathbf{r}_1\epsilon v$ and $\mathbf{r}_2\epsilon v$).

We thus see that for any geometrically well-defined element of volume in a colloidal solution we can always evaluate, in principle, the probability after-effect factor P in terms of the physical parameters of the problem, namely, the geometry of the volume v, the radius a of the colloidal particles, and the coefficient of viscosity η of the surrounding liquid. On the other hand, this factor P can also be determined empirically from a direct evaluation of the mean square of the differences in the numbers of particles observed on consecutive occasions in a long sequence of observations made at constant intervals τ apart and using the formula [Eq. (361)]

$$\langle\Delta^2\rangle_{\text{Av}}=2\nu P, \tag{382}$$

where ν is the average of all the numbers observed. A comparison of the predictions of the theory with the data of colloid statistics therefore becomes possible. Once P has been determined [either theoretically according to Eq. (381) or empirically from Eq. (382)] we can predict the frequency of occurrence, $H(n, m)$, of the *pair* (n, m) in the observed sequence of numbers. For, clearly;

$$H(n, m)=W(n)W(n; m), \tag{383}$$

where $W(n)$ is the frequency of occurrence of n according to Eq. (333) and $W(n; m)$ is the transition probability from the state n to the state m according to Smoluchowski's formulae (344) and (345). Again a comparison between the predictions of the theory with the results of observations becomes possible.

Comparisons of the kind indicated in the preceding paragraph were first made by Smoluchowski himself who used for this purpose the data provided by Svedberg's experiments. However, later experiments by Westgren carried out with the expressed intention of verifying Smoluchowski's theory provide a more stringent comparison between the predictions of the theory and the results of observations. We shall therefore limit ourselves to describing the results of Westgren's experiments only.

Westgren conducted two series of experiments with the arrangements shown in Figs. 3 and 4. In the first of the two arrangements (Fig. 3) the particles under observation are confined to a long rectangular parallelepiped (see the shaded portions in Fig. 3). Under the conditions of this arrangement it is clear that the variation in the number of particles observed is predominantly due to diffusion at right angles to the lengthwise edge. Consequently, the formula for P appropriate to this arrangement is [cf. Eq. (381)]

$$1-P=\frac{1}{h(4\pi D\tau)^{\frac{1}{2}}}\int_0^h\int_0^h\times\exp\left[-(x_1-x_2)^2/4D\tau\right]dx_1dx_2, \tag{384}$$

where h denotes the width of the element of

volume under observation (see Fig. 3). Introducing $2(D\tau)^{\frac{1}{2}}$ as the unit of length, Eq. (384) becomes

$$1-P=\frac{1}{\alpha\pi^{\frac{1}{2}}}\int_0^\alpha\int_0^\alpha \exp\left[-(\xi_1-\xi_2)^2\right]d\xi_1 d\xi_2, \quad (385)$$

where we have written

$$\alpha=h/2(D\tau)^{\frac{1}{2}}. \quad (386)$$

We readily verify that Eq. (385) is equivalent to

$$1-P=\frac{2}{\alpha\pi^{\frac{1}{2}}}\int_0^\alpha d\xi_1\int_0^{\xi_1} d\eta\exp(-\eta^2), \quad (387)$$

or, after an integration by parts we find

$$P=1-\frac{2}{\pi^{\frac{1}{2}}}\int_0^\alpha \exp(-\xi^2)d\xi + \frac{1}{\alpha\pi^{\frac{1}{2}}}\left[1-\exp(-\alpha^2)\right]. \quad (388)$$

For the second of Westgren's arrangements (Fig. 4) the element under observation is a cylindrical volume and the variations in the numbers observed are in this case due to the diffusion of particles in all directions at right angles to the line of sight. Accordingly we have

$$P=\frac{4}{\alpha^2\pi}\int_\alpha^\infty d\xi_1\xi_1\int_0^\alpha d\xi_2\xi_2\int_0^\pi \times\exp(-\xi_1^2-\xi_2^2+2\xi_1\xi_2\cos\vartheta)d\vartheta, \quad (389)$$

where

$$\alpha=r_0/2(D\tau)^{\frac{1}{2}}, \quad (390)$$

r_0 denoting the radius of the cylindrical element under observation. The integrals in (389) can be evaluated in terms of Bessel functions with imaginary arguments and we find

$$P=e^{-2\surd\alpha}\left[I_0(2\surd\alpha)+I_1(2\surd\alpha)\right]. \quad (391)[6]$$

Westgren has made several series of counts with both of his experimental arrangements. We give below a sample extract from one of his sequences:

$$\begin{matrix} 2\ 1\ 1\ 1\ 1\ 1\ 0\ 2\ 2\ 1\ 1\ 1\ 2\ 3\ 2\ 3\ 0\ 0\ 0\ 0\ 0\ 0\ 1\ 1\ 0\ 0\ 1\ 0\ 1\ 1\ 1\ 2\ 2\ 3\ 3\ 4\ 5 \\ 3\ 4\ 2\ 2\ 1\ 2\ 1\ 3\ 2\ 0\ 2\ 2\ 1\ 0\ 2\ 2\ 2\ 1\ 2\ 3\ 2\ 2\ 2\ 3\ 2\ 2\ 2\ 2\ 2\ 2\ 1\ 3\ 3\ 4\ 2\ 2 \end{matrix} \quad (392)$$

The foregoing counts were obtained with the first of the two experimental arrangements described with the following values for the various physical parameters:

$$\begin{aligned} &h=6.56\mu; && D=3.95\times10^{-8}; \\ &\tau=1.39\text{ sec.}; && a=49.5\mu\mu; \\ &T=290.0^\circ\text{K}; && \nu=1.428. \end{aligned} \quad (393)$$

First of all, it is of interest to see how well the Poisson distribution (333) represents the observed frequencies of occurrence of the different values of n. Table III shows this comparison for the sequence of which (392) is an extract. It is seen that the representation is satisfactory. Also, the observed mean square deviation for this sequence is 1.35 while the value theoretically predicted is ν which is 1.43; again the agreement is satisfactory.

Turning next to questions relating to probability after-effects we may first note that each of the observed sequences can be used for several comparisons. For, by suitably selecting from a given sequence of sufficient length we can derive others with intervals between consecutive observations which are integral multiples of that characterizing the original sequence. Thus, by considering only the alternate numbers we obtain a new sequence in which the interval τ between two observations is twice that in the original sequence.

As we have already remarked, for any given sequence, we can compute theoretical values of P in terms of the physical parameters of the problem according to Eq. (388) or (391) depending on the experimental arrangement used. For the same sequences, we can also, using Eq. (382), derive values of P from the observed counts

TABLE III. The Poisson distribution for $W(n)$. $\nu=1.428$.

$n=$	0	1	2	3	4	5	6	7
$W(n)_{\text{obs}}$	381	568	357	175	67	28	5	2
$W(n)_{\text{calc}}$	380	542	384	184	66	19	5	2

[6] The functions $e^{-x}I_{0,1}(x)$ are tabulated in Watson's *Bessel functions* (Cambridge, 1922), pp. 698–713.

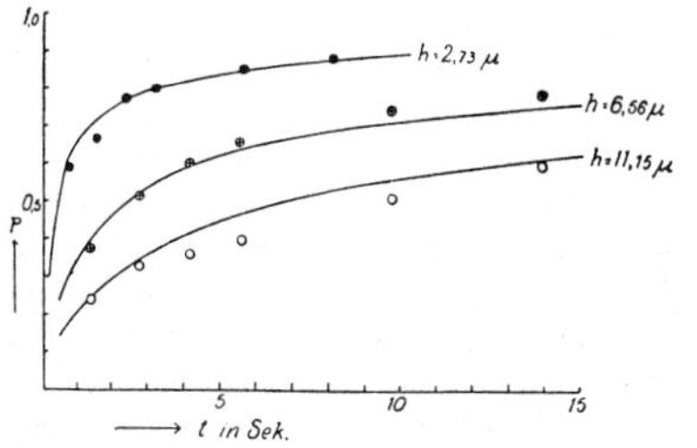

FIG. 5.

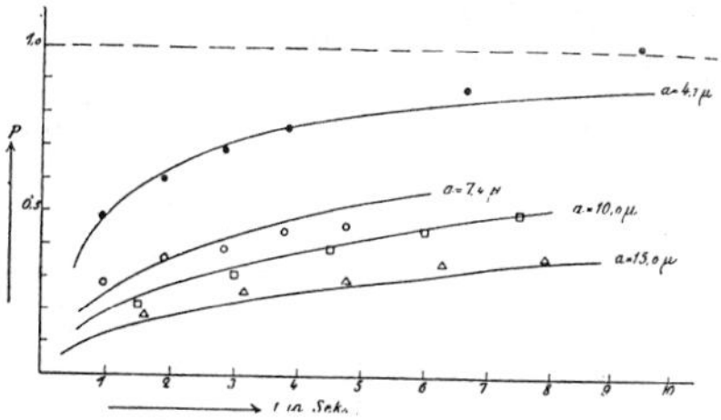

FIG. 6.

from the respective values of the mean square differences $\langle\Delta^2\rangle_{\mathrm{Av}}$. In Tables IV and V we have made, following Westgren, the comparison between the values P derived in this manner for two typical cases. The agreement is satisfactory. The confirmation of the theory is shown in a particularly striking manner in Figs. 5 and 6 where a comparison is made between the observed and the theoretical values of P in its dependence on τ for different values of h (or r_0).

It is now seen that an analysis of the data of colloid statistics actually provides us with a means of determining the Avogadro's constant N. For, from the mean square difference $\langle\Delta^2\rangle_{\mathrm{Av}}$ and the mean value of n (namely ν) we can determine P. On the other hand, according to Eqs. (380) and (381) P is determined by the geometry of the volume v *only, if* the unit of length is chosen to be $2(D\tau)^{\frac{1}{2}}$. Hence, from the empirically determined value of P we can deduce a value for this unit of length. In other words, a determination of the diffusion coefficient D is possible. But [cf. Eq. (378)]

$$D = kT/6\pi a\eta = (R/N)(T/6\pi a\eta), \quad (394)$$

where R is the gas constant and N the Avogadro's number. Thus N can be determined. With the second of his two arrangements Westgren has used this method to determine N. As a mean of 50 determinations he finds $N = 6.09 \times 10^{23}$ with a probable error of 5 percent; this is in very satisfactory agreement with other independent determinations.

TABLE IV. Comparison of the probability after-effect factor P derived from Eq. (388) and the experimental arrangement of Fig. 3 (Westgren). $h = 6.56\mu$; $a = 49.5\mu\mu$; $T = 290.0°$K; $D = 3.95 \times 10^{-8}$; $\nu = 1.428$.

τ(sec.)	$\langle\Delta^2\rangle_{\mathrm{Av}}$	P_{obs}	P_{calc}
1.39	1.068	0.374	0.394
2.78	1.452	0.513	0.517
4.17	1.699	0.600	0.587
5.56	1.859	0.656	0.634
9.73	2.125	0.744	0.713
13.90	2.265	0.793	0.760

TABLE V. Comparison of the probability after-effect factor P derived from Eq. (391) and the experimental arrangement of Fig. 4 (Westgren). $r_0 = 10.0\mu$; $a = 63.5\mu\mu$; $T = 290.1°$K; $D = 3.024 \times 10^{-8}$; $\nu = 1.933$.

τ(sec.)	$\langle\Delta^2\rangle_{\mathrm{Av}}$	P_{obs}	P_{calc}
1.50	0.836	0.217	0.238
3.00	1.200	0.310	0.332
4.50	1.512	0.391	0.401
6.00	1.718	0.444	0.456
7.50	1.939	0.502	0.503

Turning next to the frequency of occurrence $H(n, m)$ of the pair of numbers (n, m) in a given sequence, we can predict this quantity according to Eq. (383); these predicted values can again be compared with those deduced directly from the counts. Such a comparison has also been made by Westgren whose results we give in Table VI.

Finally, we shall consider the experimental basis for the formulae (368) and (376) for the mean life and the average time of recurrence of a state of fluctuation. Using the counts of Svedberg, Smoluchowski has made a comparison between the values of T_n and Θ_n derived empirically from these counts and those predicted by Eqs. (368) and (376). The results of this comparison are shown in Table VII.

The long average times of recurrence for the states of large n are to be particularly noted (see §4 below). These long times are, however, a direct consequence of the "improbable"

TABLE VI. The observed and the theoretical frequencies of occurrence of the pairs (n, m) in a given sequence ($\nu=1.428$; $P=0.374$). [In each case the top figure gives the observed number while the bottom figure (italicized), the number to be expected on the basis of Eq. (383).]

n	$m=0$	1	2	3	4	5	6
0	210 *221*	126 *119*	35 *32*	7 *6*	0 *1*	1 —	— —
1	134 *119*	281 *262*	117 *122*	29 *31*	1 *5*	1 *1*	— —
2	27 *32*	138 *122*	108 *149*	63 *63*	16 *15*	3 *3*	— —
3	10 *6*	20 *31*	76 *63*	38 *56*	24 *22*	6 *5*	0 *1*
4	2 *1*	2 *5*	14 *15*	22 *22*	13 *15*	11 *6*	3 *2*
5	— —	0 *1*	2 *3*	10 *5*	10 *6*	1 *3*	3 *1*

nature of these states. For, according to Eq. (376)

$$\Theta_n \sim (\tau/W(n)) = \tau(e^{\nu} n!/\nu^n) \quad (n \gg \nu). \quad (395)$$

which increases extremely rapidly for large values of n. For example, the number 7 was recorded only once in Svedberg's entire sequence of 518 counts; but the average time of recurrence for this state is 1105τ. Again, the number 17 (for instance) was never observed by Svedberg; and this is also understandable in view of the average time of recurrence for this state which is $\Theta_{17} \sim 10^{13}\tau$!

In concluding this discussion of the experimental verification of Smoluchowski's theory, we may remark on the inner relationships that have been disclosed to exist between the phenomena of Brownian motion, diffusion, and fluctuations in molecular concentration. But what is perhaps of even greater significance is that we have here the first example of a case in which it has been possible to follow in all its details, both theoretically and experimentally, the transition between the macroscopically irreversible nature of diffusion and the microscopically reversible nature of molecular fluctions. (These matters are further touched upon in §§3 and 4 below.)

3. Probability After-Effects for Continuous Observation

The theory of density fluctuations as developed in §1 is valid whenever the physical circumstances of the problem will permit us to introduce the probability after-effect factor P. It will be recalled that this factor $P(\tau)$ is defined as the probability that a particle, initially, somewhere inside a given element of volume will emerge from it before the elapse of a time τ. And, as we have seen in §1, we can express all the significant facts related to the phenomenon of the speed of fluctuations in terms of this single factor $P(\tau)$. But the theory as developed in §1 applies only when τ is finite, i.e., for the case of intermittent observations. We shall now show how this theory can be generalized to include the case of continuous observations.

First of all, it is clear that we should expect

$$P(\tau) \to 0 \quad \text{as} \quad \tau \to 0. \quad (396)$$

Hence, according to Eq. (364),

$$W(n;n) = e^{-\nu P}(1-P)^n + O(P^2) \quad (\tau \to 0;\ P \to 0), \quad (397)$$

or

$$W(n;n) = 1 - (n+\nu)P(\tau) + O(P^2) \quad (\tau \to 0;\ P \to 0). \quad (398)$$

From this expression for $W(n;n)$ we can derive a formula for the probability $\phi_n(t)\Delta t$ that the state n will continue to be under observation for a time t and that during t and $t+\Delta t$ there will occur a transition to a state different from n. For this purpose, we divide the interval $(0, t)$ into a very large number of subintervals of duration Δt. Then, from the definition of $\phi_n(t)\Delta t$ it follows that

$$\phi_n(t)\Delta t = [W(n;n)]^{t/\Delta t}[1 - W(n;n)], \quad (399)$$

or, using Eq. (398),

$$\phi_n(t)\Delta t = [1 - (n+\nu)P(\Delta t) + O(P^2)]^{t/\Delta t} \times (n+\nu)P(\Delta t). \quad (400)$$

This last equation suggests that to obtain consistent results it would be necessary that

$$P(\Delta t) = O(\Delta t) \quad (\Delta t \to 0). \quad (401)$$

On general physical grounds, we may expect that this would in fact be the case. But it should not be concluded that Eq. (401) will be valid for *any* arbitrary idealization of the physical problem. For example, it is *not* true that $P(\Delta t)$ is $O(\Delta t)$ for the case of Brownian motions *idealized* as a problem in pure diffusion as we have done in §2. For, according to Eq. (379)

$$\langle|\Delta \mathbf{r}|^2\rangle_{\text{Av}}=6D\Delta t; \tag{402}$$

and hence, for P defined as in Eq. (380)

$$P=O[(\Delta t)^{\frac{1}{2}}] \quad (\Delta t\rightarrow 0), \tag{403}$$

contrary to Eq. (401). However, the reason for this disagreement is that the reduction of the problem of Brownian motions to one in diffusion can be achieved only when the intervals of time we are interested in are long compared to the time of relaxation β^{-1}. When this ceases to be the case, as in the present context, Eq. (379) is no longer true and we should strictly use the general distribution derived in Chapter II, §2 [see Eq. (171)]. And, according to Eq. (175)

$$\langle|\Delta \mathbf{r}|^2\rangle_{\text{Av}}=|\mathbf{u}_0|^2(\Delta t)^2, \quad (\Delta t\rightarrow 0). \tag{404}$$

On the basis of Eq. (404) we shall naturally be led to a formula for P consistent with (401) [see Eq. (413) below]. We shall therefore assume that

$$P(\Delta t)=P_0\Delta t+O(\Delta t^2) \quad (\Delta t\rightarrow 0), \tag{405}$$

where P_0 is a constant.

Combining Eqs. (400) and (405) we have

$$\phi_n(t)\Delta t=[1-(n+\nu)P_0\Delta t+O(\Delta t^2)]^{t/\Delta t} \times(n+\nu)P_0\Delta t, \tag{406}$$

or, passing to the limit $\Delta t=0$ we obtain

$$\phi_n(t)dt=\exp[-(n+\nu)P_0t](n+\nu)P_0dt. \tag{407}$$

Equation (407) expresses *a law of decay of a state of fluctuation* quite analogous to the law of decay of radioactive substances.

According to Eq. (407), the mean life, T_n, of the state n for continuous observation can be defined by

$$T_n=\int_0^\infty t\phi_n(t)dt; \tag{408}$$

in other words

$$T_n=1/(n+\nu)P_0. \tag{409}$$

Equation (409) is our present analogue of the formula (368) valid for intermittent observations.

Again, as in §1, we can also define the average time of recurrence of a state of fluctuation for continuous observation. This can be done by introducing the probability $W(Nn; Nn)$ and proceeding exactly as in the discussion of T_n. However, without going into details, it is evident that the relation (377) between T_n and Θ_n must continue to be valid, also for the case of continuous observation. Hence

$$\Theta_n=\frac{1}{(n+\nu)P_0}\frac{1-W(n)}{W(n)}. \tag{410}$$

We shall now derive for the case of Brownian motions, an explicit formula for P_0 which we formally introduced in Eq. (405). As we have already remarked, when dealing with continuous observation, the idealization of the phenomenon of Brownian motion as pure diffusion is not tenable. Instead, we should base our discussion on the exact distribution function $W(\mathbf{r}, t; \mathbf{r}_0, \mathbf{u}_0)$ given by Eq. (171) and which is valid also for times of the order of the time of relaxation β^{-1}. However, since we are only interested in $P(\Delta t)$ for $\Delta t\rightarrow 0$ it would clearly be sufficient to consider the limiting form of the exact distribution $W(\mathbf{r}, \mathbf{u}, t; \mathbf{r}_0, \mathbf{u}_0)$ as $t\rightarrow 0$. On the other hand according to Eqs. (170)–(175) it follows that as $t\rightarrow 0$ we can regard the particles as describing linear trajectories with a Maxwellian distribution of the velocities. Hence, in our present context, $P(\Delta t)$ represents the probability that a particle initially inside a given element of volume v (with uniform probability) and with a velocity distribution governed by Maxwell's law will emerge from v before a time Δt. It is clear that formally, this is the same as the number of molecules striking the inner surface of the element of volume considered in a time Δt when the molecular concentration is $1/v$.

TABLE VII. The mean life T_n and the average time of recurrence Θ_n ($P=0.726$; $\nu=1.55$). (T_n and Θ_n are expressed in units of τ.)

n	T_n(obs.)	T_n(calc.)	Θ_n(obs.)	Θ_n(calc.)
0	1.67	1.47	6.08	5.54
1	1.50	1.55	3.13	3.16
2	1.37	1.38	4.11	4.05
3	1.25	1.23	7.85	8.07
4	1.23	1.12	18.6	20.9

Now, according to calculations familiar in the kinetic theory of gases, the number of molecules with velocities between $|\mathbf{u}|$ and $|\mathbf{u}|+d|\mathbf{u}|$ which strike unit area of any solid surface per unit time and in a direction with a solid angle $d\Omega$ at an angle ϑ with the normal to the surface is given by

$$N(m/2\pi kT)^{\frac{3}{2}} \exp\left(-m|\mathbf{u}|^2/2kT\right) \times |\mathbf{u}|^3 \cos\vartheta d\Omega d|\mathbf{u}|, \quad (411)$$

where N denotes the molecular concentration. Hence,

$$P(\Delta t) = \Delta t \frac{\sigma}{v}\left(\frac{m}{2\pi kT}\right)^{\frac{3}{2}} \int_0^\infty \int_0^\pi \exp\left(-m|\mathbf{u}|^2/2kT\right) \times |\mathbf{u}|^3 \cos\vartheta d\Omega d|\mathbf{u}|, \quad (412)$$

where σ is the total surface area of the element of volume v. On evaluating the integrals in Eq. (412) we find that

$$P(\Delta t) = (\sigma/v)(kT/2\pi m)^{\frac{1}{2}}\Delta t. \quad (413)$$

Comparing this with Eq. (405) we conclude that for the case under consideration

$$P_0 = (\sigma/v)(kT/2\pi m)^{\frac{1}{2}}. \quad (414)$$

The formulae (409) and (410) for the mean life and the average time of recurrence now take the forms

$$T_n = (v/\sigma(n+\nu))(2\pi m/kT)^{\frac{1}{2}}, \quad (415)$$

and

$$\Theta_n = (v/\sigma(n+\nu))(2\pi m/kT)^{\frac{1}{2}} \times ([1-W(n)]/W(n)). \quad (416)$$

The case of greatest interest arises when the average number of particles, ν, contained in v is a very large number and the values of n considered are relatively close to ν. Then, the Poisson distribution $W(n)$ simplifies to (see Appendix III)

$$W(n) = [1/(2\pi\nu)^{\frac{1}{2}}] \exp\left[-(n-\nu)^2/2\nu\right]. \quad (417)$$

On this approximation, Eq. (416) becomes

$$\Theta_n \simeq \pi \frac{v}{\sigma}\left(\frac{m}{\nu kT}\right)^{\frac{1}{2}} \exp\left[(n-\nu)^2/2\nu\right]. \quad (418)$$

As an illustration of Eq. (418) we shall consider, following Smoluchowski, the average time of recurrence of a state of fluctuation in which the molecular concentration of oxygen in a sphere of air of radius a will differ from the average value by 1 percent. Table VIII gives Θ_n for different values of a.

TABLE VIII. The average time of recurrence of a state of fluctuation in which the molecular concentration in a sphere of air of radius a will differ from the average value by 1 percent. $T=300°K$; $\nu=3\times 10^{19}\times(4\pi a^3/3)$.

a(cm)	1	5×10^{-5}	3×10^{-5}	2.5×10^{-5}	1×10^{-5}
Θ(sec.)	$10^{10^{14}}$	10^{68}	10^{6}	1	10^{-11}

It is seen from Table VIII that under normal conditions, for volumes which are on the edge of visual perception even appreciable fluctuations in the molecular concentrations require such colossal average times of recurrence, that for all practical purposes the phenomenon of diffusion can be regarded as an irreversible process. On the other hand, for volumes which are just on the limit of microscopic vision, fluctuations in concentrations occur to such an extent and with such frequency that there can no longer be any question of irreversibility: under such conditions the notion of diffusion very largely loses its common meaning. For example, it would scarcely occur to one to illustrate the phenomenon of diffusion by the experiments of Svedberg and Westgren on colloid statistics though it is in fact true that *on the average* the results are in perfect accord with the principles of macroscopic diffusion [as is illustrated, for example, by Eq. (356) for $\langle\Delta_n\rangle_{Av}$]. We shall return to these questions in the following section.

4. On the Reversibility of Thermodynamically Irreversible Processes, the Recurrence of Improbable States, and the Limits of Validity of the Second Law of Thermodynamics

If we formulate the second law of thermodynamics in any of its conventional forms, as, for example, that "heat cannot of itself be transferred from a colder to a hotter body" or, that "arbitrarily near to any given state there exist states which are inaccessible to the initial state by adiabatic processes" (Caratheodory), or that "the entropy of a closed system must never decrease," we, at once, get into contradiction with the kinetic molecular theory which demands the essential reversibility of all processes. Consequently, from the side of "dogmatic" thermodynamics two principal objections have been raised in the form of paradoxes and which are held to vitiate the entire outlook of the kinetic theory and statistical mechanics. We first state the two paradoxes.

(i) Loschmidt's Reversibility Paradox

Loschmidt first drew attention to the fact that in view of the essential symmetry of the laws of mechanics to the past and the future, all molecular processes must be reversible from the point of view of statistical mechanics. This is in apparent contradiction with the point of view held in thermodynamics that certain processes are irreversible.

(ii) Zermelo's Recurrence Paradox

There is a theorem in dynamics due to Poincaré which states that *in a system of material particles under the influence of forces which depend only on the spatial coordinates, a given initial state*[7] *must, in general, recur, not exactly, but to any desired degree of accuracy, infinitely often, provided the system always remains in the finite part of the phase space.* (For a proof of this theorem see Appendix V.) In other words, the trajectory described by the representative point in the phase space has a "quasi-periodic" character in the sense that after a finite interval of time (which can be specified) the system will return to the initial state to any desired degree of accuracy. Basing on this theorem of Poincaré, Zermelo has argued that the notion of irreversibility fundamental to macroscopic thermodynamics is incompatible with the standpoint of the kinetic theory.

As is well known, Boltzmann has tried to resolve these paradoxes of Loschmidt and Zermelo by probability considerations of a general nature. Thus, on the strength of certain rough estimates (see Appendix VI), Boltzmann concludes that the period of one of Poincaré's cycles is so enormously long, even for a cubic centimeter of gas, that the recurrence of an initially improbable state (i.e., the reversal to a state of lower entropy) while not strictly impossible, is yet so highly improbable that during the times normally available for observation, the chance of witnessing a thermodynamically irreversible process is *extremely* small.

Though Boltzmann's arguments and conclusions are fundamentally sound there are certain unsatisfactory features in basing on the period of a Poincaré cycle. For one thing, the period of such a cycle depends on how *nearly* we (arbitrarily) require the initial state to recur. Again, Poincaré's theorem refers to the return of the representative point in the $6N$-dimensional phase space (N denoting the number of particles in the system). Actually, in practice, we should treat two states of a gas as macroscopically distinct only if the numbers of molecules (considered indistinguishable) in the various limits of positions and velocities are different. Then, during a Poincaré cycle, the different macroscopically distinguishable states of the system will approximately recur a great many times. These recurrences of the different macroscopically distinct states, during a given Poincaré cycle, will be distributed very unequally among the states: thus, most of the recurrences will occur for the states of the system which are very close to what would be described as the thermodynamically "*normal state.*" Moreover, it can also happen that during such a cycle, states deviating by arbitrarily large amounts from the normal state are assumed by the system. In other words, during a Poincaré cycle we shall pass through many improbable states and indeed with equal frequency both in the directions of increasing and decreasing entropy.

Thus, while we may accept Boltzmann's point of view as fundamentally correct, it would clearly add to our understanding of the whole problem if we can explicitly demonstrate in a given instance how in spite of the essential reversibility of all molecular phenomena, we nevertheless get the impression of irreversibility.

Now, as we have already remarked in the preceding sections, Smoluchowski's theory of fluctuations in molecular concentrations allows us to bridge the gap between the regions of the

[7] This is defined by the positions and the velocities of all the particles, i.e., by the representative point in the phase space.

macroscopically irreversible diffusion and the microscopically reversible fluctuations. Consequently, a further discussion of this problem will enable us to follow explicitly how in this particular instance the Loschmidt and the Zermelo paradoxes resolve themselves.

(*a*) *The resolution of Loschmidt's paradox.*—Using Eqs. (333), (344), and (345) we readily verify that

$$H(n, n+k) = W(n)W(n; n+k)$$
$$= W(n+k)W(n+k; n) = H(n+k, n). \quad (419)$$

The quantity on the left-hand side in the foregoing equation represents the frequency of occurrence of the numbers n and $n+k$ on two successive occasions in a long sequence of observations; similarly, the quantity on the right-hand side gives the frequency of occurrence of the pair $(n+k, n)$. It therefore follows that under equilibrium conditions, the probability, that in a given length of time we observe a transition from the state n to the state m is equal to the probability that (in an equal length of time) we observe a transition from the state m to the state n. It is precisely the symmetry between the past and the future which guarantees this equality between $H(n; m)$ and $H(m; n)$. A glance at Table VI shows that this is amply confirmed by observations. [It may be further noted that, in accordance with Eq. (419) the numbers in italics on the opposite sides of the principal diagonal are equal.] All this, is, of course, in entire agreement with Loschmidt's requirements.

On the other hand, it is also evident from Table VI, that after a relatively large number like 5, 6, or 7 a number much smaller, generally follows; in other words, the probability that a number $n(\gg\nu)$ will further increase on the next observation is very small indeed. This circumstance illustrates how molecular concentrations differing appreciably from the average value will *almost* always tend to change in the direction indicated on the macroscopic notions concerning diffusion [cf. Eq. (356)]. This corresponds exactly to one of Boltzmann's statements that the negative entropy curve almost always decreases from any point. However this may be, in course of time, an abnormal initial state will again recur as a consequence of fluctuations, and we shall now see how in spite of this possibility for recurrence, the *apparently* irreversible nature of the phenomenon comes into being.

(*b*) *The resolution of Zermelo's paradox.*—Let us first consider the case of intermittent observations. As we have already remarked in §2, the number 17 never occurred in one of Svedberg's sequences for which ν had the value 1.55. But the average time of recurrence for this state [according to Eq. (376)] is $10^{13}\tau$; and since $\tau = 1/39$ min., for the sequence considered, $\Theta \sim 500{,}000$ years. Hence, the diffusion from the state $n=17$ will have all the *appearances* of an irreversible process simply because the average time of recurrence is so very long compared to the times during which the system is under observation.

Turning next to the case of continuous observations, we shall return to the example considered in §3. As we have seen (cf. Table VIII) the average time of recurrence of a state in which the number of molecules of oxygen contained in a sphere of radius $a \geqslant 5\times10^{-5}$ cm (and $T=300°$K and $\nu = 3\times10^{19}$ cm^{-3}) will differ from the average value by 1 percent is very long indeed ($\Theta > 10^{68}$ seconds). The factor which is principally responsible for these large values for Θ is the exponential factor in Eq. (418). Accordingly, we may say, very roughly, that *the second law of thermodynamics is valid only for those diffusion processes in which the equalization of molecular concentrations which take place are by amounts appreciably greater than the root mean square relative fluctuation* (namely, $[\langle|n-\nu|^2\rangle_{Av}/\nu^2]^{\frac{1}{2}} = \nu^{-\frac{1}{2}}$). We have thus completely reconciled (at any rate, for the processes under discussion) the notion of irreversibility which is at the base of thermodynamics and the essential reversibility of all molecular phenomena demanded by statistical mechanics. This reconciliation has become possible only because we have been able to specify the limits of validity of the second law.

Quite generally, we may conclude with Smoluchowski that *a process appears irreversible* (*or reversible*) *according as whether the initial state is characterized by a long* (*or short*) *average time of recurrence compared to the times during which the system is under observation.*

5. The Effect of Gravity on the Brownian Motion: The Phenomenon of Sedimentation

The study of the effect of gravity on the Brownian motion provides an interesting illustration of the use to which Smoluchowski's equation [Eq. (312)]

$$(\partial w/\partial t) = \mathrm{div}_r\,(q\beta^{-2}\,\mathrm{grad}_r\, w - \mathbf{K}\beta^{-1}w) \tag{420}$$

can be put. In Eq. (420) **K** represents the acceleration caused by the external field of force. If the external field is that due to gravity, we can write

$$K_x = 0;\quad K_y = 0;\quad K_z = -(1-(\rho_0/\rho))g, \tag{421}$$

provided the coordinate system has been so chosen that the z axis is in the vertical direction. In Eq. (421), g denotes the value of gravity, ρ the density of the Brownian particle and $\rho_0(\leqslant \rho)$ that of the surrounding fluid. Hence, for the case (421), Eq. (420) becomes

$$(\partial w/\partial t) = (q/\beta^2)\nabla^2 w + (1-(\rho_0/\rho))(g/\beta)(\partial w/\partial z). \tag{422}$$

It is seen that Eq. (422) is of the same general form as Eq. (126). Accordingly, we can interpret the phenomenon described by Eq. (422) as a process of diffusion in which the number of particles crossing elements of area normal to x, y, and z directions, per unit area and per unit time, are given, respectively, by [cf. Eq. (127)]

$$-D(\partial w/\partial x),\quad -D(\partial w/\partial y), \tag{423}$$

and

$$-D(\partial w/\partial z) - cw, \tag{424}$$

where

$$D = (q/\beta^2) = (kT/m\beta);\quad c = (1-(\rho_0/\rho))(g/\beta). \tag{425}$$

Thus, while the diffusion in the (x, y) plane takes exactly as in the field free case, the situation in the z direction is modified. If we, therefore, limit ourselves to considering only the distribution in the z direction, of particles uniformly distributed in the (x, y) plane, the appropriate differential equation is

$$\frac{\partial w}{\partial t} = D\frac{\partial^2 w}{\partial z^2} + c\frac{\partial w}{\partial z}. \tag{426}$$

Let us now suppose that the particle is initially at a height z_0 measured from the bottom of the vessel containing the solution. Then, the probability of occurrence of the various values of z at later times will be governed by the solution of Eq. (426) which satisfies the boundary conditions

$$\left.\begin{array}{l} w \to \delta(z - z_0) \quad \text{as} \quad t \to 0, \\ D(\partial w/\partial z) + cw = 0 \quad \text{at} \quad z = 0 \quad \text{for all} \quad t > 0. \end{array}\right\} \tag{427}$$

The second of two foregoing boundary conditions arises from the requirement that no particle shall cross the plane $z=0$ representing the bottom of the vessel [cf. Eq. (424)].

To obtain the solution of Eq. (426) satisfying the boundary conditions (427), we first introduce the following transformation of the variable [cf. Eq. (128)]

$$w = U(z, t)\exp\left[-\frac{c}{2D}(z - z_0) - \frac{c^2}{4D}t\right]. \tag{428}$$

Equation (426) reduces to the standard form

$$(\partial U/\partial t) = D(\partial^2 U/\partial z^2) \tag{429}$$

while the boundary conditions (427) become

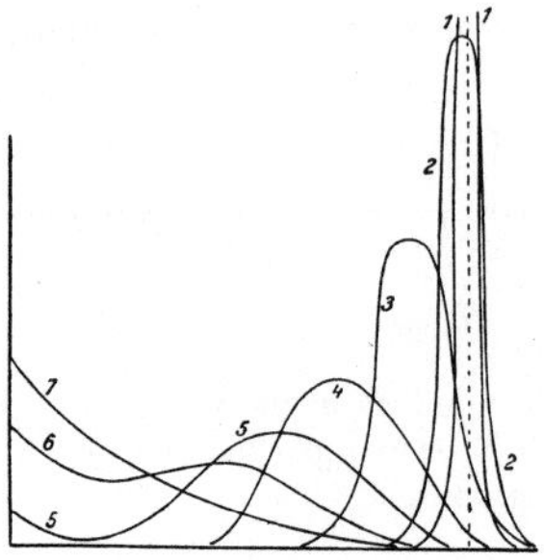

FIG. 7.

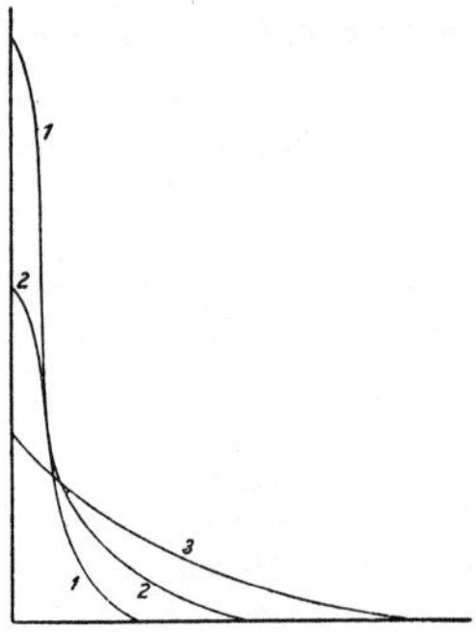

FIG. 8.

$$\left.\begin{aligned}&U\to\delta(z-z_0)\quad\text{as}\quad t\to 0,\\&D(\partial U/\partial z)+(1/2)cU=0\quad\text{at}\quad z=0\quad\text{for all}\quad t>0.\end{aligned}\right\}\tag{430}$$

Solving Eq. (429) with boundary conditions of the form (430) is a standard problem in the theory of heat conduction. We have

$$U=\frac{1}{2(\pi Dt)^{\frac{1}{2}}}\{\exp\left[-(z-z_0)^2/4Dt\right]+\exp\left[-(z+z_0)^2/4Dt\right]\}$$
$$+\frac{c}{2D(\pi Dt)^{\frac{1}{2}}}\int_{z_0}^{\infty}\exp\left[-\frac{(\alpha+z)^2}{4Dt}+\frac{c(\alpha-z_0)}{2D}\right]d\alpha.\tag{431}$$

After some elementary transformations, Eq. (431) takes the form

$$U=\frac{1}{2(\pi Dt)^{\frac{1}{2}}}\{\exp\left[-(z-z_0)^2/4Dt\right]+\exp\left[-(z+z_0)^2/4Dt\right]\}$$
$$+\frac{c}{D\sqrt{\pi}}\exp\left[\frac{c^2t}{4D}-\frac{c(z+z_0)}{2D}\right]\int_{\frac{z+z_0-ct}{2(Dt)^{\frac{1}{2}}}}^{\infty}\exp(-x^2)dx.\tag{432}$$

Returning to the variable w we have [cf. Eq. (428)]

$$w(t,z;z_0)=\frac{1}{2(\pi Dt)^{\frac{1}{2}}}\{\exp\left[-(z-z_0)^2/4Dt\right]+\exp\left[-(z+z_0)^2/4Dt\right]\}$$
$$\times\exp\left[-\frac{c}{2D}(z-z_0)-\frac{c^2}{4D}t\right]+\frac{c}{D\sqrt{\pi}}e^{-cz/D}\int_{\frac{z+z_0-ct}{2(Dt)^{\frac{1}{2}}}}^{\infty}\exp(-x^2)dx\tag{433}$$

which is the required solution. In Fig. 7 we have illustrated according to Eq. (433) the distributions $w(z, t; z_0)$ for a given value of z_0 and various values of t.

If we suppose that at time $t=0$ we have a large number of particles distributed uniformly in the plane $z=z_0$ then in the first instance diffusion takes place as in the field free case (curves 1 and 2). However, gravity makes itself felt very soon (curves 3, 4, and 5) and the maximum begins to be displaced to lower values of z with the velocity c; at the same time, the maximum becomes flatter on account of the random motions experienced by the particles. Once the probability of finding particles near enough to the bottom of the vessel becomes appreciable, the curves again begin to rise upwards (curves 5 and 6) on account of the reflection which the particles suffer at $z=0$; and, finally as $t\to\infty$ we obtain the equilibrium distribution

$$w(z,\ \infty\ ;\ z_0)=(c/D)e^{-cz/D}. \tag{434}$$

Since [cf. Eq. (425)]

$$(c/D)=(1-(\rho_0/\rho))(mg/kT), \tag{435}$$

we see that the equilibrium distribution (434) represents simply the law of isothermal atmospheres in its standard form.

The example we have just considered provides a further illustration of a case to which the conventional notions concerning entropy and the second law of thermodynamics cannot be applied. For the state of maximum entropy for the system consisting of the Brownian particles and the surrounding fluid, is that in which all the particles are at $z=0$; and, on strict thermodynamical principles we should conclude that with the continued operation of dissipative forces like dynamical friction, the state of maximum entropy will be attained. But according to Eq. (434), as $t\to\infty$ though the state of maximum entropy $z=0$ has the maximum probability, it is *not* true that the average value of the height at which the particles will be found is also zero. Actually, for the equilibrium distribution (434), we have

$$\langle z\rangle_{\text{Av}}=(D/c)=(kT/mg)[\rho/(\rho-\rho_0)], \tag{436}$$

which is the height of the equivalent homogeneous atmosphere. Moreover, even if the particles were initially at $z=0$, they will not continue to stay there. For, setting $z_0=0$ in Eq. (433) we find that

$$w(z,\ t;\ 0)=(1/(\pi Dt)^{\frac{1}{2}})\exp\left[-(z+ct)^2/4Dt\right]+(c/D\sqrt{\pi})e^{-cz/D}\int_{\frac{z-ct}{2(Dt)^{\frac{1}{2}}}}^{\infty}\exp(-x^2)dx. \tag{437}$$

Equation (437) shows that as $t\to\infty$ we are again led to the equilibrium distribution (434) (see Fig. 8). Hence, the particles do a certain amount of mechanical work *at the expense of the internal energy of the surrounding fluid;* this is of course contrary to the strict interpretation of the second law of thermodynamics. The average work done in this manner is given by [if we use Eq. (436)]

$$\langle A\rangle_{\text{Av}}=m(1-(\rho_0/\rho))g\bar{z}=kT, \tag{438}$$

per particle. Hence, on the average there is a *decrease* in entropy of amount k per particle:

$$\langle S\rangle_{\text{Av}}=S_{\max}-Nk, \tag{439}$$

where N denotes the number of Brownian particles. However, as Smoluchowski has pointed out, this work done at the expense of the internal energy of the surrounding fluid cannot be utilized to run a heat engine with an efficiency higher than that of the Carnot cycle.

We may further note that except for values of $z\lesssim D/c$, a particle has a greater probability to descend than it has to ascend. As $z\to 0$ the converse is true. We may therefore say that the tendency for the entropy to *increase* (almost always) for particles at $z\gg D/c$ is compensated by the tendency of the entropy to *decrease* for particles very near $z=0$; so that, on the average, a steady state is maintained. Of course, we have a finite probability for particles, occasionally to ascend to very great heights; but in accordance with the conclusions of §4 we should expect that the average time of recurrence for such abnormal states must be very long indeed.

6. The Theory of Coagulation in Colloids

Smoluchowski discovered a very interesting application of the theory of Brownian motion in the phenomenon of coagulation exhibited by colloidal particles when an electrolyte is added to the solution. Smoluchowski's theory of this phenomenon is based on a suggestion of Zsigmondy that coagulation results as a consequence of each colloidal particle being surrounded (on the addition of an electrolyte) by a *sphere of influence* of a certain radius R such that the Brownian motion of a particle proceeds unaffected only so long as no other particle comes within its sphere of influence and that when the particles do come within a distance R they stick to one another to form a single unit. We are not concerned here with the physico-chemical basis for Zsigmondy's suggestion except perhaps to remark that the spheres of influence are supposed to originate in the formation of electric double layers around each particle; we are here interested only in the application of the principles of Brownian motion which is possible on the acceptance of Zsigmondy's suggestion. However, we may formulate somewhat more explicitly the problem we wish to investigate:

We imagine that initially the colloidal solution contains only single particles all similar to one another and of the same spherical size. We now suppose that at time $t=0$ an (appropriate) electrolyte is added to the solution in such a way that the resulting electrolytic concentration is uniform throughout the solution. The particles are now supposed to be all instantaneously surrounded by spheres of influence of radius R. From this instant onwards, each particle will continue to describe the original Brownian motion only so long as no other particle comes within its sphere of influence. Once two particles do approach to within this distance R they will coalesce to form a "*double particle.*" This double particle will also describe Brownian motion but at a reduced rate consequent to its increased size. This double particle will, in turn, continue to remain as such only so long as it does not come within the appropriate spheres of influence of a single or another double particle: when this happens we shall have the formation of a triple or a quadruple particle; and, so on. The continuation of this process will eventually lead to the total coagulation of all the colloidal particles into one single mass.

The problem we wish to solve is the specification of the concentrations $\nu_1, \nu_2, \nu_3, \nu_4, \cdots$, of single, double, triple, quadruple, etc., particles at time t given that at time $t=0$ there were $\nu_0(=\nu_1[0])$ single particles.

As a preliminary to the discussion of the general problem formulated in the preceding paragraph we shall first consider the following more elementary situation:

A particle, assumed fixed in space, is in a medium of infinite extent in which a number of similar Brownian particles are distributed uniformly at time $t=0$. Further, if the stationary particle is assumed to be surrounded by a sphere of influence of radius R what is the rate at which particles arrive on the sphere of radius R surrounding the fixed particle?

We shall suppose that the stationary particle is at the origin of our system of coordinates. Then, in accordance with our definition of a sphere of influence, we can replace the surface $|\mathbf{r}|=R$ by a perfect absorber [cf. I, §5, see particularly Eq. (115)]. We have therefore to seek a solution of the diffusion equation [cf. Eqs. (173) and (306)]

$$(\partial w/\partial t)=D\nabla^2 w; \quad D=(q/\beta^2)=(kT/6\pi a\eta), \tag{440}$$

which satisfies the boundary conditions

$$\left.\begin{aligned} &w\equiv\nu=\text{constant, at } t=0, \text{ for } |\mathbf{r}|>R,\\ &w\equiv 0 \text{ at } |\mathbf{r}|=R \text{ for } t>0. \end{aligned}\right\} \tag{441}$$

In the first of the two foregoing boundary conditions ν denotes the average concentration of the particles exterior to $|\mathbf{r}|=R$ at time $t=0$.

Since w can depend only on the distance r from the center, the form of the diffusion equation (440) appropriate to this case is

$$(\partial/\partial t)(rw)=D(\partial^2/\partial r^2)(rw). \tag{442}$$

The solution of this equation satisfying the boundary conditions (441) is

$$w=\nu\left[1-\frac{R}{r}+\frac{2R}{r\sqrt{\pi}}\int_0^{(r-R)/2(Dt)^{\frac{1}{2}}}\exp(-x^2)dx\right]. \tag{443}$$

From Eq. (443) it follows that the rate at which particles arrive at the surface $|\boldsymbol{r}|=R$ is given by [cf. Eq. (117)]

$$4\pi D\left(r^2\frac{\partial w}{\partial r}\right)_{r=R}=4\pi DR\nu\left(1+\frac{R}{(\pi Dt)^{\frac{1}{2}}}\right). \tag{444}$$

Equation (444) gives the rate at which particles describing Brownian motion will coalesce with a stationary particle surrounded by a sphere of influence of radius R. Suppose, now, that the particle we have assumed to be stationary is also describing Brownian motion. What is the corresponding generalization of (444)? In considering this generalization we shall not suppose that the diffusion coefficients characterizing the two particles which coalesce to form a multiple particle are necessarily the same. Under these circumstances we have clearly to deal with the *relative displacements* of the two particles; and it can be readily shown that the relative displacements between two particles describing Brownian motions independently of each other and with the diffusion coefficients D_1 and D_2 also follows the laws of Brownian motion with the diffusion coefficient $D_{12}=D_1+D_2$. For, the probability that the relative displacement of two particles, initially, together at $t=0$, lies between $\boldsymbol{r}$ and $\boldsymbol{r}+d\boldsymbol{r}$ is clearly

$$\left.\begin{aligned} W(\boldsymbol{r})d\boldsymbol{r}&=d\boldsymbol{r}\int_{-\infty}^{+\infty}W_1(\boldsymbol{r}_1)W_2(\boldsymbol{r}_1+\boldsymbol{r})d\boldsymbol{r}_1\\ &=\frac{d\boldsymbol{r}}{(4\pi D_1t)^{\frac{3}{2}}(4\pi D_2t)^{\frac{3}{2}}}\int_{-\infty}^{+\infty}\exp(-|\boldsymbol{r}_1|^2/4D_1t)\exp(-|\boldsymbol{r}_1+\boldsymbol{r}|^2/4D_2t)d\boldsymbol{r}_1\end{aligned}\right\} \tag{445}$$

or, as may be readily verified [cf. the remarks following Eq. (62)]

$$W(\boldsymbol{r})=(1/[4\pi(D_1+D_2)t]^{\frac{3}{2}})\exp(-|\boldsymbol{r}|^2/4(D_1+D_2)t). \tag{446}$$

On comparing this distribution of the relative displacements with the corresponding result for the individual displacements [see for example Eq. (172)] we conclude that the relative displacements do follow the laws of Brownian motion with the diffusion coefficient (D_1+D_2).

Thus, the required generalization of Eq. (444) is

$$4\pi(D_1+D_2)R\nu\left(1+\frac{R}{[\pi(D_1+D_2)t]^{\frac{1}{2}}}\right). \tag{447}$$

More generally, let us consider two sorts of particles with concentrations ν_i and ν_k. Let the respective diffusion coefficients be D_i and D_k. Further, let R_{ik} denote the distance to which two particles (one of each sort) must approach in order that they may coalesce to form a multiple particle. Then, the rate of formation of the multiple particles by the coagulation of the particles of the kind considered is clearly given by

$$J_{i+k}dt=4\pi D_{ik}R_{ik}\nu_i\nu_k\left(1+\frac{R_{ik}}{(\pi D_{ik}t)^{\frac{1}{2}}}\right)dt \tag{448}$$

where we have written

$$D_{ik}=D_i+D_k. \tag{449}$$

In our further discussions, we shall ignore the second term in the parenthesis on the right-hand side of Eq. (447); this implies that we restrict ourselves to time intervals $\Delta t \gg R^2/D$. In most cases of practical interest, this is justifiable as $R^2/D \sim 10^{-3}-10^{-4}$ second. With this understanding we can write

$$J_{i+k}dt \simeq 4\pi D_{ik}R_{ik}\nu_i\nu_k dt. \tag{450}$$

Using Eq. (450) we can now write down the fundamental differential equations which govern the variations of $\nu_1, \nu_2, \cdots, \nu_k, \cdots$ (of single, double, $\cdots$, k-fold, $\cdots$,) particles with time:

Thus, considering the variation of the number of k-fold particles with time, we have in analogy with the equations of chemical kinetics

$$\frac{d\nu_k}{dt} = 4\pi\Big(\tfrac{1}{2}\sum_{i+j=k}\nu_i\nu_j D_{ij}R_{ij} - \nu_k\sum_{j=1}^{\infty}\nu_j D_{kj}R_{kj}\Big) \quad (k=1, \cdots). \tag{451}$$

In this equation the first summation on the right-hand side represents the increase in ν_k due to the formation of k-fold particles by the coalescing of an i-fold and a j-fold particle (with $i+j=k$), while the second summation represents the decrease in ν_k due to the formation of $(k+j)$-fold particles in which one of the interacting particles is k-fold.

A general solution of the infinite system of Eq. (451) which will be valid under all circumstances does not seem feasible. But a special case considered by Smoluchowski appears sufficiently illustrative of the general solution.

First, concerning R_{ik}, the assumption is made that

$$R_{ik} = \tfrac{1}{2}(R_i + R_k), \tag{452}$$

where R_i and R_k are the radii of the spheres of influence of the i-fold and the k-fold particles. We can, if we choose, regard the assumption (452) as equivalent to Zsigmondy's suggestion concerning the basic cause of coagulation.

Again, according to Eq. (440), the diffusion coefficient is inversely proportional to the radius of the particle; and on the basis of experimental evidence it appears that the radii of the spheres of influence of various multiple particles are proportional to the radii of the respective particles. We therefore make the additional assumption that

$$D_iR_i = DR \quad (i=1, \cdots), \tag{453}$$

where D and R denote, respectively, the diffusion coefficient and the radius of the sphere of influence of the single particles.

Combining Eqs. (449), (452), and (453) we have

$$D_{ik}R_{ik} = \tfrac{1}{2}(D_i+D_k)(R_i+R_k) = \tfrac{1}{2}DR(R_i^{-1}+R_k^{-1})(R_i+R_k) = \tfrac{1}{2}DR(R_i+R_k)^2R_i^{-1}R_k^{-1}. \tag{454}$$

Finally, for the sake of mathematical simplicity we make the (not very plausible) assumption that

$$R_i = R_k. \tag{455}$$

Thus, with all these assumptions

$$D_{ik}R_{ik} = 2DR. \tag{456}$$

In view of (456), Eq. (451) becomes

$$\frac{d\nu_k}{dt} = 8\pi DR\Big(\tfrac{1}{2}\sum_{i+j=k}\nu_i\nu_j - \nu_k\sum_{j=1}^{\infty}\nu_j\Big) \quad (k=1, \cdots). \tag{457}$$

If we now let

$$\tau = 4\pi DRt, \tag{458}$$

Eq. (457) takes the more convenient form

$$\frac{d\nu_k}{d\tau} = \sum_{i+j=k}\nu_i\nu_j - 2\nu_k\sum_{j=1}^{\infty}\nu_j \quad (k=1, \cdots). \tag{459}$$

From Eq. (459) we readily find that

$$\left.\begin{aligned}\frac{d}{dt}\Big(\sum_{k=1}^{\infty}\nu_k\Big)&=\sum_{i=1}^{\infty}\sum_{j=1}^{\infty}\nu_i\nu_j-2\sum_{k=1}^{\infty}\sum_{j=1}^{\infty}\nu_k\nu_j,\\&=-\Big(\sum_{k=1}^{\infty}\nu_k\Big)^2,\end{aligned}\right\}\tag{460}$$

or,

$$\sum_{k=1}^{\infty}\nu_k=\frac{\nu_0}{1+\nu_0\tau},\tag{461}$$

remembering that at $t=0$, $\sum\nu_k=\nu_0$.

Using the integral (461) we can successively obtain the solutions for ν_1, ν_2, etc. Thus, considering the equation for ν_1 we have [cf. Eq. (459)]

$$d\nu_1/dt=-2\nu_1\sum_{k=1}^{\infty}\nu_k=-2\nu_1\nu_0/(1+\nu_0\tau);\tag{462}$$

in other words,

$$\nu_1=\frac{\nu_0}{(1+\nu_0\tau)^2},\tag{463}$$

again using the boundary condition that $\nu_1=\nu_0$ at $t=0$. Proceeding in this manner we can prove (by induction) that

$$\nu_k=\nu_0[(\nu_0\tau)^{k-1}/(1+\nu_0\tau)^{k+1}]\quad(k=1,\,2,\,\cdots).\tag{464}$$

In Fig. 9 we have illustrated the variations of $\sum\nu_k$, ν_1, ν_2, $\cdots$ with time. We shall not go into the details of the comparison of the predictions of this theory with the data of observations. Such comparisons have been made by Zsigmondy and others and the general conclusion is that Smoluchowski's theory gives a fairly satisfactory account of the broad features of the coagulation phenomenon.

7. The Escape of Particles over Potential Barriers

As a final illustration of the application of the principles of Brownian motion we shall consider, following Kramers, the problem of the escape of particles over potential barriers. The solution to this problem has important bearings on a variety of physical, chemical, and astronomical problems.

The situation we have in view is the following:

Limiting ourselves for the sake of simplicity to a one-dimensional problem, we consider a particle moving in a potential field $\mathfrak{V}(x)$ of the type shown in Fig. 10; more generally, we may consider an

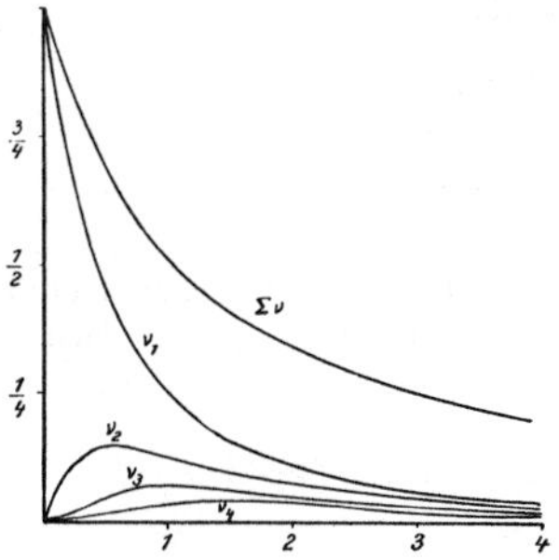

Fig. 9.

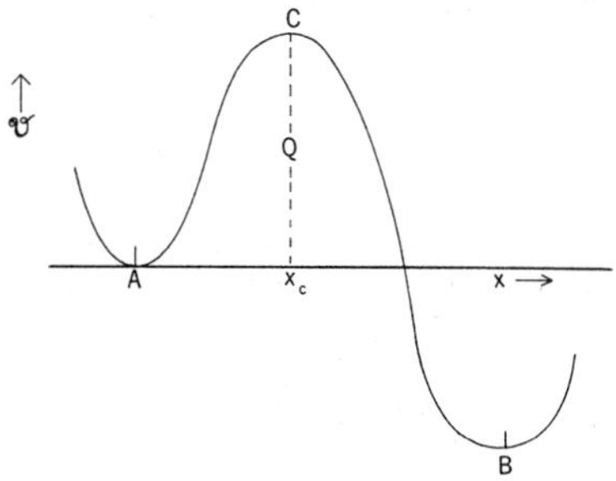

Fig. 10.

ensemble of particles moving in the potential field $\mathfrak{B}(x)$ without any mutual interference. We suppose that the particles are initially caught in the potential hole at A. The general problem we wish to solve concerns the rate at which particles will escape over the potential barrier in consequence of Brownian motion.

In the most general form, the solution to the problem formulated in the foregoing paragraph is likely to be beset with considerable difficulties. But a special case of interest arises when the height of the potential barrier is large compared to the energy of the thermal motions:

$$mQ \gg kT. \tag{465}$$

Under these circumstances, the problem can be treated as one in which the conditions are *quasi-stationary*. More specifically, we may suppose that to a high degree of accuracy a Maxwell-Boltzmann distribution obtains in the neighborhood of A. But the equilibrium distribution will not obtain for all values of x. For, by assumption, the density of particles beyond C is very small compared to the equilibrium values; and in consequence of this there will be a slow diffusion of particles (across C) tending to restore equilibrium conditions throughout. If the barrier were sufficiently high, this diffusion will take place as though stationary conditions prevailed.

Assuming first that we are interested only in time intervals that are long compared to the time of relaxation β^{-1} we can use Smoluchowski's Eq. (312). Under stationary conditions, Smoluchowski's equation predicts a current density j given by [cf. Eq. (316)]

$$\mathbf{j}\cdot\int_A^B \beta e^{m\mathfrak{B}/kT}d\mathbf{s}=(kT/m)we^{m\mathfrak{B}/kT}\Big|_B^A, \tag{466}$$

where, in the integral on the right-hand side, the path of integration of $\mathbf{s}$ from A to B is arbitrary. In our present case β is a constant and, since further we are dealing with a one-dimensional problem, we can express Eq. (466) in the form

$$j=\frac{kT}{m\beta}\frac{we^{m\mathfrak{B}(x)/kT}\Big|_B^A}{\int_A^B e^{m\mathfrak{B}(x)/kT}dx}. \tag{467}$$

Now, the number of particles ν_A in the vicinity of A can be calculated; for, in accordance with our earlier remarks we shall be justified in assuming that the Maxwell-Boltzmann distribution

$$d\nu_A=w_Ae^{-m\mathfrak{B}/kT}dx \tag{468}$$

is valid in the neighborhood of A. If we now further suppose that

$$\mathfrak{B}\simeq\tfrac{1}{2}\omega_A{}^2x^2 \quad (\omega_A=\text{constant};\ x\sim 0), \tag{469}$$

we obtain from Eq. (468)

$$\nu_A=w_A\int_{-\infty}^{+\infty}\exp\left(-m\omega_A{}^2x^2/2kT\right)dx, \tag{470}$$

where the range of integration over x has been extended from $-\infty$ to $+\infty$ in view of the fact that the main contribution to the integral for ν_A must arise only from a small region near $x=0$. Hence,

$$\nu_A=(w_A/\omega_A)(2\pi kT/m)^{\frac{1}{2}}. \tag{471}$$

Returning to Eq. (467), we can write with sufficient accuracy [cf. Eq. (469)]:

$$j\simeq\frac{kT}{m\beta}w_A\left\{\int_A^B e^{m\mathfrak{B}/kT}dx\right\}^{-1}. \tag{472}$$

In writing Eq. (472) we have assumed that the density of particles near B is very small: this is true to begin with anyway.

From Eqs. (471) and (472) we directly obtain for the rate at which a particle, initially caught in the potential hole at A, will escape over the barrier at C, the expression

$$P=\frac{j}{\nu_A}=\frac{\omega_A}{\beta}\left(\frac{kT}{2\pi m}\right)^{\frac{1}{2}}\left\{\int_A^B e^{m\mathfrak{B}/kT}dx\right\}^{-1}. \tag{473}$$

The principal contribution to the integral in the curly brackets in the foregoing equation arises from only a very small region near C [on account of the strong inequality (465)]. The value of the integral will therefore depend, very largely, only on the shape of the potential curve in the immediate neighborhood of C. If we now suppose that near $x=x_C$, $\mathfrak{B}(x)$ has a continuous curvature, we may write

$$\mathfrak{B}\simeq Q-\tfrac{1}{2}\omega_C^2(x-x_C)^2 \quad (\omega_C=\text{constant};\ x\sim x_C). \tag{474}$$

On this assumption, to a sufficient degree of accuracy we have

$$\left.\begin{aligned}\int_A^B e^{m\mathfrak{B}/kT}dx&\simeq e^{mQ/kT}\int_{-\infty}^{+\infty}\exp\left[-m\omega_C^2(x-x_C)^2/2kT\right]dx,\\ &=e^{mQ/kT}(2\pi kT/m\omega_C^2)^{\frac{1}{2}}.\end{aligned}\right\} \tag{475}$$

Combining Eqs. (473) and (475) we obtain

$$P=(\omega_A\omega_C/2\pi\beta)e^{-mQ/kT}, \tag{476}$$

which gives the probability, per unit time, that a particle originally in the potential hole at A, will escape to B crossing the barrier at C.

The formula (476) has been derived on the basis of Eq. (467) and this implies, as we have already remarked, that we are ignoring effects which take place in intervals of the order β^{-1}. Alternatively, we may say that the validity of Eq. (476) depends on how large the coefficient of dynamical friction β is: if β were sufficiently large, the formula (476) for P may be expected to provide an adequate approximation [see Eqs. (507) and (508) below]. On the other hand, if this should not be the case, we must, in accordance with our remarks in Chapter II, §4, subsection (*vi*), base our discussion of the generalized Liouville Eq. (249) in phase space; and in one dimension this equation has the form

$$\frac{\partial W}{\partial t}+u\frac{\partial W}{\partial x}+K\frac{\partial W}{\partial u}=\beta u\frac{\partial W}{\partial u}+\beta W+q\frac{\partial^2 W}{\partial u^2}, \tag{477}$$

where it may be recalled that

$$q=\beta(kT/m);\quad K=-(\partial\mathfrak{B}/\partial x). \tag{478}$$

In II, §5 we have shown that the Maxwell-Boltzmann distribution identically satisfies Eq. (249). Accordingly,

$$W=C\exp\left[-(mu^2+2m\mathfrak{B})/2kT\right], \tag{479}$$

where C is a constant, satisfies Eq. (477). However, under the conditions of our problem the equilibrium distribution (479) cannot be valid for *all* values of x; for, if it were, there would be no diffusion across the barrier at C and the conditions of the problem would not be met. On the other hand, we do expect the distribution (479) to be realized to a high degree of accuracy in the neighborhood of A. We, therefore, look for a stationary solution of Eq. (477) of the form

$$W=CF(x, u)\exp\left[-m(u^2+2\mathfrak{B})/2kT\right], \tag{480}$$

where $F(x, u)$ is very nearly unity in the neighborhood of $x=0$. Since we have further supposed that the density of particles in the region B is quite negligible, we should also require that $F(x, u)\rightarrow 0$ for values of x appreciably greater than $x=x_C$. We may express these conditions formally in the form

$$\begin{aligned} F(x, u)&\simeq 1 \quad \text{at} \quad x\sim 0,\\ F(x, u)&\simeq 0 \quad \text{for} \quad x\gg x_C. \end{aligned} \tag{481}$$

We shall now show how such a function $F(x, u)$ can be determined.

First of all it is evident that for the purposes of determining the rate of escape of particles across the barrier at C it is particularly important to determine F accurately in this region. Assuming that in the vicinity of C, $\mathfrak{B}$ has the form (474) and that stationary conditions prevail throughout, the equation for W in the neighborhood of $x=x_C$ becomes [cf. Eq. (477)]:

$$u\frac{\partial W}{\partial X}+\omega_C^2 X\frac{\partial W}{\partial u}=\beta u\frac{\partial W}{\partial u}+\beta W+q\frac{\partial^2 W}{\partial u^2}, \tag{482}$$

where for the sake of brevity, we have used

$$X=x-x_C. \tag{483}$$

According to Eqs. (474), (480), and (483) the appropriate form for W valid in the region C, is

$$W=Ce^{-mQ/kT}F(X, u)\exp\left[-m(u^2-\omega_C^2X^2)/2kT\right]. \tag{484}$$

Substituting for W according to this equation in Eq. (482), we obtain

$$u\frac{\partial F}{\partial X}+\omega_C^2 X\frac{\partial F}{\partial u}=q\frac{\partial^2 F}{\partial u^2}-\beta u\frac{\partial F}{\partial u}. \tag{485}$$

It is seen that $F=$constant satisfies this equation identically: this solution corresponds of course to the equilibrium distribution. However, the solution of Eq. (485) which we are seeking must satisfy the boundary conditions [cf. Eq. (481)]

$$\begin{aligned} F(X, u)&\rightarrow 1 \quad \text{as} \quad X\rightarrow -\infty,\\ F(X, u)&\rightarrow 0 \quad \text{as} \quad X\rightarrow +\infty. \end{aligned} \tag{486}$$

Assume for F the form

$$F\equiv F(u-aX)=F(\xi) \quad \text{(say)}, \tag{487}$$

where a is, for the present, an unspecified constant. Substituting this form of F in Eq. (485) we obtain

$$-\left[(a-\beta)u-\omega_C^2X\right]\frac{dF}{d\xi}=q\frac{d^2F}{d\xi^2}. \tag{488}$$

In order that Eq. (488) be consistent it is clearly necessary that [cf. Eq. (487)]

$$\left[\omega_C^2/(a-\beta)\right]=a; \tag{489}$$

and in this case Eq. (488) becomes

$$-(a-\beta)\xi\frac{dF}{d\xi}=q\frac{d^2F}{d\xi^2}. \tag{490}$$

Equation (490) is readily integrated to give

$$F=F_0\int^{\xi}\exp\left[-(a-\beta)\xi^2/2q\right]d\xi, \tag{491}$$

where F_0 is a constant. On the other hand, according to Eq. (489) a is the root of the equation

$$a^2-a\beta-\omega_C^2=0; \tag{492}$$

i.e.,

$$a=(\beta/2)\pm([\beta^2/4]+\omega_C^2)^{\frac{1}{2}}. \tag{493}$$

If we choose for a the *positive root*, then

$$a-\beta=([\beta^2/4]+\omega_C^2)^{\frac{1}{2}}-(\beta/2) \tag{494}$$

is also positive, and as we shall show presently, the solution (491) leads to an F which satisfies the required boundary conditions (486). For, by choosing

$$F_0=[(a-\beta)/2\pi q]^{\frac{1}{2}}, \tag{495}$$

and setting the lower limit of integration in Eq. (491) as $-\infty$ we obtain the solution

$$F=\left(\frac{a-\beta}{2\pi q}\right)^{\frac{1}{2}}\int_{-\infty}^{\xi}\exp\left[-(a-\beta)\xi^2/2q\right]d\xi, \tag{496}$$

which satisfies the conditions

$$F\to 1 \quad \text{as} \quad \xi\to+\infty; \quad F\to 0 \quad \text{as} \quad \xi\to-\infty. \tag{497}$$

On the other hand, since $\xi=u-aX$ and $a(=[(\beta/2)^2+\omega_C^2]^{\frac{1}{2}}+[\beta/2])$ is positive, $\xi\to+\infty$ or $-\infty$ is the same as $X\to-\infty$ or $+\infty$; in other words, the solution (496) for F satisfies the necessary boundary conditions (486).

Combining Eqs. (484) and (496) we have, therefore, the solution

$$W=C[(a-\beta)/2\pi q]^{\frac{1}{2}}e^{-mQ/kT}\exp\left[-m(u^2-\omega_C^2X^2)/2kT\right]\int_{-\infty}^{\xi}\exp\left[-(a-\beta)\xi^2/2q\right]d\xi. \tag{498}$$

Equation (498) is, of course, valid only in the neighborhood of C.

In the vicinity of A we have the solution [cf. Eqs. (469) and (479)]

$$W=C\exp\left[-m(u^2+\omega_A^2x^2)/2kT\right]. \tag{499}$$

Accordingly, the number of particles, ν_A, in the potential hole at A is given by

$$\left.\begin{aligned}\nu_A&\simeq C\int_{-\infty}^{+\infty}\int_{-\infty}^{+\infty}\exp\left[-m(u^2+\omega_A^2x^2)/2kT\right]dxdu,\\ &=C(2\pi kT/m\omega_A).\end{aligned}\right\} \tag{500}$$

(This equation will enable us to normalize the distribution in such a way so as to correspond to one particle in the potential hole: for this purpose we need only choose $C=m\omega_A/2\pi kT$.)

Now, the diffusion current across C is given by

$$j=\int_{-\infty}^{+\infty}W(X=0;u)udu, \tag{501}$$

or, using the solution (498), we have

$$j=C[(a-\beta)/2\pi q]^{\frac{1}{2}}e^{-mQ/kT}\int_{-\infty}^{+\infty}duu\exp\left(-mu^2/2kT\right)\int_{-\infty}^{u}d\xi\exp\left[-(a-\beta)\xi^2/2q\right]. \tag{502}$$

After an integration by parts, we find

$$j=C[(a-\beta)/2\pi q]^{\frac{1}{2}}(kT/m)e^{-mQ/kT}\int_{-\infty}^{+\infty}\exp\left\{-u^2[m/2kT+(a-\beta)/2q]\right\}du. \tag{503}$$

But [cf. Eq. (478)]

$$(m/2kT)+[(a-\beta)/2q]=(a/2q). \tag{504}$$

From Eqs. (503) and (504) we now obtain

$$j=C(kT/m)[(a-\beta)/a]^{\frac{1}{2}}e^{-mQ/kT}. \tag{505}$$

Hence, the rate of escape of particles across C is given by

$$P=(j/\nu_A)=(\omega_A/2\pi)[(a-\beta)/a]^{\frac{1}{2}}e^{-mQ/kT}, \tag{506}$$

or, substituting for a and $a-\beta$ according to Eqs. (493) and (494), we find after some elementary reductions, that

$$P=(\omega_A/2\pi\omega_C)([\beta^2/4+\omega_C^2]^{\frac{1}{2}}-[\beta/2])e^{-mQ/kT}. \tag{507}$$

If

$$\beta\gg 2\omega_C \tag{508}$$

we readily verify that our present "exact" formula for P reduces to our earlier result (476) derived on the basis of the Smoluchowski equation. But (507) now provides in addition the precise condition for the approximate validity of (476). On the other hand, for $\beta\rightarrow 0$ we have

$$P=(\omega_A/2\pi)e^{-mQ/kT} \quad (\beta\rightarrow 0). \tag{509}$$

This last formula for P valid in the limit of vanishing dynamical friction, corresponds to what is sometimes called the approximation of the *transition-state method.*

CHAPTER IV

PROBABILITY METHODS IN STELLAR DYNAMICS: THE STATISTICS OF THE GRAVITATIONAL FIELD ARISING FROM A RANDOM DISTRIBUTION OF STARS

1. Fluctuations in the Force Acting on a Star; The Outline of the Statistical Method

One of the principal problems of stellar dynamics is concerned with the analysis of the nature of the force acting on a star which is a member of a stellar system. In a general way, it appears that we may broadly distinguish between the influence of the system as a whole and the influence of the immediate local neighborhood; the former will be a smoothly varying function of position and time while the latter will be subject to relatively rapid fluctuations (see below).

Considering first the influence of the system as a whole, it appears that we can express it in terms of the gravitational potential $\mathfrak{B}(\boldsymbol{r};t)$ derived from the density function $n(\boldsymbol{r}, M;t)$ which governs the average spatial distribution of the stars of different masses at time t. Thus,

$$\mathfrak{B}(\boldsymbol{r};t)=-G\int_{-\infty}^{+\infty}\int_0^{\infty}\frac{Mn(\boldsymbol{r}_1, M;t)}{|\boldsymbol{r}_1-\boldsymbol{r}|}dMd\boldsymbol{r}_1, \tag{510}$$

where G denotes the constant of gravitation. The potential $\mathfrak{B}(\boldsymbol{r};t)$ derived in this manner may be said to represent the "smoothed out" distribution of matter in the stellar system. The force per unit mass acting on a star due to the "system as a whole" is therefore given by

$$\boldsymbol{K}=-\text{grad}\,\mathfrak{B}(\boldsymbol{r};t). \tag{511}$$

However, the fluctuations in the *complexion* of the local stellar distribution will make the instantaneous force acting on a star deviate from the value given by Eq. (511). To elucidate the nature and origin of these fluctuations, we surround the star under consideration by an element of volume σ, which we may suppose is small enough to contain, on the average, only a relatively few stars. The actual number of stars, which will be found in σ at any given instant, will not in general be the average number that will be expected to be in it, namely σn; it will be subject to fluctuations. These fluctuations will naturally be governed by a Poisson distribution with the variance σn [see Eq. (333)]. It is in direct consequence of this changing complexion of the local stellar distribution that the influence of the near neighbors on a star is variable. The average period of such a fluctuation is readily estimated: for the order of magnitude of the time involved is evidently that required for two stars to separate by a distance equal to the average distance D between the stars (see Appendix VII). We may, therefore, expect that the influence of the immediate neighborhood will fluctuate with an average period of the order of

$$T \simeq (D/(\langle |V|^2 \rangle_{Av})^{\frac{1}{2}}), \tag{512}$$

where $\langle |V|^2 \rangle_{Av}^{\frac{1}{2}}$ denotes the root mean square relative velocity between two stars.

In the neighborhood of the sun, $D \sim 3$ parsecs, $\langle |V|^2 \rangle_{Av}^{\frac{1}{2}} \sim 50$ km/sec. Hence

$$T \text{ (near the sun)} \simeq 6 \times 10^4 \text{ years.} \tag{513}$$

When we compare this time with the period of galactic rotation (which is about 2×10^8 years) we observe that in conformity with our earlier remarks, the fluctuations in the force acting on a star due to the changing local stellar distribution do occur with extreme rapidity compared to the rate at which any of the other physical parameters change. Accordingly we may write for the force per unit mass acting on a star, the expression

$$\mathfrak{F} = \boldsymbol{K}(\boldsymbol{r}; t) + \boldsymbol{F}(t), \tag{514}$$

where $\boldsymbol{K}$ is derived from the smoothed out distribution [as in Eqs. (510) and (511)] and $\boldsymbol{F}$ denotes the fluctuating force due to the near neighbors. Moreover, if Δt denotes an interval of time long compared to (512), we may write

$$\mathfrak{F}\Delta t = \boldsymbol{K}\Delta t + \boldsymbol{\delta}(t+\Delta t; t), \tag{515}$$

where

$$\boldsymbol{\delta}(t+\Delta t; t) = \int_t^{t+\Delta t} \boldsymbol{F}(\xi) d\xi \quad (\Delta t \gg T). \tag{516}$$

Under the circumstances stated (namely, $\Delta t \gg T$) the accelerations $\boldsymbol{\delta}(t+\Delta t; t)$ and $\boldsymbol{\delta}(t+2\Delta t; t+\Delta t)$ suffered during two successive intervals $(t+\Delta t, t)$ and $(t+2\Delta t, t+\Delta t)$ will not be expected to show any correlation. We may, therefore, anticipate the existence of a definite law of distribution which will govern the probability of occurrence of the different values of $\boldsymbol{\delta}(t+\Delta t; t)$. We thus see that the acceleration which a star suffers during an interval $\Delta t \gg T$ can be formally expressed as the sum of two terms: a *systematic* term $\boldsymbol{K}\Delta t$ due to the action of the gravitational field of the smoothed out distribution, and a *stochastic* term $\boldsymbol{\delta}(t+\Delta t; t)$ representing the influence of the near neighbors. Stated in this fashion, we recognize the similarity[8] between our present problems in stellar dynamics and those in the theory of Brownian motion considered in Chapters II and III. One important difference should however be noted: Under our present circumstances it is possible, as we shall presently see, to undertake an analysis of the statistical properties of $\boldsymbol{F}(t)$ and $\boldsymbol{\delta}(t+\Delta t; t)$ based on first principles and without appealing to any "intuitive" or *a priori* considerations as in the discussions of Brownian motion [see the remarks at the end of II, §1 and also those following Eq. (318)].

We shall now outline a general method which appears suitable for analyzing the statistical properties of $\boldsymbol{F}$.

The force $\boldsymbol{F}$ acting on a star, per unit mass, is given by

$$\boldsymbol{F} = G \sum_i \frac{M_i}{|\boldsymbol{r}_i|^3} \boldsymbol{r}_i, \tag{517}$$

where M_i denotes the mass of a typical "field" star and $\boldsymbol{r}_i$ its position vector relative to the star under consideration; further, in Eq. (517) the summation is to be extended over all the neighboring stars. The actual value of $\boldsymbol{F}$ given by Eq. (517) at any particular instant of time will depend on the instantaneous complexion of the local stellar distribution; it is in consequence subject to fluctuations. We can therefore ask only for the probability of occurrence,

$$W(\boldsymbol{F}) dF_x dF_y dF_z = W(\boldsymbol{F}) d\boldsymbol{F}, \tag{518}$$

of $\boldsymbol{F}$ in the range $\boldsymbol{F}$ and $\boldsymbol{F}+d\boldsymbol{F}$. In evaluating this probability distribution, we shall (consistent with the physical situations we have in view) suppose that fluctuations subject only to the restriction of a constant average density occur.

The probability distribution $W(\boldsymbol{F})$ of $\boldsymbol{F}$ can be obtained by a direct application of Markoff's method outlined in Chapter I, §3. We shall obtain the explicit form of this distribution (sometimes called the Holtsmark distribution) in §2 below, but we should draw attention, already at this stage, to the fact that the specification of $W(\boldsymbol{F})$ does *not* provide us with all the

[8] Cf. particularly Eq. (317) and Eq. (515) above.

necessary information concerning the fluctuating force $\boldsymbol{F}$ for an equally important aspect of $\boldsymbol{F}$ concerns the *speed of fluctuations.*

According to Eq. (517) the rate of change of $\boldsymbol{F}$ with time is given by

$$f=\frac{d\boldsymbol{F}}{dt}=G\sum_i M_i\left\{\frac{\boldsymbol{V}_i}{|\boldsymbol{r}_i|^3}-3\boldsymbol{r}_i\frac{(\boldsymbol{r}_i\cdot\boldsymbol{V}_i)}{|\boldsymbol{r}_i|^5}\right\}, \quad (519)$$

where $\boldsymbol{V}_i$ denotes the velocity of a typical field star *relative* to the star under consideration. It is now clear that the speed of fluctuations in $\boldsymbol{F}$ can be specified in terms of the bivariate distribution

$$W(\boldsymbol{F}, \boldsymbol{f}) \quad (520)$$

which governs the probability of the simultaneous occurrence of prescribed values for both $\boldsymbol{F}$ and $\boldsymbol{f}$. It is seen that this distribution function $W(\boldsymbol{F}, \boldsymbol{f})$ will depend on the assignment of *a priori* probability in the *phase space* in contrast to the distribution $W(\boldsymbol{F})$ of $\boldsymbol{F}$ which depends only on a similar assignment in the *configuration space.* Again, it is possible by an application of Markoff's method *formally* to write down a general expression for $W(\boldsymbol{F}, \boldsymbol{f})$; but it does not appear feasible to obtain the required distribution function in an explicit form. However, as Chandrasekhar and von Neumann have shown, explicit formulae for *all* the first and the second moments of $\boldsymbol{f}$ for a given $\boldsymbol{F}$ can be obtained; and it appears possible to make some progress in the specification of the statistical properties of $\boldsymbol{F}$ in terms of these moments.

2. The Holtsmark distribution $W(\boldsymbol{F})$

We shall now obtain the stationary distribution $W(\boldsymbol{F})$ of the force $\boldsymbol{F}$ acting on a star, per unit mass, due to the gravitational attraction of the neighboring stars.

Without loss of generality we can suppose that the star under consideration is at the origin O of our system of coordinates. About O describe a sphere of radius R and containing N stars. In the first instance we shall suppose that

$$\boldsymbol{F}=G\sum_{i=1}^{N}\frac{M_i}{|\boldsymbol{r}_i|^3}\boldsymbol{r}_i=\sum_{i=1}^{N}\boldsymbol{F}_i. \quad (521)$$

But we shall subsequently let R and N tend to infinity simultaneously in such a way that

$$(4/3)\pi R^3 n = N$$
$$(R\to\infty;\ N\to\infty;\ n=\text{constant}). \quad (522)$$

This limiting process is permissible, in view of what we shall later show to be the case, namely, that the dominant contribution to $\boldsymbol{F}$ is made by the nearest neighbor [cf. Eqs. (560) and (564) below]; consequently, the formal extrapolation to infinity of the density of stars obtaining only in a given region of a stellar system can hardly affect the results to any appreciable extent.

Considering first the distribution $W_N(\boldsymbol{F})$ at the center of a finite sphere of radius R and containing N stars, we seek the probability that

$$\boldsymbol{F}_0 \leqslant \boldsymbol{F} \leqslant \boldsymbol{F}_0+d\boldsymbol{F}_0. \quad (523)$$

Applying Markoff's method to this problem we have [cf. Eqs. (51) and (52)]

$$W_N(\boldsymbol{F}_0)=\frac{1}{8\pi^3}\int_{-\infty}^{+\infty}\exp(-i\boldsymbol{\varrho}\cdot\boldsymbol{F}_0)A_N(\boldsymbol{\varrho})d\boldsymbol{\varrho}, \quad (524)$$

where

$$A_N(\boldsymbol{\varrho})=\prod_{i=1}^{N}\int_{M_i=0}^{\infty}\int_{|\boldsymbol{r}_i|=0}^{R}\exp(i\boldsymbol{\varrho}\cdot\boldsymbol{F}_i)\times\tau_i(\boldsymbol{r}_i, M_i)d\boldsymbol{r}_i dM_i. \quad (525)$$

In Eq. (525) $\tau_i(\boldsymbol{r}_i, M_i)$ governs the probability of occurrence of the ith star at the position $\boldsymbol{r}_i$ with a mass M_i. If we now suppose that only fluctuations which are compatible with a constant average density occur, then

$$\tau_i(\boldsymbol{r}_i, M_i)=(3/4\pi R^3)\tau(M), \quad (526)$$

where $\tau(M)$ now governs the frequency of occurrence of the different masses among the stars. With the assumption (526) concerning the τ_i's Eq. (525) reduces to

$$A_N(\boldsymbol{\varrho})=\left[\frac{3}{4\pi R^3}\int_{M=0}^{\infty}\int_{|\boldsymbol{r}|=0}^{R}\exp(i\boldsymbol{\varrho}\cdot\boldsymbol{\phi})\times\tau(M)d\boldsymbol{r}dM\right]^N, \quad (527)$$

where we have written

$$\boldsymbol{\phi}=GM\boldsymbol{r}/|\boldsymbol{r}|^3. \quad (528)$$

We now let R and N tend to infinity according to Eq. (522). We thus obtain

$$W(F)=\frac{1}{8\pi^3}\int_{-\infty}^{+\infty}\exp(-i\varrho\cdot F)A(\varrho)d\varrho, \quad (529)$$

where

$$A(\varrho)=\lim_{R\to\infty}\left[\frac{3}{4\pi R^3}\int_{M=0}^{\infty}\int_{|r|=0}^{R}\exp(i\varrho\cdot\phi)\times\tau(M)drdM\right]^{4\pi R^3 n/3}. \quad (530)$$

Since,

$$\frac{3}{4\pi R^3}\int_{M=0}^{\infty}\int_{|r|=0}^{R}\tau(M)dMdr=1, \quad (531)$$

we can rewrite our expression for $A(\rho)$ in the form

$$A(\varrho)=\lim_{R\to\infty}\left[1-\frac{3}{4\pi R^3}\int_{-0}^{\infty}\int_{|r|=0}^{R}\tau(M)\times[1-\exp(i\varrho\cdot\phi)]drdM\right]^{4\pi R^3 n/3}. \quad (532)$$

The integral over r which occurs in Eq. (532) is seen to be absolutely convergent when extended over *all* $|r|$, i.e., also for $|r|\to\infty$. We can accordingly write

$$A(\varrho)=\lim_{R\to\infty}\left[1-\frac{3}{4\pi R^3}\int_{M=0}^{\infty}\int_{|r|=0}^{\infty}\tau(M)\times[1-\exp(i\varrho\cdot\phi)]drdM\right]^{4\pi R^3 n/3}, \quad (533)$$

or

$$A(\varrho)=\exp[-nC(\varrho)], \quad (534)$$

where

$$C(\varrho)=\int_{M=0}^{\infty}\int_{|r|=0}^{\infty}\tau(M)[1-\exp(i\varrho\cdot\phi)]drdM. \quad (535)$$

In the integral defining $C(\varrho)$ we shall introduce ϕ as the variable of integration instead of r. We readily verify that

$$dr=-\tfrac{1}{2}(GM)^{3/2}|\phi|^{-9/2}d\phi. \quad (536)$$

Hence,

$$C(\varrho)=\tfrac{1}{2}G^{3/2}\int_0^{\infty}dM\,M^{3/2}\tau(M)\int_{-\infty}^{+\infty}d\phi|\phi|^{-9/2}\times[1-\exp(i\varrho\cdot\phi)], \quad (537)$$

or, in an obvious notation

$$C(\varrho)=\tfrac{1}{2}G^{3/2}\langle M^{3/2}\rangle_{Av}\int_{-\infty}^{+\infty}[1-\exp(i\varrho\cdot\phi)]\times|\phi|^{-9/2}d\phi. \quad (538)$$

The foregoing expression is clearly unaffected if we replace ϕ by $-\phi$. But this replacement changes $\exp(i\varrho\cdot\phi)$ into $\exp(-i\varrho\cdot\phi)$ under the integral sign; taking the arithmetic mean of the two resulting integrals, we obtain

$$C(\varrho)=\tfrac{1}{2}G^{3/2}\langle M^{3/2}\rangle_{Av}\int_{-\infty}^{+\infty}[1-\cos(\varrho\cdot\phi)]|\phi|^{-9/2}d\phi. \quad (539)$$

Choosing polar coordinates with the z axis in the direction of ϱ Eq. (539) can be transformed to

$$C(\varrho)=\tfrac{1}{2}G^{3/2}\langle M^{3/2}\rangle_{Av}\int_0^{\infty}\int_{-1}^{+1}\int_0^{2\pi}\times[1-\cos(|\varrho||\phi|t)]|\phi|^{-5/2}d\omega dt d|\phi|, \quad (540)$$

or, introducing further the variable $z=|\varrho||\phi|$, we have

$$C(\varrho)=\tfrac{1}{2}G^{3/2}\langle M^{3/2}\rangle_{Av}|\varrho|^{3/2}\times\int_0^{\infty}\int_{-1}^{+1}\int_0^{2\pi}[1-\cos(zt)]z^{-5/2}d\omega dt dz. \quad (541)$$

After performing the integrations over ω and t we obtain

$$C(\varrho)=2\pi G^{3/2}\langle M^{3/2}\rangle_{Av}|\varrho|^{3/2}\times\int_0^{\infty}(z-\sin z)z^{-7/2}dz, \quad (542)$$

or after several integrations by parts

$$C(\varrho)=\frac{16}{15}\pi G^{3/2}\langle M^{3/2}\rangle_{Av}|\varrho|^{3/2}\int_0^{\infty}z^{-1/2}\cos z\,dz =\frac{4}{15}(2\pi G)^{3/2}\langle M^{3/2}\rangle_{Av}|\varrho|^{3/2}. \quad (543)$$

Combining Eqs. (529), (534), and (543) we now obtain

$$W(F)=\frac{1}{8\pi^3}\int_{-\infty}^{+\infty}\exp(-i\varrho\cdot F-a|\varrho|^{3/2})d\varrho, \quad (544)$$

where we have written

$$a=(4/15)(2\pi G)^{3/2}\langle M^{3/2}\rangle_{Av}n. \quad (545)$$

Using a frame of reference in which one of the principal axes is in the direction of F and changing to polar coordinates, the formula (544) for $W(F)$ can be reduced to

$$W(F)=\frac{1}{4\pi^2}\int_0^\infty\int_{-1}^{+1}\exp(-i|\varrho||F|t-a|\varrho|^{3/2})\times|\varrho|^2dtd|\varrho|. \quad (546)$$

The integration over t is readily effected, and we obtain

$$W(F)=\frac{1}{2\pi^2|F|}\int_0^\infty\exp(-a|\varrho|^{3/2})\times|\varrho|\sin(|\varrho||F|)d|\varrho|. \quad (547)$$

If we now put

$$x=|\varrho||F|, \quad (548)$$

Eq. (547) becomes

$$W(F)=\frac{1}{2\pi^2|F|^3}\int_0^\infty\exp(-ax^{3/2}/|F|^{3/2})\times x\sin xdx. \quad (549)$$

We can rewrite the foregoing formula for $W(F)$ more conveniently if we introduce the *normal field* Q_H defined by

$$\left.\begin{aligned}Q_H&=a^{2/3}=(4/15)^{2/3}(2\pi G)((\langle M^{3/2}\rangle_{\text{Av}}n)^{2/3},\\&=2.6031G(\langle M^{3/2}\rangle_{\text{Av}}n)^{2/3}\end{aligned}\right\}\quad(550)$$

and express $|F|$ in terms of this unit:

$$|F|=\beta Q_H=\beta a^{2/3}. \quad (551)$$

Equation (549) now takes the form

$$W(F)=H(\beta)/4\pi a^2\beta^2, \quad (552)$$

where we have introduced the function $H(\beta)$ defined by

$$H(\beta)=\frac{2}{\pi\beta}\int_0^\infty\exp[-(x/\beta)^{3/2}]x\sin xdx. \quad (553)$$

Since,

$$W(|F|)=4\pi|F|^2W(F), \quad (554)$$

we obtain from Eqs. (551) and (552)

$$W(|F|)=H(\beta)/Q_H; \quad (555)$$

accordingly $H(\beta)$ defines the probability distribution of $|F|$ when it is expressed in units of Q_H. The function $H(\beta)$ has been evaluated numerically and is tabulated in Table IX.

The asymptotic behavior of the distribution $W(|F|)$ can be obtained from the formulae:

$$H(\beta)=4\beta^2/3\pi+O(\beta^4)\quad(\beta\to 0), \quad (556)$$

and

$$H(\beta)=(15/8)(2/\pi)^{1/2}\beta^{-5/2}+O(\beta^{-4})\quad(\beta\to\infty). \quad (557)$$

We find [cf. Eqs. (551) and (555)]

$$W(|F|)\simeq(4/3\pi Q_H^3)|F|^2\quad(|F|\to 0), \quad (558)$$

and

$$W(|F|)\simeq(15/8)(2/\pi)^{1/2}Q_H^{3/2}|F|^{-5/2}\quad(|F|\to\infty). \quad (559)$$

Substituting for Q_H from Eq. (550) in Eq. (559) we obtain

$$W(|F|)\simeq 2\pi G^{3/2}\langle M^{3/2}\rangle_{\text{Av}}n|F|^{-5/2}\quad(|F|\to\infty). \quad (560)$$

It is seen that while the frequency of occurrence of both the weak and the strong fields is quite small, it is only the fields of average intensity which have appreciable probabilities. In particular, the value of $|F|$ which has the maximum probability of occurrence is found to be (see Table IX) $\sim 1.6Q_H$.

Equations (552) and (553) provide, of course, the *exact* formula for the distribution of F for an *ideally* random distribution of stars. But an elementary treatment which leads to an approximate formula for $W(F)$ is of some interest and illuminates certain points in the theory. The treatment we refer to is based on the assumption that the force acting on a star is entirely due to its *nearest* neighbor.

Now, the law of distribution of the nearest neighbor is given by [see Appendix VII, Eq. (671)]

$$w(r)dr=\exp(-4\pi r^3n/3)4\pi r^2ndr, \quad (561)$$

and, since on the first neighbor approximation

$$|F|=GMr^{-2}, \quad (562)$$

we readily obtain the formula

$$W(|F|)d|F|=\exp[-4\pi(GM)^{3/2}n/3|F|^{3/2}]\times 2\pi(GM)^{3/2}n|F|^{-5/2}d|F|. \quad (563)$$

TABLE IX. The function $H(\beta)$.

β	$H(\beta)$	β	$H(\beta)$
0.0		5.0	0.04310
0.1	0.004225	5.2	0.03790
0.2	0.016666	5.4	0.03357
0.3	0.036643	5.6	0.02993
0.4	0.063084	5.8	0.02683
0.5	0.094601	6.0	0.02417
0.6	0.129598	6.2	0.02188
0.7	0.166380	6.4	0.01988
0.8	0.203270	6.6	0.01814
0.9	0.238704	6.8	0.01660
1.0	0.271322	7.0	0.01525
1.1	0.30003	7.2	0.01405
1.2	0.32402	7.4	0.01297
1.3	0.34281	7.6	0.01201
1.4	0.35620	7.8	0.01115
1.5	0.36426	8.0	0.01038
1.6	0.36726	8.2	0.00967
1.7	0.36566	8.4	0.00903
1.8	0.36004	8.6	0.00846
1.9	0.35101	8.8	0.00793
2.0	0.33918	9.0	0.00745
2.1	0.32519	9.2	0.00701
2.2	0.30951	9.4	0.00660
2.3	0.29266	9.6	0.00622
2.4	0.27485	9.8	0.00588
2.5	0.25667	10.0	0.00556
2.6	0.238	15.0	0.00188
2.7	0.222	20.0	0.00089
2.8	0.206	25.0	0.00050
2.9	0.190	30.0	0.00031
3.0	0.176	35.0	0.00021
3.2	0.150	40.0	0.00015
3.4	0.128	45.0	0.00011
3.6		50.0	0.00009
3.8		60.0	0.00005
4.0		70.0	0.00004
4.2		80.0	0.00003
4.4	0.06734	90.0	0.00002
4.6	0.05732	100.0	0.00002
4.8	0.04944		

According to the distribution (563)

$$W(|\boldsymbol{F}|) \simeq 2\pi(GM)^{3/2} n |\boldsymbol{F}|^{-5/2} \quad (|\boldsymbol{F}| \to \infty), \quad (564)$$

which is seen to be in *exact* agreement with the formula (560) derived from the Holtsmark distribution (555). The physical meaning of this agreement, for $|\boldsymbol{F}| \to \infty$ in the results derived from an exact and an approximate treatment of the same problem, is simply that the highest fields are in reality produced only by the nearest neighbor. More generally, it is found that the two distributions (555) and (563) agree over most of the range of $|\boldsymbol{F}|$. Thus, the field which has the maximum frequency of occurrence on the basis of (563) is seen to differ from the corresponding value on the Holtsmark distribution by less than five percent. The region in which the two distributions (555) and (563) differ most markedly is when $|\boldsymbol{F}| \to 0$: on the Holtsmark distribution $W(|\boldsymbol{F}|)$ tends to zero as $|\boldsymbol{F}|^2$ while on the nearest neighbor approximation $W(|\boldsymbol{F}|)$ tends to zero as $\exp(-\text{const.}\ |\boldsymbol{F}|^{-\frac{3}{2}})$ [cf. Eqs. (558) and (564)]. However, the fact that the nearest neighbor approximation should be seriously in error for the weak fields is, of course, to be expected: for, a weak field arises from a more or less symmetrical, average, complexion of the stars around the one under consideration and consequently $\boldsymbol{F}$ under these circumstances is the result of the action of several stars and not due to any one single star.

Finally, we may draw attention to one important difficulty in using the Holstmark distribution for *all* values of $|\boldsymbol{F}|$: It predicts relatively too high probabilities for $|\boldsymbol{F}|$ as $|\boldsymbol{F}| \to \infty$. Thus, on the basis of the distribution (555), $\langle |\boldsymbol{F}|^2 \rangle_{Av}$ is divergent. [The same remark also applies to the distribution (564).] These relatively high probabilities for the high field strengths is a consequence of our assumption of complete randomness in stellar distribution for *all* elements of volume. It is, however, apparent that this assumption cannot be valid for the regions in the *very* immediate neighborhoods of the individual stars. For, if $\boldsymbol{V}$ denotes the relative velocity between two stars when separated by distances of the order of the average distance between the stars, the two stars cannot come closer together (on the approximation of linear trajectories) than a certain critical distance $r(|\boldsymbol{V}|)$ such that

$$|\boldsymbol{V}|^2/2 = [G(M_1+M_2)/r(|\boldsymbol{V}|)], \quad (565)$$

or

$$r(|\boldsymbol{V}|) = [2G(M_1+M_2)/|\boldsymbol{V}|^2]. \quad (566)$$

Otherwise the two stars should be strictly regarded as the components of a binary system and this is inconsistent with our original premises. This restriction therefore leads us to infer that departures from true randomness exist for $r \sim r(|\boldsymbol{V}|)$. However, under the conditions we normally encounter in stellar systems, $r(|\boldsymbol{V}|)$ is very small compared to the average distance between the stars. Thus, in our galaxy, in the general neighborhood of the sun, $r(|\boldsymbol{V}|) \sim 2 \times 10^{-5}$ parsec, and this is to be compared with an average distance between the stars of about three parsecs. Accordingly, we may expect the

Holtsmark distribution to be very close to the true distribution, except for the very highest values of $|F|$. More particularly, the deviations from the Holtsmark distribution are to be expected for field strengths of the order of

$$|F| \sim (GM_2/[r(|V|)]^2) \simeq (M_2[\langle |V|^2\rangle_{Av}]^2/4G(M_1+M_2)^2). \quad (567)$$

When $|F|$ becomes much larger than the quantity on the right-hand side of Eq. (567), the true frequencies of occurrence will very rapidly tend to zero as compared to what would be expected on the Holtsmark distribution, namely (560). A rigorous treatment of these deviations from the distribution (555) will require a reconsideration of the whole problem in *phase space* and is beyond the scope of the present investigation.

3. The Speed of Fluctuations in F

As we have already remarked the speed of fluctuations can be specified in terms of the distribution function $W(F, f)$ which gives the simultaneous probability of a given field strength F and an associated rate of change of F of amount f [cf. Eqs. (517) and (519)]. The general expression for this probability distribution can be readily written down using Markoff's method [I, §3, Eqs. (51), (52), and (53)]. We have [cf. Eqs. (529) and (530)]

$$W(F, f) = \frac{1}{64\pi^6} \int_{|\varrho|=0}^{\infty} \int_{|\sigma|=0}^{\infty} \exp\left[-i(\varrho \cdot F + \sigma \cdot f)\right] A(\varrho, \sigma) d\varrho d\sigma, \quad (568)$$

where

$$A(\varrho, \sigma) = \underset{R\to\infty}{\text{limit}} \left[\frac{3}{4\pi R^3} \int_{0<M<\infty} \int_{|r|<R} \int_{|V|<\infty} \exp\left[i(\varrho\cdot\phi + \sigma\cdot\psi)\right] \tau dr dV dM\right]^{4\pi R^3 n/3}. \quad (569)$$

In Eqs. (568) and (569) ϱ and σ are two auxiliary vectors, n denotes the number of stars per unit volume, and

$$\phi = GM\frac{r}{|r|^3}; \quad \psi = GM\left\{\frac{V}{|r|^3} - 3\frac{r(r\cdot V)}{|r|^5}\right\}. \quad (570)$$

Further,

$$\tau dV dM = \tau(V; M) dV dM \quad (571)$$

gives the probability that a star with a relative velocity in the range $(V, V+dV)$ and with a mass between M and $M+dM$ will be found. It should also be noted that in writing down Eqs. (568) and (569) we have supposed (as in §2) that the fluctuations in the local stellar distribution which occur are subject only to the restriction of a constant average density.

Since

$$\frac{3}{4\pi R^3} \int_{M=0}^{\infty} \int_{|r|<R} \int_{|V|<\infty} \tau dr dV dM = 1, \quad (572)$$

we can rewrite (569) as

$$A(\varrho, \sigma) = \underset{R\to\infty}{\text{limit}} \left\{1 - \frac{3}{4\pi R^3} \int_{M=0}^{\infty} \int_{|r|<R} \int_{|V|<\infty} \{1 - \exp\left[i(\varrho\cdot\phi + \sigma\cdot\psi)\right]\} \tau dr dV dM\right\}^{4\pi R^3 n/3}. \quad (573)$$

The integral over r which occurs in Eq. (573) is seen to be conditionally convergent when extended over all $|r|$, i.e., also for $|r| \to \infty$. Hence, we can write

$$A(\varrho, \sigma) = \underset{R\to\infty}{\text{limit}} \left\{1 - \frac{3}{4\pi R^3} \int_{M=0}^{\infty} \int_{|r|=0}^{\infty} \int_{|V|=0}^{\infty} \{1 - \exp\left[i(\varrho\cdot\phi + \sigma\cdot\psi)\right]\} \tau dr dV dM\right\}^{4\pi R^3 n/3}, \quad (574)$$

or

$$A(\varrho, \sigma) = \exp\left[-nC(\varrho, \sigma)\right] \quad (575)$$

where

$$C(\varrho, \sigma) = \int_0^\infty \int_{-\infty}^{+\infty} \int_{-\infty}^{+\infty} \{1 - \exp[i(\varrho \cdot \phi + \sigma \cdot \psi)]\} \tau d\tau dV dM. \tag{576}$$

This formally solves the problem. It does not, however, appear that the integral representing $C(\varrho, \sigma)$ can be evaluated explicitly in terms of any known functions. But if we are interested only in the moments of f for a given F and of F for a given f we need only the behavior of $A(\varrho, \sigma)$and, therefore, also of $C(\varrho, \sigma)$ for $|\sigma|$, respectively, $|\varrho|$ tending to zero. For, considering the first and the second moments of the components f_ξ, f_η, and f_ζ of f along three directions ξ, η, and ζ at right angles to each other, we have

$$W(F)\langle f_\mu \rangle_{Av} = \int_{|f|=0}^\infty W(F, f) f_\mu df \quad (\mu = \xi, \eta, \zeta), \tag{577}$$

and

$$W(F)\langle f_\mu f_\nu \rangle_{Av} = \int_{|f|=0}^\infty W(F, f) f_\mu f_\nu df \quad (\mu, \nu = \xi, \eta, \zeta), \tag{578}$$

where $W(F)$ denotes the distribution of F for which we have already obtained an explicit formula in §2. Substituting now for $W(F, f)$ from Eq. (568) in the foregoing formulae for the moments we obtain

$$W(F)\langle f_\mu \rangle_{Av} = \frac{1}{64\pi^6} \int_{|f|=0}^\infty \int_{|\varrho|=0}^\infty \int_{|\sigma|=0}^\infty \exp[-i(\varrho \cdot F + \sigma \cdot f)] A(\varrho, \sigma) f_\mu d\varrho d\sigma df, \tag{579}$$

and

$$W(F)\langle f_\mu f_\nu \rangle_{Av} = \frac{1}{64\pi^6} \int_{|f|=0}^\infty \int_{|\varrho|=0}^\infty \int_{|\sigma|=0}^\infty \exp[-i(\varrho \cdot F + \sigma \cdot f)] A(\varrho, \sigma) f_\mu f_\nu d\varrho d\sigma df. \tag{580}$$

But

$$\left.\begin{aligned} &\frac{1}{8\pi^3} \int_{|f|=0}^\infty \exp(-i\sigma \cdot f) f_\xi df = i\delta'(\sigma_\xi)\delta(\sigma_\eta)\delta(\sigma_\zeta), \\ &\frac{1}{8\pi^3} \int_{|f|=0}^\infty \exp(-i\sigma \cdot f) f_\xi^2 df = -\delta''(\sigma_\xi)\delta(\sigma_\eta)\delta(\sigma_\zeta), \\ &\frac{1}{8\pi^3} \int_{|f|=0}^\infty \exp(-i\sigma \cdot f) f_\xi f_\eta df = -\delta'(\sigma_\xi)\delta'(\sigma_\eta)\delta(\sigma_\zeta), \end{aligned}\right\} \tag{581}$$

etc. In Eq. (581) δ denotes Dirac's δ-function and δ' and δ'' its first and second derivatives; and remembering also that

$$\int_{-\infty}^{+\infty} f(x)\delta(x)dx = f(0); \quad \int_{-\infty}^{+\infty} f(x)\delta'(x)dx = -f'(0); \quad \int_{-\infty}^{+\infty} f(x)\delta''(x)dx = f''(0), \tag{582}$$

Eqs. (579) and (580) for the moments reduce to

$$W(F)\langle f_\mu \rangle_{Av} = -\frac{i}{8\pi^3} \int_{|\varrho|=0}^\infty \exp(-i\varrho \cdot F) \left[\frac{\partial}{\partial \sigma_\mu} A(\varrho, \sigma)\right]_{|\sigma|=0} d\varrho, \tag{583}$$

and

$$W(F)\langle f_\mu f_\nu \rangle_{Av} = -\frac{1}{8\pi^3} \int_{|\varrho|=0}^\infty \exp(-i\varrho \cdot F) \left[\frac{\partial^2}{\partial \sigma_\mu \partial \sigma_\nu} A(\varrho, \sigma)\right]_{|\sigma|=0} d\varrho. \tag{584}$$

We accordingly see that the first and the second moments of f can be evaluated from a series expansion of $A(\varrho, \sigma)$ or of $C(\varrho, \sigma)$ which is correct up to the *second order* in $|\sigma|$. Such a series expansion has been found by Chandrasekhar and von Neumann and, quoting their final result, we have

$$C(\boldsymbol{\varrho},\boldsymbol{\sigma})=\frac{4}{15}(2\pi)^{3/2}G^{3/2}\langle M^{3/2}\rangle_{\text{Av}}|\boldsymbol{\varrho}|^{3/2}+\frac{2}{3}\pi iG(\sigma_1\langle MV_1\rangle_{\text{Av}}+\sigma_2\langle MV_2\rangle_{\text{Av}}-2\sigma_3\langle MV_3\rangle_{\text{Av}})$$
$$+\frac{3}{28}(2\pi)^{1/2}G^{1/2}|\boldsymbol{\varrho}|^{-1/2}[(5\sigma_1^2+4\sigma_2^2-2\sigma_3^2)\langle M^{1/2}V_1^2\rangle_{\text{Av}}+(4\sigma_1^2+5\sigma_2^2-2\sigma_3^2)\langle M^{1/2}V_2^2\rangle_{\text{Av}}$$
$$+(4\sigma_3^2-2\sigma_1^2-2\sigma_2^2)\langle M^{1/2}V_3^2\rangle_{\text{Av}}-8\sigma_2\sigma_3\langle M^{1/2}V_2V_3\rangle_{\text{Av}}-8\sigma_3\sigma_1\langle M^{1/2}V_3V_1\rangle_{\text{Av}}$$
$$+2\sigma_1\sigma_2\langle M^{1/2}V_1V_2\rangle_{\text{Av}}]+O(|\boldsymbol{\sigma}|^3)\quad(|\boldsymbol{\sigma}|\to 0),\tag{585}$$

where $\langle\ \rangle_{\text{Av}}$ indicates that the corresponding quantity has been averaged with the weight function $\tau(\boldsymbol{V};M)$ [cf. Eq. (571)]; further, in Eq. (585) $(\sigma_1, \sigma_2, \sigma_3)$ and (V_1, V_2, V_3) are the components of $\boldsymbol{\sigma}$ and $\boldsymbol{V}$ in a system of coordinates in which the z axis is in the direction of $\boldsymbol{\varrho}$.

In Eq. (585) $\boldsymbol{V}=(V_1, V_2, V_3)$ is of course the velocity of a field star relative to the one under consideration. If we now let $\boldsymbol{u}$ and $\boldsymbol{v}$ denote the velocities of the field star and the star under consideration in an appropriately chosen local standard of rest, then

$$\boldsymbol{V}=\boldsymbol{u}-\boldsymbol{v}.\tag{586}$$

In their further discussion, Chandrasekhar and von Neumann introduce the assumption that the distribution of the velocities $\boldsymbol{u}$ among the stars is *spherical*, i.e., the distribution function $\Psi(\boldsymbol{u})$ has the form

$$\Psi(\boldsymbol{u})\equiv\Psi(j^2(M)|\boldsymbol{u}|^2),\tag{587}$$

where Ψ is an arbitrary function of the argument specified and the parameter j (of the dimensions of [velocity]$^{-1}$) can be a function of the mass of the star. This assumption for the distribution of the peculiar velocities $\boldsymbol{u}$ implies that the probability function $\tau(\boldsymbol{V};M)$ must be expressible as

$$\tau(\boldsymbol{V};M)\equiv\Psi[j^2(M)|\boldsymbol{u}|^2]\chi(M),\tag{588}$$

where $\chi(M)$ governs the distribution over the different masses. For a function τ of this form we clearly have

$$\begin{gathered}\langle MV_i\rangle_{\text{Av}}=-\langle M\rangle_{\text{Av}}v_i;\quad\langle M^{1/2}V_i^2\rangle_{\text{Av}}=\tfrac{1}{3}\langle M^{1/2}|\boldsymbol{u}|^2\rangle_{\text{Av}}+\langle M^{1/2}\rangle_{\text{Av}}v_i^2\quad(i=1,2,3),\\ \langle M^{1/2}V_iV_j\rangle_{\text{Av}}=\langle M^{1/2}\rangle_{\text{Av}}v_iv_j\quad(i,j=1,2,3,\ i\neq j).\end{gathered}\tag{589}$$

Substituting these values in Eq. (577) we find after some minor reductions that

$$C(\boldsymbol{\varrho},\boldsymbol{\sigma})=\frac{4}{15}(2\pi)^{3/2}G^{3/2}\langle M^{3/2}\rangle_{\text{Av}}|\boldsymbol{\varrho}|^{3/2}-\frac{2}{3}\pi iG\langle M\rangle_{\text{Av}}(\sigma_1v_1+\sigma_2v_2-2\sigma_3v_3)+\frac{1}{4}(2\pi)^{1/2}G^{1/2}\langle M^{1/2}|\boldsymbol{u}|^2\rangle_{\text{Av}}|\boldsymbol{\varrho}|^{-1/2}(\sigma_1^2+\sigma_2^2)$$
$$+\frac{3}{28}(2\pi)^{1/2}G^{1/2}\langle M^{1/2}\rangle_{\text{Av}}|\boldsymbol{\varrho}|^{-1/2}[\sigma_1^2(5v_1^2+4v_2^2-2v_3^2)+\sigma_2^2(5v_2^2+4v_1^2-2v_3^2)$$
$$+\sigma_3^2(4v_3^2-2v_1^2-2v_2^2)+2\sigma_1\sigma_2v_1v_2-8\sigma_2\sigma_3v_2v_3-8\sigma_3\sigma_1v_3v_1]+O(|\boldsymbol{\sigma}|^3).\tag{590}$$

With a series expansion of this form we can, as we have already remarked, evaluate all the first and the second moments of f for a given $\boldsymbol{F}$.

Considering first the moment of f, Chandrasekhar and von Neumann find that

$$\langle \boldsymbol{f}\rangle_{\text{Av}}=\overline{\left(\frac{d\boldsymbol{F}}{dt}\right)}_{\boldsymbol{F},\boldsymbol{v}}=-\frac{2}{3}\pi G\langle M\rangle_{\text{Av}}nB\left(\frac{|\boldsymbol{F}|}{Q_H}\right)\left(\boldsymbol{v}-3\frac{\boldsymbol{v}\cdot\boldsymbol{F}}{|\boldsymbol{F}|^2}\boldsymbol{F}\right),\tag{591}$$

where Q_H is the "normal field" introduced in §2 [Eqs. (550) and (551)] and

$$B(\beta)=3\left(\int_0^\beta H(\beta)d\beta \Big/ \beta H(\beta)\right)-1. \tag{592}$$

We shall examine certain formal consequences of Eq. (592).

Multiplying Eq. (591) scalarly with F we obtain

$$\overline{F\cdot\left(\frac{dF}{dt}\right)_{F,\,v}}=\frac{4}{3}\pi G\langle M\rangle_{\text{Av}}nB\left(\frac{|F|}{Q_H}\right)(v\cdot F); \tag{593}$$

but

$$\overline{F\cdot\left(\frac{dF}{dt}\right)_{F,\,v}}=|F|\overline{\left(\frac{d|F|}{dt}\right)_{F,\,v}}. \tag{594}$$

Hence,

$$\overline{\left(\frac{d|F|}{dt}\right)_{F,\,v}}=\frac{4}{3}\pi G\langle M\rangle_{\text{Av}}nB\left(\frac{|F|}{Q_H}\right)\frac{v\cdot F}{|F|}. \tag{595}$$

On the other hand, if F_j denotes the component of F in an arbitrary direction at right angles at the direction of v then according to Eq. (591)

$$\overline{\left(\frac{dF_j}{dt}\right)_{F,\,v}}=2\pi G\langle M\rangle_{\text{Av}}nB\left(\frac{|F|}{Q_H}\right)\frac{v\cdot F}{|F|^2}F_j. \tag{596}$$

Combining Eqs. (595) and (596) we have

$$\frac{1}{F_j}\overline{\left(\frac{dF_j}{dt}\right)_{F,\,v}}=\frac{3}{2}\frac{1}{|F|}\overline{\left(\frac{d|F|}{dt}\right)_{F,\,v}}. \tag{597}$$

Equation (597) is clearly equivalent to

$$\overline{\frac{d}{dt}(\log F_j-\tfrac{3}{2}\log|F|)_{F,\,v}}=0. \tag{598}$$

We have thus proved that

$$\overline{\left[\frac{d}{dt}\left(\frac{F_j}{|F|^{\frac{3}{2}}}\right)\right]_{F,\,v}}=0. \tag{599}$$

We shall now examine the physical consequences of Eq. (591) more closely. In words, the meaning of this equation is that the component of

$$-\frac{2}{3}\pi G\langle M\rangle_{\text{Av}}nB\left(\frac{|F|}{Q_H}\right)\left(v-3\frac{v\cdot F}{|F|^2}F\right) \tag{600}$$

along any particular direction gives the average value of the rate of change of F that is to be expected in the specified direction when the star is moving with a velocity v. Stated in this manner we at once see the essential difference in the stochastic variations of F with time in the two cases $|v|=0$ and $|v|\neq 0$. In the former case $\langle F\rangle_{\text{Av}}\equiv 0$; but this is not generally true when $|v|\neq 0$. Or expressed differently, when $|v|=0$ the changes in F occur with equal probability in all directions while this is not the case when $|v| \neq 0$. The true nature of this difference is brought out very clearly when we consider

$$\overline{\left(\frac{d|F|}{dt}\right)}_{F,v} \tag{601}$$

according to Eq. (595). Remembering that $B(\beta) \geqslant 0$ for $\beta \geqslant 0$, we conclude from Eq. (595) that

$$\overline{\left(\frac{d|F|}{dt}\right)}_{F,v} > 0 \quad \text{if} \quad (v \cdot F) > 0, \tag{602}$$

and

$$\overline{\left(\frac{d|F|}{dt}\right)}_{F,v} < 0 \quad \text{if} \quad (v \cdot F) < 0. \tag{603}$$

In other words, if F has a positive component in the direction of v, $|F|$ increases on the average, while if F has a negative component in the direction of v, $|F|$ decreases on the average. This essential asymmetry introduced by the direction of v may be expected to give rise to the phenomenon of *dynamical friction.*

Considering next the second moments of f Chandrasekhar and von Neumann find that

$$\langle |f|^2{}_{F,v}\rangle_{Av} = 2ab\frac{\beta^{1/2}}{H(\beta)}\{2G(\beta)+7k[G(\beta)\sin^2\alpha - I(\beta)(3\sin^2\alpha-2)]\}$$
$$+\frac{g^2}{\beta H(\beta)}\{\beta H(\beta)(4-3\sin^2\alpha)+3K(\beta)(3\sin^2\alpha-2)\}, \tag{604}$$

where, α denotes the angle between the directions of F and v

$$a=\frac{4}{15}(2\pi)^{1/2}G^{3/2}\langle M^{3/2}\rangle_{Av}n;\quad b=\frac{1}{4}(2\pi)^{1/2}G^{1/2}\langle M^{1/2}|u|^2\rangle_{Av}n,\quad g=\frac{2}{3}\pi G\langle M\rangle_{Av}|v|n;\quad k=\frac{3}{7}\frac{\langle M^{3/2}\rangle_{Av}|v|^2}{\langle M^{1/2}|u|^2\rangle_{Av}}, \tag{605}$$

and

$$\left.\begin{aligned} &H(\beta)=\frac{2}{\pi\beta}\int_0^\infty \exp[-(x/\beta)^{3/2}]\beta\sin\beta d\beta,\\ &G(\beta)=\frac{3}{2}\int_0^\beta \beta^{-1/2}H(\beta)d\beta,\qquad I(\beta)=\beta^{-3/2}\int_0^\beta \beta^{1/2}G(\beta)d\beta,\qquad K(\beta)=\int_0^\beta H(\beta)d\beta.\end{aligned}\right\} \tag{606}$$

Averaging Eq. (604) for all possible mutual orientations of the two vectors F and v we readily find that

$$\langle\langle |f|^2{}_{F,|v|}\rangle\rangle_{Av} = 4ab\left\{\frac{\beta^{1/2}G(\beta)}{H(\beta)}\left(1+\frac{7}{3}k\right)+\frac{g^2}{2ab}\right\}, \tag{607}$$

or, substituting for k and $g^2/2ab$ from (605) we find

$$\langle\langle |f|^2{}_F\rangle\rangle_{Av} = 4ab\left\{\frac{\beta^{1/2}G(\beta)}{H(\beta)}\left(1+\frac{\langle M^{3/2}\rangle_{Av}|v|^2}{\langle M^{1/2}|u|^2\rangle_{Av}}\right)+\frac{5}{12\pi}\frac{\langle M\rangle_{Av}{}^2|v|^2}{\langle M^{3/2}\rangle_{Av}\langle M^{1/2}|u|^2\rangle_{Av}}\right\}. \tag{608}$$

In terms of Eq. (608) we can define an approximate formula for the mean life of the state F according to the equation

$$T_{|F|,\,|v|} = |F|/(\langle\langle |f|^2{}_F\rangle\rangle_{Av})^{1/2}. \tag{609}$$

Combining Eqs. (608) and (609) we find that

$$T_{|F|,\,|v|}=T_{|F|,\,0}\frac{1}{\left[1+\dfrac{\langle M^{\frac{1}{3}}\rangle_{\text{Av}}|v|^2}{\langle M^{\frac{1}{3}}|u|^2\rangle_{\text{Av}}}+\dfrac{5}{12\pi}\dfrac{\langle M\rangle_{\text{Av}}{}^2|v|^2}{\langle M^{\frac{2}{3}}\rangle_{\text{Av}}\langle M^{\frac{1}{3}}|u|^2\rangle_{\text{Av}}}\dfrac{H(\beta)}{\beta^{\frac{1}{2}}G(\beta)}\right]^{\frac{1}{2}}}, \tag{610}$$

where $T_{|F|,\,0}$ denotes the mean life when $|v|=0$:

$$T_{|F|,\,0}=\left[\frac{a^{\frac{1}{2}}}{4b}\frac{\beta^{\frac{3}{2}}H(\beta)}{G(\beta)}\right]^{\frac{1}{2}}. \tag{611}$$

From Eq. (610) we derive that

$$T\propto|F|\quad\text{as}\quad|F|\to 0;\quad T\propto|F|^{-\frac{1}{2}}\quad\text{as}\quad|F|\to\infty; \tag{612}$$

in other words the mean life tends to zero for both weak and strong fields.

I am greatly indebted to Mrs. T. Belland for her assistance in preparing the manuscript for the press. My thanks are also due to Dr. L. R. Henrich for his careful revision of the entire manuscript.

APPENDIXES

I. THE MEAN AND THE MEAN SQUARE DEVIATION OF A BERNOULLI DISTRIBUTION

Consider the Bernoulli distribution

$$w(x)=\frac{n!}{x!(n-x)!}p^x(1-p)^{n-x}\quad(p<1;\ x\text{ a positive integer}\leqslant n). \tag{613}$$

An alternative form for $w(x)$ is

$$w(x)=C_x{}^np^xq^{n-x}, \tag{614}$$

where $C_x{}^n$ denotes the binomial coefficient and

$$q=1-p. \tag{615}$$

From Eq. (614) it is apparent that $w(x)$ is the coefficient of u^x in the expansion of $(pu+q)^n$:

$$w(x)=\text{coefficient of } u^x \text{ in } (pu+q)^n. \tag{616}$$

That $\sum w_x=1$ follows immediately from this remark:

$$\left.\begin{aligned}\sum_{x=1}^{n}w(x)&=\sum_{x=1}^{n}\text{coefficient of } u^x \text{ in } (pu+q)^n,\\ &=[(pu+q)^n]_{u=1}=1.\end{aligned}\right\} \tag{617}$$

Consider now the mean and the mean square deviation of x. By definition

$$\langle x\rangle_{\text{Av}}=\sum_{x=1}^{n}xw(x) \tag{618}$$

and

$$\delta^2=\langle(x-\langle x\rangle_{\text{Av}})^2\rangle_{\text{Av}}=\langle x^2\rangle_{\text{Av}}-\langle x\rangle_{\text{Av}}{}^2=\sum_{x=1}^{n}x^2w(x)-\langle x\rangle_{\text{Av}}{}^2. \tag{619}$$

We have

$$\left.\begin{aligned} \langle x\rangle_{\text{Av}} &= \sum_{x=1}^{n} x\times\{\text{coefficient of } u^x \text{ in } (pu+q)^n\}, \\ &= \sum_{x=1}^{n} \text{coefficient of } u^x \text{ in } \frac{d}{du}(pu+q)^n, \\ &= \left[\frac{d}{du}(pu+q)^n\right]_{u=1} = np(p+q). \end{aligned}\right\} \tag{620}$$

Hence

$$\langle x\rangle_{\text{Av}} = np. \tag{621}$$

Similarly,

$$\left.\begin{aligned} \langle x^2\rangle_{\text{Av}} &= \sum_{x=1}^{n} x^2\times\{\text{coefficient of } u^x \text{ in } (pu+q)^n\}, \\ &= \sum_{x=1}^{n} \text{coefficient of } u^x \text{ in } \frac{d}{du}\left(u\frac{d}{du}[pu+q]^n\right), \\ &= \left\{\frac{d}{du}\left(u\frac{d}{du}[pu+q]^n\right)\right\}_{u=1}, \end{aligned}\right\} \tag{622}$$

or,

$$\langle x^2\rangle_{\text{Av}} = np+n(n-1)p^2. \tag{623}$$

Combining Eqs. (619), (621) and (623) we obtain

$$\delta^2 = np-np^2 = np(1-p) = npq. \tag{624}$$

II. A PROBLEM IN PROBABILITY: MULTIVARIATE GAUSSIAN DISTRIBUTIONS

In Chapter I (§4, subsection [a]) we considered the special case of the problem of random flights in which the N displacements which the particle suffers are all governed by Gaussian distributions but with different variances. We shall now consider a generalization of this problem which has important applications to the theory of Brownian motion (see Chapter II, §2, lemma II).

Let

$$\boldsymbol{\Psi} = \sum_{j=1}^{N} \psi_j \boldsymbol{r}; \quad \boldsymbol{\Phi} = \sum_{j=1}^{N} \phi_j \boldsymbol{r}, \tag{625}$$

where the ψ_j's and the ϕ_j's are two arbitrary sets of N real numbers each, and where further $\boldsymbol{r}$ is a stochastic variable the probability distribution of which is governed by

$$\tau(\boldsymbol{r}) = (1/(2\pi l^2)^{3/2}) \exp\left(-|\boldsymbol{r}|^2/2l^2\right), \tag{626}$$

where l is a constant. We require the probability $W(\boldsymbol{\Psi}, \boldsymbol{\Phi})d\boldsymbol{\Psi}d\boldsymbol{\Phi}$ that $\boldsymbol{\Psi}$ and $\boldsymbol{\Phi}$ shall lie, respectively, in the ranges $(\boldsymbol{\Psi}, \boldsymbol{\Psi}+d\boldsymbol{\Psi})$ and $(\boldsymbol{\Phi}, \boldsymbol{\Phi}+d\boldsymbol{\Phi})$. Applying Markoff's method to this problem, we have [cf. Eqs. (51) and (52)]

$$W(\boldsymbol{\Psi}, \boldsymbol{\Phi}) = \frac{1}{64\pi^6}\int_{-\infty}^{+\infty}\int_{-\infty}^{+\infty} \exp\left[-i(\boldsymbol{\varrho}\cdot\boldsymbol{\Psi}+\boldsymbol{\sigma}\cdot\boldsymbol{\Phi})\right] A_N(\boldsymbol{\varrho}, \boldsymbol{\sigma})\, d\boldsymbol{\varrho}\, d\boldsymbol{\sigma}, \tag{627}$$

where $\boldsymbol{\varrho}$ and $\boldsymbol{\sigma}$ are two auxiliary vectors and

$$A_N(\boldsymbol{\varrho}, \boldsymbol{\sigma}) = \prod_{j=1}^{N} \frac{1}{(2\pi l^2)^{3/2}} \int_{-\infty}^{+\infty} \exp\left[i(\boldsymbol{\varrho}\cdot\psi_j\boldsymbol{r}+\boldsymbol{\sigma}\cdot\phi_j\boldsymbol{r})\right] \exp\left(-|\boldsymbol{r}|^2/2l^2\right) d\boldsymbol{r}. \tag{628}$$

To evaluate $A_N(\boldsymbol{\varrho}, \boldsymbol{\sigma})$ we need the value of the typical integral

$$J = \frac{1}{(2\pi l^2)^{\frac{3}{2}}} \int_{-\infty}^{+\infty} \exp\left[i\mathbf{r}\cdot(\psi_j\boldsymbol{\varrho}+\phi_j\boldsymbol{\sigma}) - (|\mathbf{r}|^2/2l^2)\right] d\mathbf{r}. \tag{629}$$

We have

$$\left.\begin{aligned} J &= \prod_{x,y,z} \frac{1}{(2\pi l^2)^{\frac{1}{2}}} \int_{-\infty}^{+\infty} \exp\{-[x^2+2il^2x(\rho_1\psi_j+\sigma_1\phi_j)]/2l^2\} dx, \\ &= \exp\{-l^2[(\rho_1\psi_j+\sigma_1\phi_j)^2+(\rho_2\psi_j+\sigma_2\phi_j)^2+(\rho_3\psi_j+\sigma_3\phi_j)^2]/2\}. \end{aligned}\right\} \tag{630}$$

Hence

$$\left.\begin{aligned} A_N(\boldsymbol{\varrho}, \boldsymbol{\sigma}) &= \exp\{-l^2 \sum_{j=1}^{N} [(\rho_1\psi_j+\sigma_1\phi_j)^2+(\rho_2\psi_j+\sigma_2\phi_j)^2+(\rho_3\psi_j+\sigma_3\phi_j)^2]/2\} \\ &= \exp[-(P|\boldsymbol{\varrho}|^2+2R\boldsymbol{\varrho}\cdot\boldsymbol{\sigma}+Q|\boldsymbol{\sigma}|^2)/2], \end{aligned}\right\} \tag{631}$$

where we have written

$$P = l^2 \sum_{j=1}^{N} \psi_j^2; \quad R = l^2 \sum_{j=1}^{N} \phi_j\psi_j; \quad Q = l^2 \sum_{j=1}^{N} \phi_j^2. \tag{632}$$

Substituting for $A_N(\boldsymbol{\varrho}, \boldsymbol{\sigma})$ from Eq. (632) in the formula for $W(\boldsymbol{\Psi}, \boldsymbol{\Phi})$ [Eq. (627)] we obtain

$$W(\boldsymbol{\Psi}, \boldsymbol{\Phi}) = \frac{1}{64\pi^6} \prod_{i=1}^{3} \int_{-\infty}^{+\infty}\int_{-\infty}^{+\infty} \exp\{-[P\rho_i^2+2R\rho_i\sigma_i+Q\sigma_i^2+2i(\rho_i\Psi_i+\sigma_i\Phi_i)]/2\} d\rho_i d\sigma_i. \tag{633}$$

To evaluate the integrals occurring in the foregoing formula, we first perform a translation of the coordinate system according to

$$\rho_i = \xi_i + \alpha_i; \quad \sigma_i = \eta_i + \beta_i \quad (i = 1, 2, 3), \tag{634}$$

where α_i and β_i are so chosen that

$$P\alpha_i + R\beta_i = -i\Psi_i; \quad R\alpha_i + Q\beta_i = -i\Phi_i \quad (i = 1, 2, 3). \tag{635}$$

With this transformation of the variables we have

$$\left.\begin{aligned} P\rho_i^2+2R\rho_i\sigma_i+Q\sigma_i^2+2i(\rho_i\Psi_i+\sigma_i\Phi_i) &= P\xi_i^2+2R\xi_i\eta_i+Q\eta_i^2+i(\alpha_i\Psi_i+\beta_i\Phi_i), \\ &= P\xi_i^2+2R\xi_i\eta_i+Q\eta_i^2+\frac{1}{PQ-R^2}(P\Phi_i^2-2R\Phi_i\Psi_i+Q\Psi_i^2). \end{aligned}\right\} \tag{636}$$

Hence,

$$W(\boldsymbol{\Psi}, \boldsymbol{\Phi}) = \frac{1}{64\pi^6} \prod_{i=1}^{3} \exp[-(P\Phi_i^2-2R\Phi_i\Psi_i+Q\Psi_i^2)/2(PQ-R^2)] \times \int_{-\infty}^{+\infty}\int_{-\infty}^{+\infty} \exp[-(P\xi_i^2+2R\xi_i\eta_i+Q\eta_i^2)/2] d\xi_i d\eta_i. \tag{637}$$

From this equation we readily find that

$$W(\boldsymbol{\Psi}, \boldsymbol{\Phi}) = [1/8\pi^3(PQ-R^2)^{\frac{3}{2}}] \exp[-(P|\boldsymbol{\Phi}|^2-2R\boldsymbol{\Psi}\cdot\boldsymbol{\Phi}+Q|\boldsymbol{\Psi}|^2)/2(PQ-R^2)], \tag{638}$$

which gives the required probability distribution.

III. THE POISSON DISTRIBUTION AS THE LAW OF DENSITY FLUCTUATIONS

Consider an element of volume v which is a part of a larger volume V. Let there be N particles distributed in a random fashion inside the volume V. Under these conditions the probability that a particular particle will be found in the element of volume v is clearly v/V; similarly, the probability that it will *not* be found inside v is $(V-v)/V$. Hence, the probability $W_N(n)$ that *some* n particles will be found inside v is given by the Bernoulli distribution

$$W_N(n)=\frac{N!}{n!(N-n)!}\left(\frac{v}{V}\right)^n\left(1-\frac{v}{V}\right)^{N-n} \tag{639}$$

The average value of n is therefore given by [cf. Eq. (621)]

$$\langle n\rangle_{\text{Av}}=N(v/V)=\nu \quad \text{(say)}. \tag{640}$$

In terms of ν Eq. (639) can be expressed in the form

$$W_N(n)=\frac{N!}{n!(N-n)!}\left(\frac{\nu}{N}\right)^n\left(1-\frac{\nu}{N}\right)^{N-n}. \tag{641}$$

The case of greatest practical interest arises when both N and V tend to infinity but in such a way that ν remains constant [see Eq. (640)]. To obtain the corresponding limiting form of the distribution (641) we first rewrite it as

$$\left.\begin{aligned} W_N(n)&=\frac{1}{n!}N(N-1)(N-2)\cdots(N-n+1)\left(\frac{\nu}{N}\right)^n\left(1-\frac{\nu}{N}\right)^{N-n},\\ &=\frac{\nu^n}{n!}1\left(1-\frac{1}{N}\right)\left(1-\frac{2}{N}\right)\cdots\left(1-\frac{n-1}{N}\right)\left(1-\frac{\nu}{N}\right)^{N-n}, \end{aligned}\right\} \tag{642}$$

and then let $N\rightarrow\infty$ keeping both ν and n fixed. We have

$$\left.\begin{aligned} W(n)&=\lim_{N\rightarrow\infty} W_N(n),\\ &=\frac{\nu^n}{n!}\lim_{N\rightarrow\infty}\left\{\left(1-\frac{1}{N}\right)\left(1-\frac{2}{N}\right)\cdots\left(1-\frac{n-1}{N}\right)\left(1-\frac{\nu}{N}\right)^{N-n}\right\},\\ &=\frac{\nu^n}{n!}\lim_{N\rightarrow\infty}\left(1-\frac{\nu}{N}\right)^{N}. \end{aligned}\right\} \tag{643}$$

Hence,

$$W(n)=\nu^n e^{-\nu}/n!, \tag{644}$$

which is the required Poisson distribution.

In some applications of Eq. (644) (e.g., III, §3) ν is a very large number; and when this is the case, interest is attached to only those values of n which are relatively close to ν. We shall now show that under these conditions the Poisson distribution specializes still further to a Gaussian distribution.

Rewriting Eq. (644) in the form

$$\log W(n)=n\log\nu-\nu-\log n! \tag{645}$$

and adopting Stirling's approximation for $\log n$ [cf. Eq. (7)] we obtain

$$\log W(n)=n\log\nu-\nu-(n+\tfrac{1}{2})\log n+n-\tfrac{1}{2}\log 2\pi+O(n^{-1}). \tag{646}$$

Let

$$n=\nu+\delta. \tag{647}$$

Equation (646) becomes

$$\log W(n)=-(\nu+\delta+\tfrac{1}{2})\log\left(1+\frac{\delta}{\nu}\right)+\delta-\tfrac{1}{2}\log(2\pi\nu)+O(n^{-1}). \tag{648}$$

If we now suppose that $\delta/\nu \ll 1$ we can expand the logarithmic term in Eq. (648) as a power series in δ/ν. Retaining only the dominant term, we find

$$\log W(n) = -(\delta^2/2\nu) - \tfrac{1}{2}\log(2\pi\nu) \quad (\nu \to \infty;\ \delta/\nu \to 0). \tag{649}$$

Thus,

$$W(n) = [1/(2\pi\nu)^{\frac{1}{2}}] \exp[-(n-\nu)^2/2\nu], \tag{650}$$

which is the required Gaussian form.

IV. THE MEAN AND THE MEAN SQUARE DEVIATION OF THE SUM OF TWO PROBABILITY DISTRIBUTIONS

Let $w_1(x)$ and $w_2(y)$ represent two probability distributions. For the sake of definiteness we shall suppose that x and y take on only discrete values. A probability distribution which is said to be the *sum* of the two distributions is defined by

$$w(z) = \sum_{x+y=z} w_1(x)w_2(y), \tag{651}$$

where in the summation on the right-hand side we include all pairs of values of x and y (each in their respective domains) which satisfy the relation $x+y=z$. We may first verify that $w(z)$ defined according to Eq. (651) does in fact represent a probability distribution. To see this we have only to show that $\sum w(z)=1$. Now,

$$\sum_z w(z) = \sum_z \sum_{x+y=z} w_1(x)w_2(y); \tag{652}$$

accordingly, in the summation on the right-hand side, x and y can now run through their respective ranges of values *independently* of each other. Hence,

$$\textstyle\sum_z w(z) = [\sum_x w_1(x)][\sum_y w_2(y)] = 1. \tag{653}$$

We shall now prove that *the mean and the mean square deviation of the sum of two probability distributions is the sum of the means and the mean square deviations of the component distributions.*

To prove this theorem, we observe that by definitions

$$\langle z\rangle_{\text{Av}} = \sum_z z w(z) = \sum_z \sum_{x+y=z} (x+y) w_1(x)w_2(y), \tag{654}$$

or

$$\langle z\rangle_{\text{Av}} = \textstyle\sum_x \sum_y [x w_1(x)w_2(y) + y w_1(x)w_2(y)], \tag{655}$$

where in the summations on the right-hand side we can again let x and y run their respective ranges of values independently of each other. Hence,

$$\langle z\rangle_{\text{Av}} = [\textstyle\sum_x x w_1(x)][\sum_y w_2(y)] + [\sum_x w_1(x)][\sum_y y w_2(y)], \tag{656}$$

or

$$\langle z\rangle_{\text{Av}} = \langle x\rangle_{\text{Av}} + \langle y\rangle_{\text{Av}}. \tag{657}$$

Similarly,

$$\left.\begin{aligned}
\langle (z-\langle z\rangle_{\text{Av}})^2\rangle_{\text{Av}} &= \textstyle\sum_z (z-\langle z\rangle_{\text{Av}})^2 w(z),\\
&= \sum_z \sum_{x+y=z} (x+y-\langle x\rangle_{\text{Av}} - \langle y\rangle_{\text{Av}})^2 w_1(x)w_2(y).\\
&= \textstyle\sum_x \sum_y [(x-\langle x\rangle_{\text{Av}})^2 + 2(x-\langle x\rangle_{\text{Av}})(y-\langle y\rangle_{\text{Av}}) + (y-\langle y\rangle_{\text{Av}})^2] w_1(x)w_2(y),\\
&= [\textstyle\sum_x (x-\langle x\rangle_{\text{Av}})^2 w_1(x)][\sum_y w_2(y)] + [\sum_x w_1(x)][\sum_y (y-\langle y\rangle_{\text{Av}})^2 w_2(y)]\\
&\qquad + 2[\textstyle\sum_x (x-\langle x\rangle_{\text{Av}}) w_1(x)][\sum_y (y-\langle y\rangle_{\text{Av}}) w_2(y)].
\end{aligned}\right\} \tag{658}$$

Hence,

$$\langle(z-\langle z\rangle_{Av})^2\rangle_{Av}=\langle(x-\langle x\rangle_{Av})^2\rangle_{Av}+\langle(y-\langle y\rangle_{Av})^2\rangle_{Av}. \tag{659}$$

The theorem is now proved.

The extension of the foregoing results to include the case when x and y are continuously variable is, of course, obvious. Similarly the definitions and results can be further extended to include the sums of more than two probability distributions.

V. ZERMELO'S PROOF OF POINCARÉ'S THEOREM CONCERNING THE QUASI-PERIODIC CHARACTER OF THE MOTIONS OF A CONSERVATIVE DYNAMICAL SYSTEM

Consider a conservative dynamical system of n degrees of freedom and which is described by a Hamiltonian function H of the generalized coordinates $q_1, \cdots, q_n$ and momenta $p_1, \cdots, p_n$. The state of such a dynamical system can be represented by a point in the $2n$ dimensional phase space of the q's and p's. Similarly, the trajectory described by the representative point will describe the evolution of the dynamical system.

Through each point in the phase space there passes a unique trajectory which can be derived from the canonical equations of motion

$$\dot{q}_s=\frac{\partial H}{\partial p_s};\quad \dot{p}_s=-\frac{\partial H}{\partial q_s}\quad (s=1, \cdots, n). \tag{660}$$

More generally, consider any arbitrary continuous domain of points g_0 (of finite measure) in the phase space. Let the points g_0 be the representatives at time $t=0$ of an ensemble of dynamical systems all described by the same Hamiltonian function $H(p_1, \cdots, p_n; q_1, \cdots, q_n)$. At a later time t the representatives of the ensemble will occupy a continuous domain of points g_t which can be obtained by tracing through each point of g_0 the corresponding trajectory and following along the various trajectories for a time t. Because of the uniqueness, in general, of the trajectories passing through a given point in the phase space, the construction of the domain g_t from an initial domain g_0 is a unique process. We shall accordingly refer to g_t as the *future phase* (at time t) of the *initial phase* g_0 (at time $t=0$) of the given dynamical system.

Now, according to Liouville's theorem of classical dynamics, the density of any element of phase space remains constant during its motion according to the canonical Eqs. (660). Hence, if ω_t denotes the volume extension of the domain of points g_t introduced in the preceding paragraph, it follows from Liouville's theorem that ω_t remains constant as t varies.

We have already described how from an initial phase g_0 we can derive the future phase g_t at time t. The domain of points g_0 together with *all* its future phases g_t, $(0<t<\infty)$ clearly form a continuous domain of points which we shall denote by Γ_0: Γ_0 is accordingly the class of all states which at some finite past occupied states belonging to g_0. The extension of Γ_0 will be finite if we are considering a dynamical system which is enclosed—for, then, none of the coordinates or momenta can take on infinite values and the entire accessible region of the phase space remains finite. We shall suppose that this is the case and denote by Ω_0 the extension of Γ_0. Clearly $\Omega_0 \geqslant \omega_0$. In a similar manner we can, quite generally, define the domain of points Γ_t which includes all the future phases of g_t. Let Ω_t denote the extension of Γ_t. It is evident that

$$\Omega_{t_1} \geqslant \Omega_{t_2} \quad \text{whenever} \quad t_1<t_2. \tag{661}$$

For, Ω_{t_1} denoting the extension of *all* the future phases of g_{t_1} must therefore necessarily include also the future phases of g_{t_2} if $t_1<t_2$. On the other hand, considering Γ_0 itself as a domain of points, we can construct *its* future phases in exactly the same way as the future phases g_t of g_0 were constructed. But the future phase of Γ_0 after a time t is clearly Γ_t. And therefore applying Liouville's theorem to the domain Γ_0 and its future phases Γ_t we conclude that

$$\Omega_t=\text{constant}. \tag{662}$$

Comparing this result with the inequality (661) we infer that *the domain of points* Γ_t *can differ from* Γ_0 *by at most a set of points of measure zero*. Hence, the future phases of $g_t(t>0$ but arbitrary otherwise) must include g_0 apart, possibly, from a set of points of measure zero. But the points of g_t are themselves future phases of the points of g_0. Hence, the states belonging to g_0 (again, with the possible exception of a set of zero measure) must recur after the elapse of a sufficient length of time; and this is true no matter how small the extension ω_0 of g_0 is, provided it is only finite. From this, the deduction of Poincaré's theorem is immediate. (For a formal statement of Poincaré's theorem see Chapter III, §4).

VI. BOLTZMANN'S ESTIMATE OF THE PERIOD OF A POINCARÉ CYCLE

To estimate the order of magnitude of the period of a Poincaré cycle, Boltzmann has considered the following typical example:

A cubic centimeter of air containing 10^{18} molecules is considered in which all the molecules are initially supposed to have a speed of 500 meters per second. With a concentration of 10^{18} molecules, the average distance between the neighboring ones is of the order of 10^{-6} cm. Also, under normal conditions, each molecule will suffer something like 4×10^9 collisions per second so that on the whole there will occur

$$b=2\times10^{27} \text{ collisions per second.} \tag{663}$$

Since Poincaré's theorem asserts only the quasi-periodic character of the motions (see Chapter III, §4 and Appendix V) the period to be estimated clearly depends on the closeness to which we require the initial conditions to recur. For the case under discussion Boltzmann supposes that a molecule can be said to have approximately returned to its initial state if the differences in position (x, y, z) and velocity (u, v, w) in the initial and the final states are such that

$$|\Delta x|,\ |\Delta y|,\ |\Delta z| \leqslant 10^{-7} \text{ cm,} \tag{664}$$

and

$$|\Delta u|,\ |\Delta v|,\ |\Delta w| \leqslant 1 \text{ m/sec.} \tag{665}$$

In other words, we shall require the positions to agree to within 10 percent of the average distance between the molecules and the velocities to agree within one part in 500.

We shall first estimate the order of magnitude of the time required for the recurrence of an initial "abnormal" distribution in the velocities. According to Poincaré's theorem, an initial state need not recur earlier than the time necessary for all the molecules to take on all the possible values for the velocity. We can readily determine the number N of such possibilities with the understanding that we agree to distinguish between two velocities only if at least one of the components differ by more than 1 m/sec.

The first molecule can have all velocities ranging from zero to $a=500\times10^9$ m/sec.—since we have supposed that in the initial state all the molecules have the same speed of 500 m/sec and that there are 10^{18} molecules in the system. Again, if the first molecule has a speed v_1 the second one can have speeds only in range 0 to $(a^2-v_1^2)^{\frac{1}{2}}$. Similarly, if the first and the second molecules have speeds v_1 and v_2, respectively, the third molecule can have speeds only in the range 0 to $(a^2-v_1^2-v_2^2)^{\frac{1}{2}}$; and so on. Accordingly, the required number of combinations N is

$$\left.\begin{aligned} N&=(4\pi)^{n-1}\int_0^a dv_1v_1^2\int_0^{(a^2-v_1^2)^{\frac{1}{2}}} dv_2v_2^2\int_0^{(a^2-v_1^2-v_2^2)^{\frac{1}{2}}} dv_3v_3^2\cdots\int_0^{(a^2-v_1^2\cdots-v^2_{n-2})^{\frac{1}{2}}} dv_{n-1}v^2_{n-1},\\ &=(\pi^{(3n-3)/2}/2\cdot3\cdot4\cdots[3(n-1)/2])a^{3(n-1)} \quad (n,\text{ odd}),\\ &=(2(2\pi)^{(3n-4)/2}/3\cdot5\cdot7\cdots3(n-1))a^{3(n-1)} \quad (n,\text{ even}),\end{aligned}\right\} \tag{666}$$

where

$$a=500\times10^9 \quad\text{and}\quad n=10^{18}. \tag{667}$$

Since each of these N combinations occurs on the average in a time $1/b$ seconds [cf. Eq. (663)] the total time required for the velocities to run through all the possible values is

$$N/b. \tag{668}$$

After this length of time we may expect the initial distribution of the velocities to recur to within the limits of accuracy specified except for one single molecule the direction of whose motion has been left unrestricted. On the other hand we have still left unspecified the positions of the centers of gravity of all the molecules. But in order that we may say that the initial state has recurred to a sufficient approximation, we must require the positions of the molecules in the final state also to agree with the initial values to some stated degree of accuracy. This would clearly require the time (668) to be multiplied by another number of order similar to N. However, the extremely large value already of N/b gives some indication of the enormous times which are involved Moreover, comparing these times with the time of relaxation of a gas which is of the order 10^{-8} second under normal conditions, we get an idea as to how extremely small the fraction of the total number of complexions is for which appreciable departures from a Maxwellian distribution occur. (For a further discussion of these and related matters see Chapter III, §4.)

VII. THE LAW OF DISTRIBUTION OF THE NEAREST NEIGHBOR IN A RANDOM DISTRIBUTION OF PARTICLES

This problem was first considered by Hertz (see reference 71 in the Bibliographical Notes for Chapter IV).

Let $w(r)dr$ denote the probability that the nearest neighbor to a particle occurs between r and $r+dr$. This probability must be clearly equal to the probability that no particles exist interior to r times the probability that a particle does exist in the spherical shell between r and $r+dr$. Accordingly, the function $w(r)$ must satisfy the relation

$$w(r)=\left[1-\int_0^r w(r)dr\right]4\pi r^2 n, \tag{669}$$

where n denotes the average number of particles per unit volume. From Eq. (669) we derive:

$$\frac{d}{dr}\left[\frac{w(r)}{4\pi r^2 n}\right]=-4\pi r^2 n\frac{w(r)}{4\pi r^2 n}. \tag{670}$$

Hence

$$w(r)=\exp\left(-4\pi r^3 n/3\right)4\pi r^2 n, \tag{671}$$

since, according to Eq. (669)

$$w(r)\rightarrow 4\pi r^2 n \quad \text{as} \quad r\rightarrow 0. \tag{672}$$

Equation (671) gives then the required law of distribution of the nearest neighbor.

Using the distribution (671) we can derive an *exact* formula for the "average distance" D between the particles. For, by definition

$$D=\int_0^\infty rw(r)dr, \tag{673}$$

or, if we use Eq. (671)

$$D=\int_0^\infty \exp\left(-4\pi r^3 n/3\right)4\pi r^3 n dr. \tag{674}$$

After some elementary reductions, Eq. (674) becomes

$$\left.\begin{aligned} D&=\frac{1}{(4\pi n/3)^{\frac{1}{3}}}\int_0^\infty e^{-x}x^{\frac{1}{3}}dx,\\ &=\Gamma(4/3)/(4\pi n/3)^{\frac{1}{3}}. \end{aligned}\right\} \quad (675)$$

Substituting for $\Gamma(4/3)$, we find

$$D=0.55396n^{-\frac{1}{3}}. \quad (676)$$

BIBLIOGRAPHICAL NOTES

Chapter I

§*1*.—We may briefly record here the history of the problem of random flights considered in this chapter:

Karl Pearson appears to have been the first to explicitly formulate a problem of this general type:

1. K. Pearson, Nature **77**, 294 (1905). Pearson's formulation of the problem was in the following terms: "A man starts from a point O and walks l yards in a straight line; he then turns through any angle whatever and walks another l yards in a second straight line. He repeats this process n times. I require the probability that after these n stretches he is at a distance between r and $r+dr$ from his starting point O." After Pearson had formulated this problem Lord Rayleigh pointed out that the problem is formally "the same as that of the composition of n isoperiodic vibrations of unit amplitude and of phases distributed at random" which he had considered as early as in 1880:

2. Lord Rayleigh, Phil. Mag. **10**, 73 (1880); see also ibid. **47**, 246 (1889). These papers are reprinted in *Scientific Papers of Lord Rayleigh*, Vol. I, p. 491, and Vol. IV, p. 370. In the foregoing papers Rayleigh obtains the asymptotic form of the solution as $n\to\infty$. But for finite values of n the general solution of Pearson's problem was given by

3. J. C. Kluyver, Konink. Akad. Wetenschap. Amsterdam **14**, 325 (1905). The general solution of the problem of random walk in one dimension was obtained by Smoluchowski apparently independently of the earlier investigators.

4. M. von Smoluchowski, Bull. Acad. Cracovie, p. 203 (1906). In its most general form the problem of random flights was formulated by A. A. Markoff who also outlined the method for obtaining the general solution.

5. A. A. Markoff, *Wahrscheinlichkeitsrechnung* (Leipzig, 1912), §§16 and 33.

§*2*.—The problem of the random walk with reflecting and absorbing barriers was first considered by Smoluchowski:

6. M. v. Smoluchowski, (a) Wien Ber. **124**, 263 (1915); also (b) "Drei Vortrage uber Diffusion, Brownsche Bewegung und Koagulation von Kolloidteilchen," Physik. Zeits. **17**, 557, 585 (1916). See also

7. R. von Mises, *Wahrscheinlichkeitsrechnung* (Leipzig and Wien), pp. 479–518.

§*3*.—Markoff's method described in this section is a somewhat generalized version of what is given in Markoff (reference 5). See also

8. M. von Laue, Ann. d. Physik **47**, 853 (1915).

§*4*.—See A. A. Markoff (reference 5). The case of finite N considered in subsection (b) follows the treatment of

9. Lord Rayleigh, Phil. Mag. **37**, 321 (1919) (or *Scientific Papers*, Vol. VI, p. 604).

§*5*.—The passage to a differential equation for the case of the one-dimensional problem of the random walk was achieved by Rayleigh:

10. Lord Rayleigh, Phil. Mag. **47**, 246 (1899) (or *Scientific Papers*, Vol. IV, p. 370). See also Smoluchowski (reference 6). But the general treatment given in this section appears to be new.

We may also note the following further reference:

11. W. H. McCrea, Proc. Roy. Soc. Edinburgh **60**, 281 (1939).

Chapter II

The following general references may be noted.

12. The Svedberg, *Die Existenz der Molekule* (Leipzig, 1912).

13. G. L. de Haas-Lorentz, *Die Brownsche-Bewegung und einige verwandte Erscheinungen*, (Braunschweig, 1913).

14. M. v. Smoluchowski, see reference 6(b).

15. J. Perrin, *Atoms* (Constable, London, 1916).

16. R. Fürth, *Schwankungserscheinungen in der Physik* (Sammlung Vieweg, Braunschweig, 1920), Vol. 48.

§*1*.—As is well known the modern theory of Brownian motion was initiated by Einstein and Smoluchowski:

17. A. Einstein, Ann. d. Physik **17**, 549 (1905); also, ibid. **19**, 371 (1906).

18. M. v. Smoluchowski, Ann. d. Physik **21**, 756 (1906). In Einstein's and in Smoluchowski's treatment of the problem, Brownian motion is idealized as a problem in random flights; but as we have seen, this idealization is valid only when we ignore effects which occur in time intervals of order β^{-1}. For the general treatment of the problem we require to base our discussion on an equation of the type first introduced by Langevin:

19. P. Langevin, Comptes rendus **146**, 530 (1908). In this connection see

20. F. Zernike, *Handbuch der Physik* (Berlin, 1928), Vol. 3, p. 456.

§*2*.—The treatment of the Brownian motion of a free particle given in this section is derived from:

21. L. S. Ornstein and W. R. van Wijk, Physica **1**, 235 (1933). See also

22. W. R. van Wijk, Physica **3**, 1111 (1936). Earlier, but somewhat less general treatment along the same lines is contained in

23. G. E. Uhlenbeck and L. S. Ornstein, Phys. Rev. **36**, 823 (1930). In the foregoing papers the discussion has been carried out only for the case of one-dimensional motion. In the text we have treated the general three-dimensional problem; further, the arguments in references 21 and 22 have been rearranged considerably to make the presentation more direct and straightforward.

§*3*.—See Ornstein and Wijk (reference 21); also

24. G. E. Uhlenbeck and S. Goudsmidt, Phys. Rev. **34**, 145 (1929).

25. G. A. van Lear and G. E. Uhlenbeck, Phys. Rev. **38**, 1583 (1931).

§*4*.—The passage to a differential equation for the description of the Brownian motion of a free particle in the velocity space was achieved by

26. A. D. Fokker, Ann. d. Physik **43**, 812 (1914). A more general discussion of this problem is due to

27. M. Planck, Sitz. der preuss. Akad. p. 324 (1917). See also references 21 and 23; further,

28. Lord Rayleigh, *Scientific Papers*, Vol. III, p. 473.

29. L. S. Ornstein, Versl. Acad. Amst. **26**, 1005 (1917); also Konink. Akad. Wetenschap. Amsterdam **20**, 96 (1917).

30. H. C. Burger, Versl. Acad. Amst. **25**, 1482 (1917).

31. L. S. Ornstein and H. C. Burger, Versl. Acad. Amst. **27**, 1146 (1919); **28**, 183 (1919); also Konink. Akad. Wetenschap. Amsterdam **21**, 922 (1918).

Earlier attempts to generalize Liouville's equation of classical dynamics to include Brownian motion are contained in

32. O. Klein, Arkiv for Matematik, Astronomi, och Fysik **16**, No. 5 (1921); and

33. H. A. Kramers, Physica **7**, 284 (1940).

The passage to a differential equation in configuration space was first achieved by

34. M. v. Smoluchowski, Ann. d. Physik **48**, 1103 (1915); see also,

35. R. Fürth, Ann. d. Physik **53**, 177 (1917).

In the text the discussion of the various differential equations has been carried out more generally and more completely than in the references given above; this is particularly true of the discussion relating to the generalization of the Liouville equation of classical dynamics (subsections, ii–v).

§*5*.—See H. A. Kramers (reference 33).

Approaches to the problem of the Brownian motion somewhat different to the one we have adopted are contained in

36. G. Krutkov, Physik. Zeits. der Sowjetunion **5**, 287 (1934). See also the various articles by the same author in C. R. Acad. Sci. USSR during the years (1934) and (1935).

37. S. Bernstein, C. R. Acad. Sci. USSR, p. 1 (1934), and p. 361 (1934). A more particularly mathematical discussion of the problems of Brownian motion has been given by

38. J. L. Doob, Ann. Math. **43**, 351 (1942); see also the references given in this paper.

Chapter III

The following general references may be noted.

39. M. v. Smoluchowski, reference 6(b).

40. A. Sommerfeld, "Zum Andenken an Marian von Smoluchowski," Physik. Zeits. **18**, 533 (1917).

41. R. Fürth, Physik. Zeits. **20**, 303, 332, 350, 375 (1919); also reference 16.

42. H. Freundlich, Kapillarchemie (Leipzig, 1930–1932), Vols. **I** and **II**; see particularly pp. 485–510 in Vol. **I** and pp. 140–162 in Vol. **II**.

43. The Svedberg, *Die Existenz der Molekule* (Leipzig, 1912).

In reference 39 we have an extremely valuable account of the entire subject of Brownian motion and molecular fluctuations; there exists no better introduction to this subject than these lectures of Smoluchowski. In reference 40 Sommerfeld gives a fairly extensive bibliography of Smoluchowski's writings.

§*1*.—The theory of density fluctuations as developed by Smoluchowski represents one of the most outstanding achievements in molecular physics. Not only does it quantitatively account for and clarify a wide range of physical and physico-chemical phenomena, it also introduces such fundamental notions as the "probability after-effect" which are of very great significance in other connections (see Chapter IV).

44. M. v. Smoluchowski, Wien. Ber. **123**, 2381 (1914); see also Physik. Zeits. **16**, 321 (1915) and Kolloid Zeits. **18**, 48 (1916). For discussions of the problem of density fluctuations prior to the introduction of the notion of the "speed of fluctuations" see

45. M. v. Smoluchowski, *Boltzmann Festschrift* (1904), p. 626; *Bull. Acad. Cracovie*, p. 1057, 1907; Ann. d. Physik **25**, 205 (1908). Also

46. R. Lorenz and W. Eitel, Zeits. f. physik. Chemie **87**, 293, 434 (1914).

It is of some interest to recall that referring to his deviation of the formulae for $\langle\Delta_n\rangle_{Av}$ and $\langle\Delta_n^2\rangle_{Av}$ [Eqs. (356) and (358)] Smoluchowski says, "Aus diesem komplizierten Formeln [referring to the formula for $W(n; m)$] lassen sich mittels verwickelter summationen merkwurdigerweise recht einfache resultate fur die durschschnittliche Änderung der Teilchenzahl ableiten. . . . So wie fur das Anderungsquadrat bei unbestimmter Anfangszahl n [Eq. (363)]." This led to some heated discussion whether these formulae cannot be derived more simply; for example, see

47. L. S. Ornstein, Konink. Akad. Wetenschap. Amsterdam **21**, 92 (1917). But neither Ornstein nor Smoluchowski seems to have noticed that the formulae for $\langle\Delta_n\rangle_{Av}$ and $\langle\Delta_n^2\rangle_{Av}$ can be derived very directly from the fact that the transition probability $W(n; m)$ is the sum (in a technical sense) of a Bernoulli and a Poisson dis-

tribution; it is to this fact that the simplicity of the results are due.

§*2*.—Comparisons between the predictions of his theory with the data of colloid statistics were first made by Smoluchowski himself (reference 44). The experiments which were used for these first comparisons were those of

48. The Svedberg, Zeits. f. physik. Chemie **77**, 147 (1911); see also references 43 and 46. But precision experiments carried out with expressed intention of verifying Smoluchowski's theory are those of

49. A. Westgren, Arkiv for Matematik, Astronomi, och Fysik **11**, Nos. 8 and 14 (1916) and **13**, No. 14 (1918).

An interesting application of Smoluchowski's theory to a problem of rather different sort has been made by Fürth:

50. R. Fürth, Physik. Zeits. **19**, 421 (1918); **20**, 21 (1919). Fürth made systematic counts of the number of pedestrians in a block every five seconds. This interval of five seconds was chosen because the length of the block was such that a pedestrian observed in the block on one occasion has an appreciable probability of remaining in the same block when the next observation is made five seconds later. We can, accordingly, define a probability after-effect factor P $(=v\tau/a$, where v is the average speed of a pedestrian, τ the chosen interval of time and a the length of the block), and Smoluchowski's theory applies. A statistical analysis of this data showed that the agreement with the theory is excellent. It is amusing that by systematic counts of the kind made by Fürth it is possible actually to determine the average speed of a pedestrian!

§*3*.—The theory outlined in this section is derived from

51. M. v. Smoluchowski, Wien. Ber. **124**, 339 (1915); see also references 39 and 41.

§*4*.—Among the early discussions on the compatibility between the notions of conventional thermodynamics and the then new standpoint of the kinetic molecular theory, we may refer to

52. J. Loschmidt, Wien. Ber. **73**, 139 (1876); **75**, 67 (1877).

53. L. Boltzmann, Wien. Ber. **75**, 62 (1877); **76**, 373 (1877); also Nature **51**, 413 (1895) and *Vorlesungen über Gas Theorie* (Leipzig, 1895) Vol. **I**, p. 42 (or the reprinted edition of 1923).

54. E. Zermelo, Ann. d. Physik **57**, 485 (1896); **59**, 793 (1896).

55. L. Boltzmann, Ann. d. Physik **57**, 773 (1896); **60**, 392 (1897).

Smoluchowski's fundamental discussions of the limits of validity of the second law of thermodynamics are contained in

56. M. v. Smoluchowski, Physik. Zeits. **13**, 1069 (1912); **14**, 261 (1913). See also references 39 and 51.

It is somewhat disappointing that the more recent discussions of the laws of thermodynamics contain no relevant references to the investigations of Boltzmann and Smoluchowski [e.g., P. W. Bridgman, *The Nature of Thermodynamics* (Harvard University Press, 1941)]. The absence of references, particularly to Smoluchowski, is to be deplored since no one has contributed so much as Smoluchowski to a real clarification of the fundamental issues involved.

For an exhaustive discussion of the foundations of statistical mechanics, see

57. P. and T. Ehrenfest, *Begriffliche Grundlagen der Statistischen Auffassung in der Mechanik, Encyklopadie der Mathematischen Wissenschaften* (1911), Vol. 4, p. 4. And for Carathéodory's version of thermodynamics see

57a. S. Chandrasekhar, *An Introduction to the Study of Stellar Structure* (University of Chicago Press, 1939), Chap. I, pp. 11–37.

§*5*.—See Smoluchowski, reference 39; also

58. M. v. Smoluchowski, Ann. d. Physik **48**, 1103 (1915).

59. R. Fürth, Ann. d. Physik **53**, 177 (1917).

§*6*.—See Smoluchowski reference 39; also

60. M. v. Smoluchowski, Zeits. f. physik. Chemie **92**, 129 (1917).

61. R. Zsigmondy, Zeits. f. physik. Chemie **92**, 600 (1917). The papers **60** and **61** contain references to the earlier literature on the subject of coagulation. For the more recent literature see Freundlich (reference 42, particularly Vol. **II**, pp. 140–162).

§*7*.—See

62. H. A. Kramers, Physica **7**, 284 (1940). Also,

63. H. Pelzer and E. Wigner, Zeits. f. physik. Chemie, **B15**, 445 (1932).

An aspect of the theory of Brownian motion we have not touched upon concerns the natural limit set by it to all measuring processes. But an excellent review of this entire field exists:

64. R. B. Barnes and S. Silverman, Rev. Mod. Phys. **6**, 162 (1934).

Chapter IV

The ideas developed in this chapter are in the main taken from

65. S. Chandrasekhar, Astrophys. J. **94**, 511 (1941).

66. S. Chandrasekhar and J. von Neumann, Astrophys. J. **95**, 489 (1942).

67. S. Chandrasekhar and J. von Neumann, Astrophys. J. **97**, 1, (1943).

§*1*.—See references 65, 66, and 67; also

68. S. Chandrasekhar, *Principles of Stellar Dynamics* (University of Chicago Press, 1942), Chapters II and V.

§*2*.—The problem considered in this section is clearly equivalent to finding the probability of a given *electric* field strength at a point in a gas composed of simple ions. This latter problem was first considered by Holtsmark:

69. J. Holtsmark, Ann. d. Physik **58**, 577 (1919); also Physik. Zeits. **20**, 162 (1919) and **25**, 73 (1924). Among other papers on related subjects we may refer to

70. R. Gans, Ann. d. Physik **66**, 396 (1921).

71. P. Hertz, Math. Ann. **67**, 387 (1909).

72. R. Gans, Physik. Zeits. **23**, 109 (1922).

73. C. V. Raman, Phil. Mag. **47**, 671 (1924).

§*3*.—See references 66 and 67. See also three further papers on "Dynamical Friction" by Chandrasekhar in forthcoming issues of *The Astrophysical Journal* where further applications of the Fokker-Planck equation will be found.

SEPTEMBER 1, 1930 PHYSICAL REVIEW VOLUME 36

ON THE THEORY OF THE BROWNIAN MOTION

By G. E. Uhlenbeck and L. S. Ornstein

University of Michigan, Ann Arbor and Physisch Laboratorium der R. U. Utrecht, Holland

(Received July 7, 1930)

Abstract

With a method first indicated by Ornstein the mean values of *all* the powers of the velocity u and the displacement s of a free particle in Brownian motion are calculated. It is shown that $u-u_0\exp(-\beta t)$ and $s-u_0/\beta[1-\exp(-\beta t)]$ where u_0 is the initial velocity and β the friction coefficient divided by the mass of the particle, follow the normal Gaussian distribution law. For s this gives the exact frequency distribution corresponding to the exact formula for $\overline{s^2}$ of Ornstein and Fürth. Discussion is given of the connection with the Fokker-Planck partial differential equation. By the same method exact expressions are obtained for the square of the deviation of a harmonically bound particle in Brownian motion as a function of the time and the initial deviation. Here the periodic, aperiodic and overdamped cases have to be treated separately. In the last case, when β is much larger than the frequency and for values of $t \gg \beta^{-1}$, the formula takes the form of that previously given by Smoluchowski.

I. General Assumptions and Summary

In the theory of the Brownian motion the first concern has always been the calculation of the mean square value of the displacement of the particle, because this could be immediately observed. As is well known, this problem was first solved by Einstein[1] in the case of a *free* particle. He obtained the famous formula:

$$\overline{s^2} = 2Dt = \frac{2kT}{f}t \tag{1}$$

where f is the friction coefficient, T the absolute temperature and t the time. The influence of the surrounding medium is characterized by f as well as by T. For this Einstein used the formula of Stokes, because almost always the particle is immersed in a liquid or gas at ordinary pressure. In that case the mean free path of the molecules is small compared with the particle, and we may consider the surrounding medium as continuous and may use the results hydrodynamics gives for the friction coefficient for bodies of simple form (sphere, ellipsoid etc.). This will depend on the viscosity coefficient of the medium and therefore be *independent* of the pressure.

But of course, when the surrounding medium is a rarefied gas (mean free path of the molecules great in comparison with the particle), the friction

[1] A. Einstein, Ann. d. Physik **17**, 549 (1905). This and the further articles of Einstein have been collected in a book called: "Investigations on the theory of the Brownian Movement". Edited by R. Fürth, translated by A. D. Cowper. New York, Dutton. To this we shall always refer.

will change in character. Instead of a Stokes friction, we then get what we may call a Doppler friction and this can also be calculated for simple forms of the particle. It is based on the fact that a particle moving, say to the right, will be hit by more molecules from the right than from the left. This friction coefficient will be proportional to the pressure. To cover all cases, we will always leave the friction coefficient explicitly in the formulas.

The basis of formula (1), which since Einstein has been derived in various other ways,[2] has been almost always the equation of motion:

$$m\frac{du}{dt} = -fu + F(t) \tag{2}$$

where u is the velocity of the particle. Characteristically of this equation, the influence of the surrounding medium is split into two parts:
(1) a systematic part $-fu$, which causes the friction
(2) a fluctuating part $F(t)$. Concerning this we will naturally make the following assumptions:

A: The mean of $F(t)$, at given t, over an ensemble of particles (a large number of similar, but independent particles), which have started at $t=0$, with the same velocity u_0, is zero. We will denote this by:

$$\overline{F(t)}^{u_0} = 0. \tag{3}$$

B: There will be correlation between the values of $F(t)$ at different times t_1 and t_2 only when $|t_1-t_2|$ is very small. More explicitly we shall suppose that:

$$\overline{F(t_1)F(t_2)}^{u_0} = \phi_1(t_1 - t_2) \tag{4}$$

where $\phi_1(x)$ is a function with a very sharp maximum at $x=0$. More generally, when $t_1, t_2 \ldots . t_{n+1}$ are all lying very near each other, we assume:

$$\overline{F(t_1)F(t_2)\cdots F(t_{n+1})}^{u_0} = \phi_n(r, \theta_1, \theta_2 \cdots \theta_{n-1}) \tag{5}$$

where r is the distance perpendicular to the line $t_1=t_2= \ldots . =t_{n+1}$ in the $(n+1)$ dimensional $(t_1, t_2 \ldots . t_{n+1})$ space, and $\theta_1, \theta_2 \ldots . \theta_{n-1}$ are $(n-1)$ angles to determine the position of r in the subspace perpendicular to this line. The function ϕ_n has again a very sharp maximum for $r=0$. Further, when $t_1, t_2 \ldots . t_k$ are lying near each other, and also $t_{k+1}, t_{k+2} \ldots . t_l$ but far from the group $t_1, t_2 \ldots . t_k$ and so on, then:

$$\overline{F(t_1)\cdots F(t_k)F(t_{k+1})\cdots F(t_l)F(t_{l+1})\cdots F(t_m)\cdots}^{u_0}$$
$$= \overline{F(t_1)\cdots F(t_k)}^{u_0}\cdot\overline{F(t_{k+1})\cdots F(t_l)}^{u_0}\cdot\overline{F(t_{l+1})\cdots F(t_m)}^{u_0}\cdots \tag{6}$$

The justification, or eventually the criticism, of these assumptions must come from a more precise, kinetic, theory. We will not go into that.

[2] Compare G. L. de Haas-Lorentz: Die Brownsche Bewegung (Braunschweig, Vieweg, 1913).

§3. In the later development, especially when given outside forces like gravitation were also considered, so that (2) had to be replaced by:

$$m\frac{du}{dt} = -fu + F(t) + K(x) \tag{2a}$$

the attention was fixed more on the determination of the frequency distribution of quantities like the displacement or the velocity. Given the value ϕ_0 of the quantity ϕ at $t=0$, we wish to find the probability $F(\phi_0, \phi, t)d\phi$ that after the time t the value lies between ϕ and $\phi+d\phi$. It is clear, that when we know $F(\phi_0, \phi, t)$ *all* mean values are determined. For instance:

$$\overline{\phi^k}^{\phi_0} = \int \phi^k F(\phi_0, \phi, t)d\phi.$$

The frequency distribution is the most general thing the theory can predict. In the case of a free particle, the function $F(x_0, x, t)$, which will now depend only on $x-x_0=s$, was already determined by Einstein. He found:

$$F(x_0, x, t) = \left(\frac{1}{4\pi Dt}\right)^{1/2} e^{-(x-x_0)^2/4Dt} \tag{7}$$

of which (1) is an immediate consequence. He derived this, by finding for F a partial differential equation, which in this case is the *diffusion equation:*

$$\frac{\partial z}{\partial t} = D\frac{\partial^2 z}{\partial s^2} \tag{8}$$

and of which $F(x_0, x, t)$ is then the so-called fundamental solution. This is that solution of (8) which for $t=0$ becomes $\delta(x-x_0)$, when $\delta(x)$ means the function, defined by the properties:

$$\delta(x) = 0 \quad \text{for} \quad x \neq 0$$

$$\int_{-\infty}^{+\infty} \delta(x)dx = 1$$

This is clear from the definition of $F(x_0, x, t)$ because for $t=0$, there is certainty that $x=x_0$. Further there are boundary conditions, which express the behavior of the particle at the walls; in the case of a completely free particle they are simply $F=0$ for $x=\pm\infty$. The relation between the diffusion coefficient D and the friction coefficient f, Einstein then derived very simply, using the osmotic pressure idea.

This connection between the frequency distribution function and a partial differential equation of the parabolic type like (8), has later been generlized considerably by Smoluchowski, Fokker, Planck, Ornstein, Burger,

Fürth and others.[3] The equation is generally called the Fokker-Planck equation. Especially for a particle under influence of outside forces, Smoluchowski showed that the generalization of (8) was:

$$\frac{\partial z}{\partial t} = -\frac{1}{f}\frac{\partial}{\partial x}(Kz) + D\frac{\partial^2 z}{\partial x^2} \tag{9}$$

For special forces (gravitation, elastic binding etc.) and by different boundary conditions, Smoluchowski, Fürth and others have determined the fundamental solution, and from this all sorts of mean values, which they have compared with experiments.

§4. With the results (1), (7), (8) and (9) of Einstein and Smoluchowski the problem seems completely solved. But there is one restriction, which was first stressed by Einstein. All these results hold only when t is large compared to m/f. The generalization of (1) for all times was given by Ornstein[4] and Fürth[5], independently of each other.
The result is:

$$\overline{s^2} = \frac{2mkT}{f^2}\left(\frac{f}{m}t - 1 + e^{-ft/m}\right) \tag{10}$$

For values of t large compared to m/f this becomes again Einstein's formula (1). For very short times on the other hand, we get:

$$\overline{s^2} = \frac{kT}{m}t^2 = \overline{u_0^2}t^2$$

as one would expect, because in the beginning the motion must be uniform.

The problem now arises to generalize the other results also. In part III we will do this for the frequency distribution $F(x_0, x, t)$. The result is rather complicated; for $t \gg m/f$ it goes over into (7), and (10) is an immediate consequence of it. The *method*, we used, was the momentum method. From the equation of motion (2), and using the assumptions (3) to (6), we could calculate the mean value of all the powers of

$$S = s - \frac{mu_0}{f}(1 - e^{-ft/m})$$

[3] M. v. Smoluchowski, Phys. Zeits. **17**, 557 (1916). A. Fokker, Dissertation Leiden, 1913, p. 000. M. Planck, Berl. Ber. p. 324, 1927. L. S. Ornstein, Versl. Acad. Amst. **26**, 1005 (1917). H. C. Burger, Versl. Acad. Amst; **25**, 1482 (1917); L. S. Ornstein and H. C. Burger, Versl. Acad. Amst. **27**, 1146 (1919); **28**, 183 (1919). R. Fürth, Ann. d. Physik **53**, 177 (1917). R. Fürth gives a survey in Riemann-Weber, Die Partiellen Differential-gleichungen der Mathematischen Physik (Edited by R. v. Mises and Ph. Frank, Braunschweig Vieweg 1928) Vol. II, p. 177. Comp. also the article of F. Zernike, Handbuch der Physik, Vol. III, p. 456 (Berlin, Springer, 1928).

[4] L. S. Ornstein, Versl. Acad. Amst. **26**, 1005 (1917) (=Proc. Acad. Amst. **21**, 96 (1919).

[5] R. Fürth, Zeits. f. Physik **2**, 244 (1920).

and prove that S follows the normal Gaussian distribution law. We did not succeed in generalizing the diffusion Eq. (8), and determining the distribution function by this method.

As a preparation we derive in part II the frequency distribution function $G(u_0, u, t)$ for the velocity of a free particle in Brownian motion, first with the momentum method, and then also with the Fokker-Planck equation.

This extension to short times becomes especially interesting in the case of outside periodic forces. In part IV we shall treat the problem of the Brownian motion of an elastically bound particle. By using the same method as before, we could get exact expressions for the mean square of the displacement as a function of the initial deviation and of the time. The periodic, aperiodic and overdamped cases have to be treated separately. The way in which the equipartition value is reached for $t \to \infty$ is different in the three cases. In the last case, for very strong damping and $t \gg m/f$ the formula goes over into the result of Smoluchowski, which is a consequence of the frequency distribution function following from (9).

II. The Frequency Distribution of the Velocity

§5. The problem is to determine the probability that a free particle in Brownian motion after the time t has a velocity which lies between u and $u+du$, when it started at $t=0$ with the velocity u_0.

The first method to solve the problem is by calculating all the mean values $\overline{u^k}$ for given u_0. As has first been shown by Ornstein[6] for $\bar{u}$ and $\overline{u^2}$, this is possible by integrating the equation of motion:

$$\frac{du}{dt} + \beta u = A(t)$$

when $\beta = f/m$ and $A = F/m$. Of course, the assumptions (3) to (6) hold for the fluctuating acceleration A(t), as well as for the fluctuating force F(t). Integrating we get:

$$u = u_0 e^{-\beta t} + e^{-\beta t} \int_0^t e^{\beta \xi} A(\xi) d\xi. \tag{11}$$

Taking the mean over an ensemble of particles, which have started at $t=0$ with the same velocity u_0, and using (3) we get:

$$\bar{u}^{u_0} = u_0 e^{-\beta t}. \tag{12}$$

The mean velocity goes down exponentially due to the friction. Squaring (11) and taking the mean, gives:

$$\overline{u^2}^{u_0} = u_0^2 e^{-2\beta t} + e^{-2\beta t} \int_0^t \int_0^t e^{\beta(\xi+\eta)} \overline{A(\xi)A(\eta)} d\xi d\eta.$$

[6] L. S. Ornstein, Proc. Acad. Amst. **21**, 96 (1919).

By taking $\xi+\eta=v$, $\xi-\eta=w$ as new variables and by using (4), we can write for the integral:

$$\frac{1}{2}e^{-2\beta t}\int_0^{2t} e^{\beta v}dv \int_{-\infty}^{+\infty}\phi_1(w)dw = \frac{\tau_1}{2\beta}(1-e^{-2\beta t})$$

because $\phi_1(w)$ is such a rapidly decreasing function, that we may integrate from $-\infty$ to $+\infty$. The value of the constant

$$\tau_1 = \int_{-\infty}^{+\infty}\phi_1(w)dw$$

we find with the help of the theorem of the equipartition of energy. For $t\to\infty$, we must have:

$$\lim_{t\to\infty}\overline{u^2}^{u_0} = \frac{\tau_1}{2\beta} = \frac{kT}{m}$$

so that:

$$\tau_1 = \frac{2\beta kT}{m}. \tag{13}$$

Substituting, we get:

$$\overline{u^2}^{u_0} = \frac{kT}{m} + \left(u_0^2 - \frac{kT}{m}\right)e^{-2\beta t} \tag{14}$$

which shows, how the equipartition value is reached. So we can go on. Using the assumptions (3) to (6) for A(t) and the fact that we must get the equipartition values for $t\to\infty$, we will prove in Note I, that for $u-u_0\exp(-\beta t)$ the normal Gaussian distribution law holds. For the velocity itself we get, therefore, the distribution law:

$$G(u_0, u, t) = \left(\frac{m}{2\pi kT(1-e^{-2\beta t})}\right)^{1/2} \exp\left\{\frac{m}{2kT}\frac{(u-u_0e^{-\beta t})^2}{1-e^{-\beta t}}\right\} \tag{15}$$

which shows how the Maxwell distribution is reached, when at $t=0$ all the particles started with the same velocity u_0.

§6. The second method for deriving (15) is, as we have already said, by constructing the Fokker-Planck partial differential equation for the problem, of which $G(u_0\, u, t)$ is then the fundamental solution. We will first derive the equation in general and then later specialize to our case.[7] Consider the distribution function $F(\phi_0, \phi, t)$. When t increases by Δt, ϕ will increase by a $\Delta\phi$, which will be different for every particle. Let the probability for an increase between the limits $\Delta\phi$ and $\Delta\phi+d(\Delta\phi)$ be $\psi(\Delta\phi,\phi,t)d(\Delta\phi)$. Writing $\phi'=\phi+\Delta\phi$ we have then:

$$F(\phi_0, \phi', t+\Delta t) = \int F(\phi_0, \phi'-\Delta\phi, t)\psi(\Delta\phi, \phi'-\Delta\phi, t)d(\Delta\phi) \tag{16}$$

[7] Comp. F. Zernike, Handbuch der Physik, Vol. III, p. 457.

when we may suppose that the probability of an increase $\Delta\phi$ is *independent* of the fact that for $t=0$, $\phi=\phi_0$. We now develop the integrand after powers of $\Delta\phi$:

$$F(\phi_0, \phi' - \Delta\phi, t)\psi(\Delta\phi, \phi' - \Delta\phi, t) = F(\phi_0, \phi', t)\psi(\Delta\phi, \phi', t) - \Delta\phi(F'\psi + F\psi') + \frac{\Delta\phi^2}{2}(F''\psi + 2F'\psi' + F\psi'') + \cdots$$

The resulting integrals all have simple meanings, for instance:

$$\int \psi(\Delta\phi, \phi', t)d(\Delta\phi) = 1; \quad \int \Delta\phi\psi d(\Delta\phi) = \overline{\Delta\phi}; \quad \int \Delta\phi^2\psi'' d(\Delta\phi) = \frac{\partial^2}{\partial\phi'^2}\overline{\Delta\phi^2}$$

and so on. Developing the left hand side in powers of Δt, putting:

$$\lim_{\Delta t\to 0} \frac{\overline{\Delta\phi}}{\Delta t} = f_1(\phi', t); \quad \lim_{\Delta t\to 0} \frac{\overline{\Delta\phi^2}}{\Delta t} = f_2(\phi', t) \tag{17}$$

and supposing that:

$$\lim_{\Delta t\to 0} \frac{\overline{\Delta\phi^k}}{\Delta t} = 0 \cdot \text{ for } k > 2 \tag{18}$$

we get, when we write again ϕ for ϕ':

$$\frac{\partial F}{\partial t} = \frac{1}{2}f_2\frac{\partial^2 F}{\partial\phi^2} + \left(\frac{\partial f_2}{\partial\phi} - f_1\right)\frac{\partial F}{\partial\phi} + \left(\frac{1}{2}\frac{\partial^2 f_2}{\partial\phi^2} - \frac{\partial f_1}{\partial\phi}\right)F. \tag{19}$$

We must of course in each special case determine the functions $f_1(\phi,t)$ and $f_2(\phi,t)$ and verify the supposition (18). We always can do that, when we know the equation of motion.

Let us return now to the velocity distribution. From the equation of motion we have:

$$u' - u = \Delta u = -\beta u\Delta t + \int_t^{t+\Delta t} A(\xi)d\xi$$

Using (3), we get therefore:

$$\overline{\Delta u} = -\beta u\Delta t = -\beta u'\Delta t$$

neglecting higher powers of Δt. From this:

$$\lim_{\Delta t\to 0} \frac{\overline{\Delta u}}{\Delta t} = f_1(u') = -\beta u'.$$

In the same way, we find using (4) as before:

$$\overline{\Delta u^2} = \tau_1\Delta t$$

so that:

$$f_2(u') = \tau_1 = \frac{2\beta kT}{m} = \text{const}.$$

All the higher powers of Δu become proportional to powers of Δt higher than the first, so that (18) is satisfied. We get therefore:[8]

$$\frac{\partial G}{\partial t} = \beta \frac{\partial}{\partial u}(uG) + \frac{\tau_1}{2} \frac{\partial^2 G}{\partial u^2} \cdot \qquad (20)$$

The systematic way of finding the fundamental solution of this equation is by solving the equation:

$$\frac{\partial z}{\partial t} = \beta \frac{\partial}{\partial u}(uz) + \frac{\tau_1}{2} \frac{\partial^2 z}{\partial u^2}$$

when for $t=0$, $z=f(u)$. This is an ordinary boundary value problem, which can easily be solved by the method of particular solutions. By summing the infinite series which we get, one can write the solution:

$$z(u, t) = \int_{-\infty}^{+\infty} f(u_0) G(u_0 u t) du_0$$

and G is then clearly the fundamental solution. For the details, see Note II. The result is again formula (15). One can derive the same result much more briefly when one is so clever as to substitute in (20):

$$G = (\phi)^{1/2} \exp\{-(u - u_0\chi)\phi\}$$

where ϕ and χ are functions of t only.[9] This is suggested a little by the result one ought to expect. Substituting, one sees that (18) is fulfilled, when χ and ϕ are solutions of the ordinary differential equations:

$$\frac{d\chi}{dt} = -\beta\chi$$

$$\frac{1}{\beta}\frac{d\phi}{dt} = 2\phi - 4\phi^2.$$

These can be immediately integrated, and the integration constants can be determined from the fact that for $t=0$ we must get $\delta(u-u_0)$ and for $t=\infty$ the Maxwell distribution law.

III. The Frequency Distribution of the Displacement

§7. The problem is to determine the probability that a free particle in Brownian motion which, at $t=0$ starts from $x=x_0$ with the velocity u_0 after the time t lies between x and $x+dx$. It is clear that this probability will depend only on $s=x-x_0$, and on t.

[8] This equation has been derived already by Rayleigh (Phil. Mag. **32,** 424 (1891) = Scient. Papers III, p. 473) and he gives also the fundamental solution (15). Later it has again been treated by v. Smoluchowski (Krakauer Ber. 1913, p. 418). Because Rayleigh's proof is a little artificial, and the treatment of v. Smoluchowski is not easily accessible, we thought it not superfluous to give the proof again.

[9] Comp. Lord Rayleigh, reference 8.

We will use again the momentum method, and calculate all the mean values $\overline{s^k}^{u_0}$. This goes in an analogous way as with the velocity. By integrating (11) again we find:

$$x = x_0 + \frac{u_0}{\beta}(1 - e^{-\beta t}) + \int_0^t e^{-\beta \eta} d\eta \int_0^\eta e^{\beta \xi} A(\xi) d\xi \tag{21}$$

or integrating partially:

$$s = x - x_0 = \frac{u_0}{\beta}(1 - e^{-\beta t}) - \frac{1}{\beta} e^{-\beta t} \int_0^t e^{\beta \xi} A(\xi) d\xi + \frac{1}{\beta} \int_0^t A(\xi) d\xi.$$

Taking the mean, gives:

$$\bar{s}^{u_0} = \frac{u_0}{\beta}(1 - e^{-\beta t}) \tag{22}$$

which can be interpreted as the distance travelled in the time t with the mean velocity $\bar{u} = u_0 \exp(-\beta t)$. By squaring, averaging, and calculating the double integrals in the same way as before, we get:

$$\overline{s^2}^{u_0} = \frac{\tau_1}{\beta^2} t + \frac{u_0}{\beta^2}(1 - e^{-\beta t})^2 + \frac{\tau_1}{2\beta^3}(-3 + 4e^{-\beta t} - e^{-2\beta t}) \tag{23}$$

where the constant τ_1 is known from the corresponding calculation of $\overline{u^2}^{u_0}$. This result (23) was first derived by Ornstein; for very long times t it goes over in:

$$\overline{s^2}^{u_0} = \frac{\tau_1}{\beta^2} t = \frac{2kT}{m\beta} t$$

the result of Einstein. For very short times t on the other hand, we get:

$$\bar{s}^{u_0} = u_0 t$$

$$\overline{s^2}^{u_0} = u_0^2 t^2$$

The motion is then uniform with the velocity u_0. Taking a second average over u_0, remembering that $\overline{u^2_0} = kT/m$, we get:

$$\bar{\bar{s}} = 0$$

$$\overline{\overline{s^2}} = \frac{2kT}{m\beta^2}(\beta t - 1 + e^{-\beta t})$$

which is the result quoted above (formula 10). The calculation of the higher powers goes similarly. In the result we get constants $\tau_2, \tau_3 \cdots$ which have been determined in part II in the corresponding calculation of $\overline{u^k}^{u_0}$ from the equipartition law. We can show in this way that for:

$$S = s - \frac{u_0}{\beta}(1 - e^{-\beta t})$$

again the normal Gaussian distribution law holds. For the details of the proof, see Note III. We get therefore:

$$F(x_0, x, t) = \left(\frac{m\beta^2}{2\pi kT(2\beta t - 3 + 4e^{-\beta t} - e^{-2\beta t})}\right)^{1/2} \exp\left[\frac{m\beta^2}{2kT}\frac{\{x - x_0 - u_0(1 - e^{-\beta t})/\beta\}^2}{2\beta t - 3 + 4e^{-\beta t} - e^{-2\beta t}}\right] \tag{24}$$

For large t this becomes of course the distribution law (7), already derived by Einstein. For $t \to 0$ it becomes $\delta(x - x_0)$ as it should.

§8. When we want to derive (24) in the same way as $G(u_0, u, t)$ from a partial differential equation we run into the following difficulty. According to the general Eq. (19), we have to calculate $\overline{\Delta x}$ and $\overline{\Delta x^2}$. Now it follows from the equation of motion, when the prime denotes the value of the quantities at the time $t+\Delta t$, that:

$$u' - u = -\beta(x' - x) + \int_t^{t+\Delta t} A(\xi)d\xi$$

so that:

$$-\beta(\overline{x' - x}) = -\beta\overline{\Delta x} = \overline{u'} - \overline{u} = u_0 e^{-\beta t}(e^{\beta\Delta t} - 1)$$

or:

$$\overline{\Delta x} = u_0 e^{-\beta t}\Delta t. \tag{25}$$

When one now calculates in the same way $\overline{\Delta x^2}$, then one finds that $\overline{\Delta x^2}$ becomes proportional to Δt^2, so that the function f_2 in (19) would become zero, and the differential equation would become:

$$\frac{\partial F}{\partial t} = -u_0 e^{-\beta t}\frac{\partial F}{\partial t}$$

which does not become the diffusion equation for $t \gg \beta^{-1}$. On the other hand, when we suppose $t \gg \beta^{-1}$ and Δt so large that we may apply the formula of Einstein for $\overline{\Delta x^2}$, we have:

$$\begin{aligned} \overline{\Delta x} &= 0 \\ \overline{\Delta x^2} &= 2D\Delta t \end{aligned} \tag{26}$$

and this substituted in (19), gives immediately:

$$\frac{\partial F}{dt} = D\frac{\partial^2 F}{\partial x^2}\cdot$$

It seems impossible to derive from (19) the rigorous differential equation for $F(x_0,x,t)$, which for $t \gg \beta^{-1}$ would become the diffusion equation, and of which (24) would be the fundamental solution. The reason for this, it seems to us, is that in the derivation of (19) we suppose that the change Δx in the time Δt is independent of the fact that at the time $t=0$ the particle is at $x = x_0$ and has the velocity u_0.

IV. The Brownian Motion of a Harmonically Bound Particle

§9. We will first derive, following Ornstein[10] the equation (9) first proposed by Smoluchowski from macroscopic considerations. We have to determine again $\overline{\Delta x}$ and $\overline{\Delta x^2}$. Now, when there are external forces the equation of motion is:

$$\frac{du}{dt} + \beta u = A(t) + \frac{1}{m}K(x).$$

Integrating as in §8, we get:

$$u' - u = -\beta(x' - x) + \int_t^{t+\Delta t} A(\xi)d\xi + \frac{1}{m}K\Delta t$$

from which follows, when we may neglect the influence of the initial velocity:

$$\beta\overline{\Delta x} = \frac{1}{m}K(x)\Delta t \tag{27}$$

so that:

$$f_1(x) = \frac{1}{\beta m}K(x) = \frac{1}{f}K(x).$$

When again Δt is not too small, we may put:

$$\overline{\Delta x^2} = \frac{2kT}{m\beta}\Delta t = 2D\Delta t \tag{28}$$

and substituting in the general equation (19), we get:

$$\frac{\partial F}{\partial t} = -\frac{1}{f}\frac{\partial}{\partial x}(KF) + D\frac{\partial^2 F}{\partial x^2}$$

which is (9).

Let us apply this to the case of a harmonically bound particle, for which:

$$\frac{1}{m}K(x) = -\omega^2 x$$

where ω is the frequency in 2π sec. We get then:

$$\frac{\partial F}{\partial t} = \frac{\omega^2}{\beta}\frac{\partial}{\partial x}(xF) + D\frac{\partial^2 F}{\partial x^2}.$$

This is completely similar to the equation (20) for $G(u_0, u, t)$. We find therefore for the fundamental solution

$$F(x_0, x, t) = \left(\frac{\omega^2}{2\pi\beta D(1 - e^{-(2\omega^2/\beta)t})}\right) \exp\left\{-\frac{\omega^2}{2\beta D}\cdot\frac{(x - x_0 e^{-(\omega^2/\beta)t})}{1 - e^{-(2\omega^2/\beta)t}}\right\}$$

[10] L. S. Ornstein, Proc. Acad. Amst. **21**, 96 (1919).

which gives:

$$\bar{x}^{x_0} = x_0 e^{-(\omega^2/\beta)t}$$

$$\overline{x^2}^{x_0} = \frac{kT}{m\omega^2} + \left(x_0^2 - \frac{kT}{m\omega^2}\right)e^{-(2\omega^2/\beta)t}.$$

This shows how the equipartition value is reached. For ω^2 very small we get approximately:

$$\bar{x}^{x_0} = x_0$$

$$\overline{x^2}^{x_0} = x_0^2 + \frac{2kT}{m\beta}t$$

which are the results for a free particle. We may not expect though, that the equations are generally valid. According to the derivation, there are clearly *two* limitations:

a. Because we have used (27) and (28) which correspond to (26) in §8, we must expect (30) to hold only for times $t \gg \beta^{-1}$

b. Because we have in (28) used the result for a *free* particle, we must expect (30) to hold only when β is large, the motion therefore being strongly overdamped. This is also the reason why apparently there is no distinction between the periodic, aperiodic and overdamped cases in the result for $\overline{x^2}^{x_0}$.

§10. To get exact results, we have to use the same method as before. We have first to integrate the equation of motion, and then take the average. The periodic, aperiodic and overdamped case must now be treated separately. We will indicate the calculations only for the periodic case.

The equation of motion is:

$$\frac{d^2x}{dt^2} + \beta\frac{dx}{dt} + \omega^2 x = A(t)$$

when at $t=0$, $x=x_0$ and $u=dx/dt=u_0$ we get from this:

$$u = -\frac{2\omega^2 x_0 + \beta u_0}{2\omega_1}e^{-\beta t/2}\sin\omega_1 t + u_0 e^{-(\beta/2)t}\cos\omega_1 t$$

$$+\frac{1}{\omega_1}\int_0^t A(\xi)e^{-\beta(t-\xi)/2}\left\{-\frac{\beta}{2}\sin\omega_1(t-\xi) + \omega_1\cos\omega_1(t-\xi)\right\}d\xi$$

$$x = \frac{\beta x_0 + 2u_0}{2\omega_1}e^{-(\beta/2)t}\sin\omega_1 t + x_0 e^{-(\beta/2)t}\cos\omega_1 t + \frac{1}{\omega_1}\int_0^t A(\xi)e^{-\beta(t-\xi)/2}\sin\omega_1(t-\xi)d\xi$$

where:

$$\omega_1^2 = \omega^2 - \frac{\beta^2}{4}.$$

Supposing, in correspondence with (3):

$$\overline{A(\xi)}^{x_0 u_0} = 0$$

this gives immediately, for instance:

$$\bar{x}^{x_0 u_0} = \frac{\beta x_0 + 2u_0}{2\omega_1} e^{-(\beta/2)t} \sin \omega_1 t + x_0 e^{-(\beta/2)t} \cos \omega_1 t. \tag{31}$$

The mean value here has to be understood as follows. We have a canonical ensemble of harmonic oscillators, from which at $t=0$ we pick a sub-ensemble (A) of oscillators, which have a deviation and velocity x_0, u_0, resp. and which we follow in their motion. At the time t we take an average over the x of the different members of this sub-ensemble (A), and the result is then given by (31). If we would follow a *sub-ensemble* (B), of which the members at $t=0$ had the deviation x_0 but arbitrary velocity, we would get at the time t a mean deviation, which will follow from (31) by taking the average over u_0. Since in a canonical ensemble of oscillators the deviation is not correlated with the velocity, we may put:

$$\bar{u}_0^{x_0} = 0$$
$$\overline{u^2}^{x_0} = \frac{kT}{m}. \tag{32}$$

Uisng this, we get:

$$\bar{\bar{x}}^{x_0} = x_0 e^{(-\beta/2)t}\left(\frac{\beta}{2\omega_1} \sin \omega_1 t + \cos \omega_1 t\right). \tag{33}$$

Let us now consider u^2 and x^2. Using again the assumption analogous to (4):

$$\overline{A(t_1)A(t_2)}^{x_0 u_0} = \phi(t_1 - t_2)$$

where $\phi(x)$ is an even function with a sharp maximum at $x=0$, and calculating the double integrals exactly as before, we get:

$$\overline{x^2}^{x_0 u_0} = \left(\frac{\beta x_0 + 2u_0}{2\omega_1} e^{-(\beta/2)t} \sin \omega_1 t + x_0 e^{-(\beta/2)t} \cos \omega_1 t\right)^2 + \frac{\tau_1}{2\omega_1^2 \beta}(1 - e^{-\beta t})$$
$$- \frac{\tau_2}{8\omega^2 \omega_1^2}(\beta - \beta e^{-\beta t} \cos 2\omega_1 t + 2\omega_1 e^{-\beta t} \sin 2\omega_1 t)$$

where we have put:

$$\tau_1 = \int_{-\infty}^{+\infty} \phi(w) \cos \omega_1 w dw$$
$$\tau_2 = \int_{-\infty}^{+\infty} \phi(w) dw.$$

The condition, that for $t \to \infty$ we must get the equipartition value, gives us one relation between τ_1 and τ_2. One would expect that from:

$$\lim_{t\to\infty} \overline{u^2}^{x_0 u_0} = \frac{kT}{m}.$$

we would get a second relation, but the calculation of $\overline{u^2}^{x_0u_0}$ shows that this is the same as the first. The fact that $\phi(w)$ has such a sharp maximum suggests, that in the integral for τ_1 we may replace $\cos\omega_1 w$ by unity, which would make $\tau_1=\tau_2$. We can prove this more exactly by calculating $\overline{xu}^{x_0u_0}$[11] and determining the limit for $t\to\infty$, which must be zero, because for $t\to\infty$ sub ensemble (A) must again become a canonical ensemble. We get in this way:

$$\tau_1 = \tau_2 = \frac{2\beta kT}{m}.$$

This solves the problem completely. Averaging again over u_0, using (32) we get:

$$\overline{\overline{x^2}}^{x_0} = \frac{kT}{m\omega^2} + \left(x_0^2 - \frac{kT}{m\omega^2}\right)e^{-\beta t}\left(\cos\omega_1 t + \frac{\beta}{2\omega_1}\sin\omega_1 t\right)^2 \tag{34}$$

which shows how the equipartition value is reached. So we can calculate all sorts of mean values. The further result is perhaps interesting, that:

$$\overline{\overline{xu}}^{x_0} = \frac{1}{\omega_1\omega^2}\left(\frac{kT}{m\omega^2} - x_0^2\right)e^{-\beta t}\sin\omega_1 t\left(\cos\omega_1 t + \frac{\beta}{2\omega_1}\sin\omega_1 t\right)$$

which shows how the correlation between x and u, beginning with being zero, oscillates and goes to zero again for $t\to\infty$. Of course, averaging over x_0, we get $\overline{\overline{\overline{xu}}}=0$ as it must be.

§11. In the aperiodic case we get:

$$\overline{\overline{x^2}}^{x_0} = x_0\left(1+\frac{\beta t}{2}\right)e^{-(\beta/2)t} \tag{33a}$$

$$\overline{\overline{x^2}}^{x_0} = \frac{kT}{m\omega^2} + \left(x_0^2 - \frac{kT}{m\omega^2}\right)\left(1+\frac{\beta t}{2}\right)^2 e^{-\beta t}. \tag{34a}$$

The equipartition value is now reached monotonously. The calculation goes similarly, except that instead of the integral τ_1, we have to introduce an integral:

$$\tau_1' = \int_{-\infty}^{+\infty} w^2\phi(w)dw.$$

The calculation of $\overline{xu}^{x_0u_0}$ proves then that this is zero, which could be expected.

In the overdamped case we get:

$$\overline{\overline{x}}^{x_0} = x_0 e^{-(\beta/2)t}\left(\cosh\omega' t + \frac{\beta}{2\omega'}\sinh\omega' t\right) \tag{33b}$$

$$\overline{\overline{x^2}}^{x_0} = \frac{kT}{m\omega^2} + \left(x_0^2 - \frac{kT}{m\omega^2}\right)e^{-\beta t}\left(\cosh\omega' t + \frac{\beta}{2\omega'}\sinh\omega' t\right)^2 \tag{34b}$$

[11] Here we use: $\int_{-\infty}^{+\infty}\phi(w)\sin\omega_1 w\,dw=0$, which follows from the fact that $\phi(w)$ is an even function.

where:

$$\omega'^2 = \frac{\beta^2}{4} - \omega^2 = -\omega_1^2.$$

The equipartition value is again reached monotonously. It is easy to show further, that when $\beta \gg 2\omega$ and $t \gg \beta^{-1}$ these last equations go over into the results (30) of v. Smoluchowski, as we would expect according to the remarks at the end of §9.

§12. The problem of the rotatorial Brownian motion of a small mirror suspended on a fine wire, has been treated recently by S. Goudsmit and one of us,[12] by a method analogous to the well-known treatment of the shot effect by Schottky.[13] If the displacement, registered during a time, long compared to the characteristic period of the mirror, is developed in a Fourier series, an expression was derived for the square of the amplitude of each Fourier component. It was found that this depended, besides on the temperature, on the pressure and molecular weight of the surrounding gas. This explains in principle, why the curves registered by Gerlach[14] at different pressures, though all giving the same mean square deviation, are quite different in appearance. The calculations were made under the condition that the surrounding gas is much rarified, and though they can easily be generalized, the exact comparison with the experimental data of Gerlach is very difficult.

The results (33) and (34) (when we replace m by the moment of inertia) are in this respect much better. They could be tested easily, and they hold for all pressures of the surrounding gas. They show that, though the mean square deviation depends only on the temperature, the correlation between successive values of the deviation depends in a more interesting way on the surrounding medium. Its influence is expressed by the friction coefficient β.

Notes

I. To prove that for $U = u - u_0 \exp(-\beta t)$ the normal Gaussian distribution law holds, we have to show that:

$$\begin{aligned} \overline{U^{2n+1}} &= 0 \\ \overline{U^{2n}} &= 1\cdot 3\cdot 5 \cdots (2n-1)(\overline{U^2})^n \end{aligned} \qquad (A)$$

We have from §5:

$$\bar{U} = 0$$

$$\overline{U^2} = \frac{\tau_1}{2\beta}(1 - e^{-2\beta t}).$$

From (11) we get further:

$$\overline{U^3} = e^{-3\beta t} \int_0^t \int_0^t \int_0^t e^{\beta(\xi_1+\xi_2+\xi_3)} \overline{A(\xi_1)A(\xi_2)A(\xi_3)}\, d\xi_1 d\xi_2 d\xi_3.$$

[12] G. E. Uhlenbeck and S. Goudsmit, Phys. Rev. **34,** 145 (1929).
[13] W. Schottky, Ann. d. Physik **57,** 541 (1918).
[14] W. Gerlach, Naturwiss. **15,** 15 (1927).

According to the assumptions made about $A(\xi)$ the integrand will be different from zero only in the neighborhood of the line $\xi_1=\xi_2=\xi_3$. Taking cylindrical coordinates with this line as z-axis, and using (5), we find:

$$\overline{U^3} = \frac{\tau_2}{\beta(3)^{1/2}}(1 - e^{-3\beta t})$$

where τ_2 denotes the constant:

$$\tau_2 = \int_0^\infty \int_0^{2\pi} \phi_2(r, \theta) r dr d\theta .$$

The value of τ_2 follows again from the equipartition law. For $t\to\infty$, $\overline{U^3}$ must go to zero, so that $\tau_2=0$.

Going to the fourth power we find:

$$\overline{U^4} = e^{-4\beta t} \int_0^t\int_0^t\int_0^t\int_0^t e^{\beta(\xi_1+\xi_2+\xi_3+\xi_4)} \overline{A(\xi_1)A(\xi_2)A(\xi_3)A(\xi_4)} d\xi_1 d\xi_2 d\xi_3 d\xi_4 .$$

When ξ_1 and ξ_2 are lying near each other and also ξ_3 and ξ_4 (but far from ξ_1, ξ_2), we will have according to (6):

$$\overline{A(\xi_1)A(\xi_2)A(\xi_3)A(\xi_4)} = \overline{A(\xi_1)A(\xi_2)} \cdot \overline{A(\xi_3)A(\xi_4)}$$

so that this integration region will contribute:

$$e^{-4\beta t}\frac{\tau_1^2}{4\beta^2}(e^{2\beta t} - 1)^2 .$$

We will get this 3 times because we can divide $A(\xi_1)$ $A(\xi_2)$ $A(\xi_3)$ $A(\xi_4)$ into two pairs in 3 ways. There remains the region in the neighborhood of the line $\xi_1=\xi_2=\xi_3=\xi_4$. For this we get, introducing cylindrical coordinates and using (5):

$$\frac{\tau_3}{2\beta} e^{-4\beta t}(e^{4\beta t} - 1)$$

where:

$$\tau_3 = \int_0^\infty \iint \phi_3(r, \theta, \theta_2) dr dw$$

For $t\to\infty$ we get therefore:

$$\lim_{t\to\infty} \overline{U^4} = \frac{3\tau_1^2}{4\beta^2} + \frac{\tau_3}{2\beta}$$

but according to the Maxwell distribution law, we have:

$$\lim_{t\to\infty} \overline{U^4} = \lim_{t\to\infty} \overline{u^4} = 3(\lim_{t\to\infty} \overline{u^2})^2 = \frac{3\tau_1^2}{4\beta^2}$$

so that $\tau_3 = 0$ and we get:

$$\overline{U^4} = 3(\overline{U^2}).$$

To write down the general proof for (A) is tedious, because one has more and more integration regions to consider. However, since (A) holds for $t \to \infty$, one can convince oneself of the fact that only those regions where the ξ are lying in pairs near each other give a real contribution. All the other regions give contributions proportional to constants $\tau_k (k > 1)$ which by the equipartition law prove to be zero. This gives A_1 immediately and since the number of ways in which we can divide $2n$ objects into n pairs is $1.3.5 \cdots (2n-1)$ we get A_2 also.

II. When we substitute in (20):

$$x = \beta t$$

$$y = u\left(\frac{2\beta}{\tau_1}\right)^{1/2}$$

we get:

$$\frac{\partial z}{\partial x} = z + y\frac{\partial z}{\partial y} + \frac{\partial^2 z}{\partial y^2}$$

and we have to solve this when for $x = 0$, $z = f(y)$ and for $y = \pm\infty$, $z = 0$. By separating we find as a particular solution:

$$A_n e^{-nx} D_n(y) e^{-y^2/4}$$

where D_n denotes Weber's function of the nth order[15]
We have then to determine A_n:

$$f(y) = \sum_0^\infty A_n D_n(y) e^{-y^2/4}$$

which gives:

$$A_n = \frac{1}{n!(2\pi)^{1/2}} \int_{-\infty}^{+\infty} D_n(\eta) f(\eta) e^{-\eta^2/4} d\eta$$

and we get for the solution:

$$z(x, y) = \frac{1}{(2\pi)^{1/2}} \int_{-\infty}^{+\infty} d\eta f(\eta) e^{(\eta^2 - y^2)/4} \sum_0^\infty \frac{D_n(y) D_n(\eta)}{n!} e^{-nx} \qquad (B)$$

We have now to sum the infinite series. As Professor H. A. Kramers showed to us, this can be done in the following way. Put, suppressing the arguments y and η:

$$M(x) = \sum_0^\infty \frac{D_n D_n}{n!} e^{-nx}$$

[15] Comp. Whittaker-Watson, Modern Analysis, p. 347.

then:

$$-\frac{dM}{dx} = \sum_0^\infty \frac{D_{n+1}D_{n+1}}{n!}e^{-(n+1)x}.$$

Using the recurrence formula:

$$D_{n+1}(z) = zD_n(z) - nD_{n-1}(z)$$

we get:

$$-\frac{dM}{dx} = y\eta e^{-x}M - e^{-2x}\frac{dM}{dx} - \sum_0^\infty \frac{yD_{n+1}D_n + \eta D_nD_{n+1} - D_nD_n}{n!}e^{-(n+2)x}.$$

Calling the last sum N and using again the recurrence relation, we find:

$$N = (y^2 + \eta^2 - 1)e^{-2x}M - \sum_0^\infty \frac{yD_nD_{n+1} + \eta D_{n+1}D_n}{n!}e^{-(n+3)x}$$

Again using the recurrence relation, we find for the last sum

$$L = (2y\eta e^{-3x} - e^{-4x})M - e^{-2x}N.$$

Substituting back, we get for M the differential equation:

$$-(1 - e^{-2x})^2\frac{dM}{dx} = M\{y\eta e^{-x} - (y^2 + \eta^2 - 1)e^{-2x} + y\eta e^{-3x} - e^{-4x}\}$$

This we can immediately integrate, which gives:

$$M = \frac{C(y, \eta)}{(1-e^{-2x})^{1/2}}\exp\left\{-\frac{y^2 + \eta^2 - 2y\eta e^{-x}}{2(1-e^{-2x})}\right\}.$$

The integration constant $C(y,\eta)$ can be determined from the fact that:

$$\lim_{x\to\infty} M = D_0(y)D_0(\eta) = C(y, \eta)e^{-(y^2+\eta^2)/2}$$

which gives:

$$C(y, \eta) = e^{(y^2+\eta^2)/4}.$$

Substituting in the solution (B) gives finally:

$$z(x, y) = \frac{1}{(2\pi)^{1/2}}\int_{-\infty}^{+\infty} d\eta f(\eta)\frac{1}{(1 - e^{-2x})^{1/2}}\exp\left\{-\frac{(y - \eta e^{-x})^2}{2(1 - e^{-2x})}\right\}$$

which shows that the fundamental solution ($f(\eta)$ is then $\delta(y-y_0)$) is given by:

$$G(y_0, y, x) = \frac{1}{(2\pi(1 - e^{-2x}))^{1/2}}\exp\left\{-\frac{(y - y_0e^{-x})^2}{2(1 - e^{-2x})}\right\}.$$

Introducing again t and u, we get (15).

III. To prove that for $S=s-u_0/\beta(1-e^{-\beta t})$ the Gaussian distribution law holds, we have to show again:

$$\overline{S^{2n+1}} = 0$$

$$\overline{S^{2n}} = 1\cdot 3\cdot 5 \cdots (2n-1)(\overline{S^2})^n \qquad (C)$$

We have from §7:

$$\overline{S} = 0$$

$$\overline{S^2} = \frac{\tau_1}{2\beta^3}(2\beta t - 3 + 4e^{-\beta t} - e^{-2\beta t}).$$

The calculation of the 3-fold integrals in $\overline{S^3}$ is analogous to the calculation of $\overline{U^3}$ in Note I. We find that the result is proportional to τ_2, and from Note I we know that $\tau_2=0$, so that:

$$\overline{S^3} = 0.$$

In the 4-fold integrals occurring in $\overline{S^4}$ we have to consider only the regions where ξ_1, ξ_2, ξ_3, ξ_4 are lying in pairs near each other, because the other regions will give results proportional to τ_3 which is zero, as is proved in Note I. The calculation gives:

$$\overline{S^4} = 3(\overline{S^2})^2$$

as could be expected. The factor *3* comes again from the fact that we can divide ξ_1, ξ_2, ξ_3, ξ_4 into two pairs in three ways. In the same way as in Note I then, one convinces oneself further of the truth of the general relations (*C*).

REVIEWS OF MODERN PHYSICS VOLUME 17, NUMBERS 2 AND 3 APRIL–JULY, 1945

On the Theory of the Brownian Motion II

MING CHEN WANG AND G. E. UHLENBECK
H. M. Randall Laboratory of Physics, University of Michigan, Ann Arbor, Michigan

1. INTRODUCTION

IN 1930, Ornstein and one of us[1] tried to summarize and partially extend the existing theory of the Brownian motion for simple systems like the free particle and the harmonic oscillator. Since that time the theory has been developed and clarified to a considerable extent, so that it seems worth while again to try to summarize the theory. In this we will restrict ourselves to the case of the Brownian motion of a system of coupled harmonic oscillators, or in the electrical analogy to the theory of the thermal noise in a linear, passive network.[2] It is now clear that in this case we have to do with the theory of the so-called Gaussian random process, and we shall try, therefore, to present the theory of the Brownian motion against the background of the general theory of the random process.[3] This will also allow us to show the connection with some of the mathematical literature on this subject.

There are two approaches to the theory of the Gaussian random process. In the first the attention is focused on the actual random variation in time of the displacement, or voltage, or whichever variables of the system one is especially interested in. One usually[4] develops this variable in a Fourier series in time, of which the coefficients can vary in a random fashion. A fundamental notion is the notion of the spectrum of the random process, and the connection between the spectrum and the so-called correlation function is one of the basic theorems. For many purposes, and especially in the electrical case when the "noise" passes through *non-linear* circuit elements (like rectifiers for instance), this method is the most natural. Recently the method has been applied systematically to a whole series of problems by S. O. Rice,[5] and we shall call it, therefore, the *method of Rice* or the *Fourier series method*, and in the following we shall give only a short account of it.

The second method is the *method of Fokker-Planck* or the *diffusion equation method*. Macroscopically, for an ensemble of particles or systems, the variations which occur are like a diffusion process. The distribution function of the random variables of the system will, therefore, fulfill a partial differential equation of the diffusion type, and this is the basic equation of the method.[6] We shall discuss this method in more detail, mainly because, thanks to a recent article by Kramers,[7] one is now able to derive the distribution function for any of the random variables in the Brownian motion of a system of coupled oscillators. Thus it becomes completely clear that the two methods give identical results, and the relation between the two methods can then perhaps be better appreciated.

2. THE GENERAL RANDOM PROCESS

Roughly speaking, what we mean by a random process $y(t)$ is a process in which the variable y[8]

[1] G. E. Uhlenbeck and L. S. Ornstein, Phys. Rev. **36**, 823 (1930). We will refer to this paper as I.

[2] Only in the last section, where we shall mention some unsolved problems, we shall go beyond this restriction.

[3] Recently S. Chandrasekhar has also reviewed several aspects and applications of the general theory in Rev. Mod. Phys. **15**, 1 (1943). Although there will be some overlapping, we hope that our review will complement the exposition of Chandrasekhar.

[4] It sometimes is convenient to use a development in other sets of orthogonal functions.

[5] S. O. Rice, Bell Tel. J. **23**, 282 (1944); **25**, 46 (1945). We shall refer to these papers as Rice I and II. One finds here also references to previous applications of the method.

[6] For simple examples see I, and also R. Furth, Ann. d. Physik **53**, 177 (1917) and Riemann-Weber, 3rd ed. Vol. II, p. 177. In the mathematical literature the method has been analyzed by A. Kolmogoroff, Math. Ann. **104**, 415 (1931); **108**, 149 (1933); and by W. Feller, Math. Ann. **113**, 113 (1936); Trans. Am. Math. Soc. **48**, 488 (1940).

[7] H. A. Kramers, Physica **7**, 284 (1940).

[8] It may be that y is the displacement or velocity of a particle in Brownian motion or a fluctuating voltage or current when we have thermal noise. It may also denote a combination of two or more of such quantities, and we shall speak then of two-dimensional or more dimensional random processes. In the following, everything will be written as if y and t were continuous variables. This is *not* necessary; it may happen that either y or t or both y and t can assume only discrete values. We propose to let the words continuous and discrete refer only to the dependent variable y; while the words process and series refer to continuous and discrete t respectively. The well-

does not depend in a completely definite way on the independent variable t (=time), as in a *causal* process; instead one gets in different observations different functions $y(t)$, so that only certain probability distributions are directly observable. In fact the random process $y(t)$ is completely described (or defined) by the following *set of probability distributions:*

$W_1(yt)dy$ = probability of finding y in the range $(y, y+dy)$ at time t.

$W_2(y_1t_1; y_2t_2)dy_1dy_2$ = joint probability of finding y in the range (y_1, y_1+dy_1) at time t_1 *and* in the range (y_2, y_2+dy_2) at time t_2.

$W_3(y_1t_1; y_2t_2; y_3t_3)dy_1dy_2dy_3$ = joint probability of finding a triple of values of y in the ranges dy_1, dy_2, dy_3 at times t_1, t_2, t_3. (1)

And so on! The set of functions (1) must fulfill the following obvious conditions:

(*a*) $W_n \geqq 0$.

(*b*) $W_n(y_1t_1; y_2t_2 \cdots y_nt_n)$ is a symmetric function in the set of variables $y_1t_1, y_2t_2 \cdots y_nt_n$. This is clear since W_n is a *joint* probability.

(*c*) $W_k(y_1t_1; \cdots y_kt_k)$

$$= \int \cdots \int dy_{k+1} \cdots dy_n W_n(y_1t_1 \cdots y_nt_n)$$

since each function W_n must imply all the previous W_k with $k<n$. The set of functions (1) form, therefore, a kind of hierarchy; they describe successively the random process in more detail.[9]

To determine the functions W_n experimentally, it is clear that one needs a great number of records $y(t)$ obtained from a great number of experiments on "similarly prepared" systems (an ensemble of observations). To find then, for instance, $W_1(yt)$, one determines at a definite time t how often in the different experiments y occurs in a given interval $(y, y+\Delta y)$, etc. In most applications (and especially for the Brownian motion problems) we can make, however, a simplification because the processes are *stationary in time*. This means that the underlying "mechanism" which causes the fluctuations does not change in course of time. A shift of the t-axis will then not influence the functions W_n, and as a result the set (1) becomes:

$W_1(y)dy$ = probability of finding y between y and $y+dy$.

$W_2(y_1y_2t)dy_1dy_2$ = joint probability of finding a pair of values of y in the ranges dy_1 and dy_2, which are a time interval t apart from each other (t is therefore $=t_2-t_1$).

And so on again. These functions can now be experimentally determined from *one* record $y(t)$ taken over a sufficiently long time. One can then cut the record in pieces of length T (where T is long compared to all "periods" occurring in the process), and one may consider the different pieces as the different records of an ensemble of observations. In computing average values one has *in general* to distinguish between an ensemble average and a time average. However, for a stationary process these two ways of averaging will always give the same result, and one can, therefore, use either of them.

3. CLASSIFICATION OF RANDOM PROCESSES

The set of probability distributions (1) leads immediately to a method of classifying the random processes.

(*a*) We shall call a random process a *purely random process* when the successive values of y are not correlated at all. This means that:

$$W_2(y_1t_1; y_2t_2) = W_1(y_1t_1) \cdot W_1(y_2t_2)$$

known theory of Smoluchowski for the concentration fluctuations of a colloidal suspension is in our terminology an example of the analysis of a discrete random series. The general random walk problem is an example of the theory of continuous random series, etc. One finds a complete account of these examples in the article of Chandrasekhar.

[9] So far as we know, the first attempt to give a general theory of a random process is contained in two papers by L. S. Ornstein and H. C. Burger, Versl. Kon. Acad. Amst. **27**, 1146 (1919); **28**, 183 (1919). Here the set of distributions (1) and the property (*c*) are mentioned. Compare further A. Kolmogoroff, *Grundbegriffe der Wahrscheinlichkeitsrechnung* (Berlin, 1933), p. 27; H. Wold, "A study in the analysis of stationary time series, Diss. Uppsala (1938); B. Hostinky, Ann. Inst. H. Poincaré, **3**, 1 (1933); **7**, 69 (1937). The authors are aware of the fact that in the mathematical literature (especially in papers by N. Wiener, J. L. Doob, and others; cf. for instance Doob, Ann. Math. **43**, 351 (1942), also for further references) the notion of a random (or stochastic) process has been defined in a much more refined way. This allows for instance to determine in certain cases the probability that the random function $y(t)$ is of bounded variation, or continuous or differentiable, etc. However, it seems to us that these investigations have not helped in the solution of problems of direct physical interest, and we will, therefore, not try to give an account of them.

and analogously for the higher W_n. All the information about the process is then completely contained in the first distribution function W_1. When t is discrete, it is easy to give examples, but for continuous t, the purely random process can only be considered as a kind of limiting case; in any actual example, the y_1 and y_2 will surely be correlated when the time interval t_2-t_1 is small enough.

(*b*) In the next more complicated case, all the information about the process will be contained in W_2. Such processes are called *Markoff processes*. For the more precise definition it is useful first to introduce the notion of *conditional probabilities*. We will write $P_2(y_1|y_2, t)dy_2$ for the probability that given y_1 one finds y in the range (y_2, y_2+dy_2) a time t later. Of course, one finds P_2 from W_2 according to[10]

$$W_2(y_1y_2t) = W_1(y_1)P_2(y_1|y_2, t), \tag{2}$$

P_2 must further fulfill the obvious relations (which also follow from the properties (*a*), (*b*), (*c*), of Section 2):

$$P_2(y_1|y_2, t) \geqq 0, \tag{3a}$$

$$\int dy_2 P_2(y_1|y_2, t) = 1, \tag{3b}$$

$$W_1(y_2) = \int W_1(y_1)P_2(y_1|y_2, t)dy_1, \tag{3c}$$

while in the Brownian motion problems one also always has:[11]

$$\lim_{t\to\infty} P_2(y_1|y_2, t) = W_1(y_2). \tag{4}$$

Analogously one can introduce higher order conditional probabilities, and we will use an analogous notation; a bar will always separate the variables which are given from those for which the probability has to be found. All this holds, of course, still for any stationary random process. A Markoff process can now be defined more precisely by stating that for such a process the conditional probability that y lies in the interval (y_n, y_n+dy_n) at time t_n, given that y is equal to $y_1, y_2 \cdots y_{n-1}$ at the times $t_1, t_2 \cdots t_{n-1}$ (where $t_n > t_{n-1} \cdots > t_2 > t_1$) depends besides on $y_n t_n$ *only* on the value of y at the previous time t_{n-1}. Or in a formula, a Markoff process is defined by the equation:

$$P_n(y_1t_1, y_2t_2 \cdots y_{n-1}t_{n-1}|y_nt_n) = P_2(y_{n-1}t_{n-1}|y_nt_n). \tag{5}$$

This makes it clear that all the W_n for $n>2$ can be found, when only W_2 is known. One derives for instance easily from (5) that:

$$W_3(y_1t_1, y_2t_2, y_3t_3) = \frac{W_2(y_1t_1, y_2t_2)W_2(y_2t_2, y_3t_3)}{W_1(y_2t_2)},$$

and so on. It is clear, therefore, that W_2 or P_2 completely describes the process. However, one cannot take P_2 as an arbitrary function of its variables. Besides the general relations (3) and (4), it must fulfill:

$$P_2(y_1|y_2, t) = \int dy P_2(y_1|y, t_0)P_2(y|y_2, t-t_0), \tag{6}$$

for all values of t_0 between zero and t. This follows immediately from the definition of a Markoff process and is called the *Smoluchowski equation*. It is the basic equation for the theory.

(*c*) In this way one can go on. The next class of processes will be completely described by giving W_3. However, in the physical applications there are very few examples studied of such higher order processes. Very often, when a process is *not* a Markoff process one can still consider it as a kind of "projection" of a more complicated Markoff process. Besides y, one then considers another dependent variable z (which may be, for instance, dy/dt or it may be a coordinate of another system), and it may be that for the two variables y, z combined, the process is then a Markoff process, so that:

$$P_2(y_1z_1|y_2z_2, t) = \int\int dydz P_2(y_1z_1|yz, t_0)P_2(yz|y_2z_2, t-t_0).$$

The $W_2(y_1y_2t)$ which one obtains by integrating $W_2(y_1z_1y_2z_2t)$ over z_1 and z_2 will then in general *not* be a Markoff process, and one can say that this is due to the fact that one has not given a complete enough description of the process.

[10] From now on we shall restrict ourselves to stationary processes.

[11] This property excludes, for instance, the existence of "hidden periodicities." In the theory of noise it excludes the presence of "signals."

Whether it is always possible to find the appropriate variables $z_1, z_2 \cdots$ (there may be more than one) so as to complete the given process to a Markoff process will, of course, depend on the physical "causes" of the fluctuation phenomena in question. As we shall see, for the theory of the Brownian motion, such a completion will *always* be possible, so that in some sense we will always have to do with Markoff processes.

4. THE RELATION BETWEEN THE SPECTRUM AND THE CORRELATION FUNCTION[12]

Suppose that one considers for a very long time T a stationary random process $y(t)$ whose average value is zero. Taking $y(t)=0$ outside the time interval T, one can develop the resulting function in a Fourier integral:

$$y(t)=\int_{-\infty}^{+\infty} dfA(f)e^{2\pi ift}, \tag{7}$$

where $A(f)=A^*(-f)$, since $y(t)$ is real. It is well known (Parzeval theorem) that:

$$\int_{-\infty}^{+\infty} y^2(t)dt=\int_{-\infty}^{+\infty} |A(f)|^2 df.$$

Using the fact that $|A(f)|^2$ is an even function of f and going to the limit $T\to\infty$, one can write this equation in the form:

$$\langle y^2\rangle_{\text{Av}}=\lim_{T\to\infty}\frac{1}{T}\int_{-\infty}^{+\infty} y^2(t)dt=\int_0^{\infty} dfG(f), \tag{8}$$

where

$$G(f)=\lim_{T\to\infty}\frac{2}{T}|A(f)|^2 \tag{9}$$

is the *spectral density*.

Consider next the average value:

$$\langle y(t)y(t+\tau)\rangle_{\text{Av}}=\lim_{T\to\infty}\frac{1}{T}\int_{-\infty}^{+\infty} y(t)y(t+\tau)dt. \tag{10}$$

By introducing the Fourier expansion (7) and using the Fourier integral theorem, one shows easily:

$$\langle y(t)y(t+\tau)\rangle_{\text{Av}}=\int_0^{\infty} dfG(f)\cos 2\pi f\tau, \tag{11a}$$

from which follows by inversion:

$$G(f)=4\int_0^{\infty} d\tau\langle y(t)y(t+\tau)\rangle_{\text{Av}}\cos 2\pi f\tau. \tag{11b}$$

This is the relation referred to in the title of this section. One can express it by saying that the *correlation function*

$$\rho(\tau)=\frac{\langle y(t)y(t+\tau)\rangle_{\text{Av}}}{\langle y^2\rangle_{\text{Av}}} \tag{12a}$$

and the *normalized spectrum:*

$$S(f)=\frac{G(f)}{\int_0^{\infty} dfG(f)} \tag{12b}$$

are each other's Fourier cosine transform, so that they are uniquely related to each other. For an almost pure random process, $\rho(\tau)$ is a function, which starting from unity drops very rapidly to zero, and as a result $S(f)=$constant except for very high frequencies. We call this a *white spectrum;* of course, the case that $S(f)=$constant for all f, which corresponds to a pure random process, is a limiting case, which will never occur in practice. When $S(f)$ has a sharp maximum around f_0, then $\rho(\tau)$ will look like a damped oscillation with roughly the frequency f_0.

5. SOME REMARKS ON THE THEORY OF DISCRETE RANDOM SERIES[13]

We will restrict ourselves to Markoff processes. The problem will then always be to determine $P(n|m, s\tau)$ when one knows $P(n|m, \tau)$. Here P is the analogue of the $P(y_1|y_2, t)$; y_1, y_2 can only have discrete values n, m and also the time t can only have discrete values $s\tau$ with $s=1, 2, 3, \cdots$. From now on we will drop the τ and write also $Q(n, m)$ for $P(n|m, \tau)$ in order to emphasize that

[12] This relation is contained in the paper of N. Wiener, Acta Math. **55**, 117 (1930) on generalized harmonic analysis. It was rediscovered by Khintchine, Math. Ann. **109**, 604 (1934). See also the dissertation of H. Wold for further references and for the formulation in the discrete case. In 1938, G. I. Taylor (Proc. Roy. Soc. **164**, 476 (1938)) gave a beautiful application of the theorem to the theory of turbulence. Cf. also Rice I, p. 310.

[13] The purpose of this section is only to present some of the ideas which are of importance for the understanding of the Fokker-Planck method. For a complete discussion compare Hostinky, see reference 9 and also M. Fréchet, *Traité du Calcul des Probabilités* (1938), Vol. I, Part II, Section 3. For the discussion of the important application to the theory of the concentration fluctuations, see Chandrasekhar, reference 3.

it is the basic probability which must be given from the "mechanism" or the "physical cause" of the random process. To find then $P(n|m, s)$ one can try to make successive use of the Smoluchowski equation:

$$P(n|m, s)=\sum_k P(n|k, s-1)Q(k, m). \quad (13)$$

However, for large values of s this is usually not practicable, and one has to look for other methods.

It is instructive to write (13) in a different way by remembering that:

$$\sum_m Q(k, m)=1,$$

or

$$Q(k, k)+\sum_m{}' Q(k, m)=1,$$

where the prime means that the value $m=k$ must be omitted. Using this and dropping in (13) the initial value n one can write Eq. (13) in the form:

$$\begin{aligned}P(m, s)-P(m, s-1)\\ &=-P(m, s-1)\sum_k{}' Q(m, k)\\ &\quad+\sum_k{}' P(k, s-1)Q(k, m). \quad (13a)\end{aligned}$$

One can interpret this by saying that the rate of change of $P(m, s)$ with the time ($=s$) is owing to the "gains" of P because of transitions from k to m minus the "losses" of P because of the transitions from m to all possible k. It is clear, therefore, that (13a) is completely analogous to the well known Boltzmann equation in the kinetic theory of gases.[14] One must solve such an equation for a given "initial" distribution; in our case this is the way the variable n comes in since:

$$P(m, 0)=\delta(n, m), \quad (14)$$

where $\delta(n, m)$ denotes the Kronecker symbol.

In many cases the process has the property that the dependent variable k can change in one step at most by ± 1. This means that $Q(k, m)=0$ except when $m=k$, $k\pm 1$, and Eq. (13) or (13a) becomes then a rather simple difference equation. To illustrate this we will consider two examples.

(*a*) Discrete random walk problem in one dimension. This is the simplest possible case; a point can move on a straight line in steps Δ; at each time moment s there is an equal chance that the point makes a step Δ to the right or to the left. If at $s=0$ the point is at the position $n\Delta$, what is the probability $P(n|m, s)$ that at time s the point is at the positions $m\Delta$. It is clear that the basic transition probability $Q(k, m)$ is given by:

$$Q(k, m)=\tfrac{1}{2}\delta(m, k-1)+\tfrac{1}{2}\delta(m, k+1).$$

Introducing this in (13) and dropping again the initial state n, one obtains the difference equation:

$$P(m, s)=\tfrac{1}{2}P(m+1, s-1)+\tfrac{1}{2}P(m-1, s-1), \quad (15)$$

which has to be solved with the initial condition (14). The solution is very easy to obtain; with $\nu=|n-m|$ one gets:

$$P(n|m, s)=\frac{s!}{\left(\frac{\nu+s}{2}\right)!\left(\frac{\nu-s}{2}\right)!}\left(\frac{1}{2}\right)^s. \quad (16)$$

The first probability distribution $W_1(n)$ should of course become independent of n, and will therefore be not strictly normalizable except when one limits the number of positions. This is also in accord with the general relation (3c). One sees easily that in this case $P(n|m, s)$, besides fulfilling the general Eq. (3b), fulfills in addition the special relation:

$$\sum_n P(n|m, s)=1,$$

and as a result the equation (analogous to (3c)):

$$W(m)=\sum_n W(n)P(n|m, s)$$

has the solution $W(n)=\text{constant}$.

(*b*) An example of Ehrenfest[15]—Suppose now that the fundamental transition probability $Q(k, m)$ has the form:

$$Q(k, m)=\frac{R+k}{2R}\delta(m, k-1)+\frac{R-k}{2R}\delta(m, k+1),$$

where R is a given integer. In the language of the random walk problem of the previous example, this means that there is an attractive center; the probabilities for making a step Δ to the right or left are not more equal but $\tfrac{1}{2}(1-k/R)$ and $\tfrac{1}{2}(1+k/R)$ so that the point will have the

[14] For the case that the molecules of the gas can only collide against fixed centers or against other molecules which have a *given* velocity distribution.

[15] For the description of the probability problem and for a more complete analysis see E. Schrödinger and F. Kohlrausch, Physik. Zeits. **27**, 306 (1926).

tendency to go to the position $k=0$. The difference equation now becomes:

$$P(m,s)=\frac{R+m+1}{2R}P(m+1,s-1)+\frac{R-m+1}{2R}P(m-1,s-1), \quad (17)$$

which has again to be solved with the initial condition (14). We did not succeed in finding the solution; it is possible, however, to calculate average values. For instance it is easy to show from (17):

$$\langle m(s)\rangle_{Av}=\sum_m mP(m,s)=\left(1-\frac{1}{R}\right)\langle m(s-1)\rangle_{Av}.$$

Since for $s=0$, $\langle m\rangle_{Av}=n$, one finds:

$$\langle m(s)\rangle_{Av}=n\left(1-\frac{1}{R}\right)^s, \quad (18a)$$

which shows how the average position of the point goes to zero. In the same way one gets:

$$\langle m^2(s)\rangle_{Av}=n^2\left(1-\frac{2}{R}\right)^s+\frac{1}{2}R\left[1-\left(1-\frac{2}{R}\right)^s\right]. \quad (18b)$$

$\langle m^2\rangle_{Av}$ will therefore go to $R/2$ for $s\to\infty$. This is in accord with the first probability distribution (which is also the stationary distribution, see Eq. (4)) for which one finds easily:

$$W_1(n)=\frac{(2R)!}{(R+n)!(R-n)!}\left(\frac{1}{2}\right)^{2R}$$

One can verify that $W_1(n)$ fulfills the equation:

$$W_1(n)=\sum_k W_1(k)Q(k,n)$$

which is a special case of Eq. (3c).

6. THE GAUSSIAN RANDOM PROCESS; METHOD OF RICE

(a) Assumptions

The Gaussian random process is characterized by the fact that all the basic distribution functions (1) are Gaussian distributions, and one could take this fact as the defining property of the process. However, since as we shall see, the spectrum essentially determines everything, it is more natural to start (following Rice) with the Fourier development of the Gaussian random function $y(t)$.

Consider again the stationary random function $y(t)$ over a long time T, and suppose that $y(t)$ is repeated periodically with the period T. One can then develop $y(t)$ in a Fourier series:

$$y(t)=\sum_{k=1}^{\infty}(a_k\cos 2\pi f_k t+b_k\sin 2\pi f_k t), \quad (19)$$

where $f_k=k/T$. There is no constant term, since we will assume that the average value of y is zero. The coefficients a_k and b_l are random variables, and we will *assume*, that they are all independent of each other and Gaussianly distributed with average values zero, so that one has for the probability that the a_k and b_l are in certain ranges da_k, db_l the expression:

$$W(a_1a_2\cdots;b_1b_2\cdots)=\prod_k\frac{1}{\sigma_k\sqrt{2\pi}}\exp\left[-(a_k^2+b_k^2)/2\sigma_k^2\right], \quad (20)$$

where $\sigma_k^2=\langle a_k^2\rangle_{Av}=\langle b_k^2\rangle_{Av}=G(f_k)/T$. $G(f)$ is again the spectral density (cf. Eqs. (8) and (9)), since:

$$\langle y^2(t)\rangle_{Av}=\sum_k(\langle a_k^2\rangle_{Av}\cos^2 2\pi f_k t+\langle b_k^2\rangle_{Av}\sin^2 2\pi f_k t)=\frac{1}{T}\sum_k G(f_k)\cong\int_0^\infty dfG(f). \quad (21)$$

With these assumptions one is now able to derive all possible distribution functions for the Gaussian random function $y(t)$. As a preparation one needs:

(b) A Theorem about Gaussian Distributions

Suppose the variables $x_1x_2\cdots x_n$ are distributed according to:

$$W(x_1\cdots x_n)=\prod_{i=1}^{n}\frac{1}{\sigma_i(2\pi)^{\frac{1}{2}}}\exp\left\{-\frac{x_i^2}{2\sigma_i^2}\right\}.$$

Let $y_1y_2\cdots y_s$ $(s\leqslant n)$ be s linear combinations of the x_i:

$$y_k=\sum_{i=1}^{n}a_{ki}x_i,\quad k=1,2\cdots s$$

where the a_{ki} are constants. One can prove easily[16] that the y_k will be distributed according

[16] See Note I of the appendix.

to the s-dimensional Gaussian distribution:

$$P(y_1\cdots y_s)=\frac{1}{(2\pi)^{s/2}B^{\frac{1}{2}}}\times\exp\left[-\frac{1}{2B}\sum_{k,l=1}^{s}B_{kl}y_ky_l\right]. \quad (22)$$

Here B_{kl} is the cofactor of the element b_{kl} in the matrix b_{kl}, where:

$$b_{kl}=\sum_{i=1}^{n}a_{ki}a_{li}\sigma_i^2=\langle y_ky_l\rangle_{Av}, \quad (23)$$

and B is the determinant of the matrix b_{kl}. As a special case take, for instance, $s=2$. One then gets the two-dimensional Gaussian distribution, which according to (22) can be written in the form:

$$P(y_1y_2)=\frac{1}{2\pi\sigma\tau(1-\rho^2)^{\frac{1}{2}}}\times\exp\left[-\frac{1}{2(1-\rho^2)}\left\{\frac{y_1^2}{\sigma^2}+\frac{y_2^2}{\tau^2}-\frac{2\rho}{\sigma\tau}y_1y_2\right\}\right], \quad (24)$$

where $\sigma^2=\langle y_1^2\rangle_{Av}$, $\tau^2=\langle y_2^2\rangle_{Av}$ and $\langle y_1y_2\rangle_{Av}=\sigma\tau\rho$; ρ is the correlation coefficient.

(c). Distribution Functions for $y(t)$

Using the general theorem mentioned above one can now derive from (19) any kind of distribution function referring to $y(t)$. The method is best explained by considering a few examples.

1. The distribution of y at fixed t. According to (19) y is for a given t a linear function of the basic variables a_k, b_l. We know, therefore, that the probability distribution for y will be Gaussian with a mean square value given by (21). The time t has disappeared, and this is as it should be since $W_1(y)$ must be independent of t, because the process is stationary. In the same way one can compute the distribution of the velocity $\dot{y}(t)$. This is also a linear function of the a_k, b_l, so the distribution will again be Gaussian with the mean square value:

$$\langle\dot{y}(t)^2\rangle_{Av}=4\pi^2\int_0^\infty f^2G(f)df. \quad (25)$$

2. The distribution of y and $\dot{y}$ at a fixed t. This will now be a two-dimensional Gaussian distribution, which is, however, especially simple since y and $\dot{y}$ at a given t are not correlated. One gets namely:

$$\langle y(t)\dot{y}(t)\rangle_{Av}=\sum_k 2\pi f_k\sin 2\pi f_kt\times\cos 2\pi f_kt(\langle -a_k^2\rangle_{Av}+\langle b_k^2\rangle_{Av})=0. \quad (26)$$

3. The joint distribution of $y(t_1)$ and $y(t_2)$. This will again be a two-dimensional Gaussian distribution. One gets:

$$\langle y(t_1)y(t_2)\rangle_{Av}=\frac{1}{T}\sum_k G(f_k)\cos 2\pi f_k(t_1-t_2)\cong\int_0^\infty G(f)\cos 2\pi f\tau df. \quad (27)$$

The correlation depends therefore only on $\tau=t_2-t_1$, as it should be since the process is stationary.

4. In this way one can go on. One can consider for instance the third distribution function $W_3(y_1t_1, y_2t_2, y_3t_3)$, which will be a three-dimensional Gaussian distribution depending only on t_2-t_1 and t_3-t_2. One can find the four-dimensional Gaussian distribution $W(y_1\dot{y}_1, y_2\dot{y}_2, \tau)$. One can bring in the acceleration $\ddot{y}(t)$ and its distribution functions, and so on.

7. FURTHER REMARKS ON THE METHOD OF RICE

1. We have seen that for a Gaussian random process all the distribution functions can be determined when one only knows the spectrum or the correlation function. In the actual problems of the Brownian motion this spectrum can be found from the so-called *Langevin equations* or in the electrical analogy from the *circuit equations with thermal noise sources.* For examples see Sections 9 and 10. It should be emphasized, however, that for many applications it is an advantage that one can leave open the question of the actual shape of the spectrum.

2. A disadvantage of not knowing the spectrum is that it does not allow a classification of the Gaussian processes, so that one does not know which distribution function describes the process completely. The different type of processes correspond to different type of spectra. For instance one can show[17] that a one-dimensional Gaussian process will be Markoffian only when the correlation function $\rho(t)=\exp(-\beta t)$ so

[17] This was first pointed out by J. L. Doob, Ann. Math. 43, 351 (1942).

that according to (11) the spectrum must be $\sim 1/(\beta^2+\omega^2)$. To prove this one determines with the method of Rice the distribution functions $W_3(y_1y_2y_3)$ and $W_2(y_1y_2)$. One, therefore, knows also the conditional probability $P_3(y_1y_2|y_3)$. For a Markoff process this must be identical with $P_2(y_2|y_3)$ and one finds that this can only be the case when the correlation function $\rho(\tau)$ fulfills the functional equation:

$$\rho(t_3-t_1)=\rho(t_2-t_1)\rho(t_3-t_2). \tag{28}$$

The only non-singular solution of this equation is

$$\rho(\tau)=\exp(-\beta\tau).$$

3. This theorem can be generalized to n-dimensional Gaussian processes. The dependent variable y now denotes an n-dimensional vector with components $x_1, x_2, \cdots x_n$. Instead of a correlation *function* one gets a correlation *matrix:*

$$\mathbf{R}(\tau)=\begin{pmatrix} \langle x_1(t)x_1(t+\tau)\rangle_{Av}\cdots\langle x_1(t)x_n(t+\tau)\rangle_{Av} \\ \vdots \\ \langle x_n(t)x_1(t+\tau)\rangle_{Av}\cdots\langle x_n(t)x_n(t+\tau)\rangle_{Av}\end{pmatrix}. \tag{29}$$

From the stationarity of the process, follows:

$$\mathbf{R}(\tau)=\tilde{\mathbf{R}}(-\tau), \tag{30}$$

where $\tilde{\mathbf{R}}$ denotes the transposed matrix. Following the same reasoning as for the one-dimensional process one can show[18] that the n-dimensional Gaussian process is Markoffian only when $R(\tau)$ fulfills the matrix functional equation:

$$\mathbf{R}(t_3-t_1)=\mathbf{R}(t_2-t_1)\mathbf{R}(t_3-t_2), \tag{31}$$

where we have still assumed that $\mathbf{R}(0)=\mathbf{I}$, the unit matrix.[19] The only non-singular solution is

$$\mathbf{R}(\tau)=e^{\mathbf{Q}\tau}, \tag{32}$$

for $\tau>0$, $\mathbf{Q}$ is a constant matrix, which is in general not symmetric,[20] so that its eigenvalues may be complex. There is now of course a greater variety of possible spectra, corresponding to the different forms $\mathbf{Q}$ may have.

4. One should point out, that sometimes the distribution functions which one derives with Rice's method will have no meaning since some of the integrals over the spectrum are divergent. For instance when $G(f)\sim 1/(\alpha^2+f^2)$, the distribution functions in which the velocity $\dot{y}(t)$ appear have no meaning since (see Eq. (25)) $\langle \dot{y}^2\rangle_{Av}$ will not exist. In this case one may call the process *non-differentiable.* The degree of differentiability will be characteristic for the process and will depend again on the spectrum.

[18] This result seems to be contained in a recent paper by J. L. Doob, Ann. Am. Stat. 15, 229 (1944). See also Note II of the appendix, where we give some details of a more direct proof which we owe to Dr. M. Kac.

[19] This is no loss in generality, since it can always be achieved by using the proper linear combinations of the components of the n-dimensional vector y. In the physical language this means that we have used such coordinates that the energy is a sum of squares.

[20] For $\tau<0$ $\mathbf{R}(\tau)=\exp(-\tilde{\mathbf{Q}}\tau)$ in accordance with (30).

8. THE GAUSSIAN RANDOM PROCESS; METHOD OF FOKKER-PLANCK

(a) Basic Ideas

It is best to start with the discrete random series (cf. Section 5). Suppose that the basic transition probability $P(n|m,\tau)$ or $Q(n,m)$ has the property that in the time τ n can only change by zero or by ± 1. This was, for instance, the case in the examples (*a*) and (*b*) discussed in Section 5. Consider now for this case the limit in which n and the time $s\tau$ become continuous. The Smoluchowski equation will then become a partial differential equation of the first order in the time coordinate and of the second order in the space coordinate. For instance, in example (*a*), the Smoluchowski equation becomes Eq. (15) which may be written:

$$P(m,s)-P(m,s-1)=\tfrac{1}{2}[P(m+1,s-1) -2P(m,s-1)+P(m-1,s-1)].$$

In the limit that $s\tau=t$ and $m\Delta=x$ become continuous variables, this clearly goes over into:

$$(\partial P/\partial t)=D(\partial^2P/\partial x^2), \tag{33}$$

when $D=\operatorname{Lim}\Delta^2/2\tau$. One gets, therefore, the well-known heat conduction or diffusion equation. In the same way one shows that in example (*b*) one gets from (17) in the limit the equation:

$$\frac{\partial P}{\partial t}=\beta\frac{\partial}{\partial x}(xP)+D\frac{\partial^2P}{\partial x^2}, \tag{34}$$

where $\beta=\operatorname{Lim}1/\tau R$.

In this limit the problem of finding the probability distribution $P(x_0|x,t)$ becomes then the problem of finding the *fundamental solution* of the partial differential equation of the diffusion (or parabolic) type into which the Smoluchowski equation has degenerated. We mean by this the solution which for $t=0$ becomes the Dirac singular function $\delta(x-x_0)$. This corresponds to the condition (14) in the discrete case and expresses again the fact that for $t=0$ one is certain that $x=x_0$. For (33) this solution is given by:

$$P(x_0|x,t)=\frac{1}{(4\pi Dt)^{\frac{1}{2}}}\exp\left[-(x-x_0)^2/4Dt\right]. \quad (35)$$

It is easy to show that this is the limit into which the solution (16) of the discrete case goes over. For Eq. (34) the fundamental solution is given by:[21]

$$P(x_0|x,t)=\frac{1}{(2\pi\sigma^2)^{\frac{1}{2}}}\exp\left[-(x-\bar{x})^2/2\sigma^2\right], \quad (36)$$

where $\langle x\rangle_{\mathrm{Av}}=x_0\exp(-\beta t)$ and $\sigma^2=\langle(x-\bar{x})^2\rangle_{\mathrm{Av}}=(D/\beta)[1-\exp(-2\beta t)]$. It is clear that these average values follow also in the limit from (18a) and (18b).

One should point out that one gets in the limit a diffusion equation *only* when $P(n|m,\tau)$ is such that in the time τ n can only change by zero or ± 1, or, less precisely, when in small times the space coordinate can only change with small amounts. *In the general case*, the Smoluchowski equation will become in the limit an integro-differential equation which is of the same type as the Boltzmann equation in the kinetic theory of gases.

(b) Assumptions

In the continuous case we will start from the Smoluchowski equation in the form:[22]

$$P(x|y,t+\Delta t)=\int dz P(x|z,t)P(z|y,\Delta t). \quad (37)$$

This *assumes*, therefore, that the process is a Markoff process. The moments of the change in the space coordinate in a small time Δt are given by:

$$a_n(z,\Delta t)=\int dy(y-z)^n P(z|y,\Delta t),$$

and we shall *assume* that for $\Delta t\to 0$, only the first and second moments become proportional to Δt so that the limits

$$A(z)=\mathrm{Lim}\frac{1}{\Delta t}a_1(z,\Delta t),$$
$$B(z)=\mathrm{Lim}\frac{1}{\Delta t}a_2(z,\Delta t), \quad (38)$$

exist. This assumption expresses the fact that for these processes in small times the space coordinate can only change with small amounts. In the actual problems of the Brownian motion this assumption can be proved and the average values $A(z)$ and $B(z)$ can be calculated from the Langevin equations or in the electrical analogy from the circuit equations with thermal noise sources.[23] Just as in the method of Rice, these equations are, therefore, the real basis for the theory of the Brownian motion.

(c) Derivation of the Fokker-Planck Equation

Consider the integral

$$\int dyR(y)\frac{\partial P(x|y,t)}{\partial t},$$

where $R(y)$ is an arbitrary function, which goes to zero for $y\to\pm\infty$ sufficiently fast. Replacing the differential quotient by the limit of the difference quotient and using the Smoluchowski equation in the form (37) one can write:

$$\int dyR(y)\frac{\partial P}{\partial t}=\mathrm{Lim}\frac{1}{\Delta t}\int dyR(y)$$
$$\times[P(x|y,t+\Delta t)-P(x|y,t)]$$
$$=\mathrm{Lim}\frac{1}{\Delta t}\left[\int dyR(y)\int dzP(x|z,t)P(z|y,\Delta t)\right.$$
$$\left.-\int dzR(z)P(x|z,t)\right].$$

In the double integral, interchange the order of integration and develop $R(y)$ in a Taylor series

[21] See, for instance, I, Section II. Equation (36) is also a special case of the solution derived in Note IV of the appendix.

[22] For simplicity we consider the process to be one dimensional, since the generalization to the n-dimensional case is obvious. We follow the notation and the exposition of Kolmogoroff, Math. Ann. **104**, 415 (1931).

[23] For examples see Sections 9 and 10.

in $(z-y)$. Because of (38) one can stop at the term with $(z-y)^2$ and one gets:

$$\int dy R(y) \frac{\partial P}{\partial t} = \int dz P(x|z,t) \times [R'(z)A(z) + \tfrac{1}{2}R''(z)B(z)].$$

Integrating partially and writing y for z one obtains:

$$\int dy R(y) \left[\frac{\partial P}{\partial t} + \frac{\partial}{\partial y}(AP) - \frac{1}{2}\frac{\partial^2}{\partial y^2}(BP) \right] = 0.$$

Since this must hold for any function $R(y)$, the expression in the square brackets must be zero, which gives the general Fokker-Planck equation:

$$\frac{\partial P}{\partial t} = -\frac{\partial}{\partial y}[A(y)P] + \frac{1}{2}\frac{\partial^2}{\partial y^2}[B(y)P], \quad (39)$$

of which, of course, (33) and (34) are special cases. For an n-dimensional process one gets analogously:

$$\frac{\partial P}{\partial t} = -\sum_i \frac{\partial}{\partial y_i}[A_i(\mathbf{y})P] + \frac{1}{2}\sum_{k,l}\frac{\partial^2}{\partial y_k \partial y_l}[B_{kl}(\mathbf{y})P], \quad (39a)$$

where again the A_i and the B_{kl} are the first and second moments defined analogously as (38).

9. THE BROWNIAN MOTION OF A FREE PARTICLE[24]

(a) Assumptions; The Langevin Equation

For a free particle (mass m, velocity v) the equation of motion will be:

$$m(dv/dt) + fv = K(t), \quad (40a)$$

where f is the friction coefficient, and $K(t)$ is the fluctuating force, of which the average value is zero and which has a very sharp correlation function and therefore, a practically white spectrum. The spectral density of $K(t)$ is $4fkT$ where k is the Boltzmann constant and T the temperature of the surrounding medium. The analogous electrical problem is of course the (L, R) circuit, and the circuit equation is:

$$L(di/dt) + Ri = E(t), \quad (40b)$$

where $E(t)$ is a purely random fluctuating e.m.f. (the thermal noise source), which has a spectral density $4RkT$. We will combine these cases by writing (40a) and (40b) in the form:

$$(dy/dt) + \beta y = F(t) \quad (40c)$$

and by taking $4D$ as the spectral density of the purely random $F(t)$. This means that we *assume*:[25]

$$\langle F(t)\rangle_{\text{Av}} = 0, \quad (41a)$$

$$\langle F(t_1)F(t_2)\rangle_{\text{Av}} = 2D\delta(t_1 - t_2). \quad (41b)$$

This is, however, not enough; besides being purely random, we must *assume* that $F(t)$ is *Gaussian*. This can be expressed in different ways. Either one can postulate the Gaussian distribution of the Fourier coefficients (see Eq. (20) where now $\sigma_k^2 = \text{const.} = 4D/T$) or one can assume the two properties:[26]

$$\langle F(t_1)F(t_2)\cdots F(t_{2n+1})\rangle_{\text{Av}} = 0, \quad (42a)$$

$$\langle F(t_1)F(t_2)\cdots F(t_{2n})\rangle_{\text{Av}} = \sum_{\text{all pairs}} \langle F(t_i)F(t_j)\rangle_{\text{Av}} \cdot \langle F(t_k)F(t_l)\rangle_{\text{Av}} \cdots \quad (42b)$$

where the sum has to be taken over all the different ways in which one can divide the $2n$ time points $t_1 \cdots t_{2n}$ into n pairs. It is easy to show the equivalence of these two definitions (see Note III of the appendix).

(b) The Spectrum of $y(t)$

Since $F(t)$ is Gaussian it is clear that $y(t)$ will also be a Gaussian random process with a spectrum:

$$G_y(f) = \frac{4D}{\beta^2 + (2\pi f)^2}. \quad (43)$$

[24] Compare I, Sections II and III. The first complete derivation of the distribution functions obtained in Sections 9 and 10 was given by L. S. Ornstein and W. R. van Wyk, Physica 1, 235 (1934). The derivation from the Fokker-Planck or Kramers equation was found independently by Ming Chen Wang, Dissertation, Ann Arbor (1942) and by Chandrasekhar, reference 3.

[25] The second equation follows from (11b) since:

$$G_F(f) = 4D = 2\int_{-\infty}^{+\infty} d\tau \cos 2\pi f\tau \langle F(t)F(t+\tau)\rangle_{\text{Av}}.$$

[26] This is the starting point of some of the work of N. Wiener on the theory of the Brownian motion. The physical justification of the assumptions (41) and (42) comes from the Maxwell-Boltzmann distribution law, which in the theory of the Brownian motion is always *postulated* and not derived. Compare also I.

This corresponds to a correlation function $\rho(t) = \exp(-\beta t)$ and the second probability distribution is, therefore, the two-dimensional Gaussian distribution:

$$W_2(y_1y_2t) = \frac{\beta}{2\pi D(1-\rho^2)^{\frac{1}{2}}} \times \exp\left[-\frac{\beta}{2D(1-\rho^2)}\{y_1^2+y_2^2-2\rho y_1y_2\}\right], \quad (44)$$

since:

$$\langle y^2\rangle_{\text{Av}} = \int_0^\infty G_y(f)df = D/\beta. \quad (45)$$

According to the theorem of Doob $y(t)$ will be a Markoff process, so that $W_2(y_1y_2t)$ gives the complete description of the process.

(c) The Fokker-Planck Equation

The average values $A(y)$ and $B(y)$ can now be computed by means of (41) and one finds:

$$A(y) = -\beta y, \quad B(y) = 2D. \quad (46)$$

The proof is simple; integrating (40c) over a short time Δt one gets:

$$\Delta y = -\beta y \Delta t + \int_t^{t+\Delta t} d\xi F(\xi).$$

Therefore:

$$A(y) = \operatorname*{Lim}_{\Delta t\to 0} \frac{\langle \Delta y\rangle_{\text{Av}}}{\Delta t} = -\beta y,$$

since $\langle F\rangle_{\text{Av}} = 0$. Further:

$$\langle \Delta y^2\rangle_{\text{Av}} = \beta^2 y^2 \Delta t^2 + \int\int_t^{t+\Delta t} d\xi d\eta \langle F(\xi)F(\eta)\rangle_{\text{Av}},$$

and from (41b) one shows easily that the double integral is $2D\Delta t$, so that:

$$B(y) = \operatorname*{Lim}_{\Delta t\to 0} \frac{\langle \Delta y^2\rangle_{\text{Av}}}{\Delta t} = 2D.$$

In the same way it follows from (42) that all the higher moments of Δy go to zero in the limit $\Delta t \to 0$, so that all the assumptions of §8, b are fulfilled.[27] With the values given by (46) the Fokker-Planck equation becomes:

$$\frac{\partial P}{\partial t} = \beta \frac{\partial}{\partial y}(yP) + D\frac{\partial^2 P}{\partial y^2}. \quad (47)$$

This is identical with (34), so that the fundamental solution $P(y_0|y, t)$ is given by (36) (for the proof, cf. Note IV). For $t\to\infty$ one gets:

$$W_1(y) = \operatorname*{Lim}_{t\to\infty} P(y_0|y, t) = \left(\frac{\beta}{2\pi D}\right)^{\frac{1}{2}} \exp\left(-\frac{\beta y^2}{2D}\right),$$

in accordance with (45). For the second probability distribution:

$$W_2(y_1y_2t) = W_1(y_1)P(y_1|y_2, t)$$

one gets again Eq. (44). That $y(t)$ is a Markoff process has now, of course, been assumed from the beginning.[28]

10. THE BROWNIAN MOTION OF A SIMPLE HARMONIC OSCILLATOR

(a) The Langevin Equation

Suppose now that instead of (40c) we have the second-order differential equation:

$$\frac{d^2y}{dt^2} + \beta\frac{dy}{dt} + \omega_0^2 y = F(t). \quad (48)$$

This describes clearly the Brownian motion of a simple harmonic oscillator or the thermal noise in a (R, L, C) circuit. For the $F(t)$ we assume again the basic properties (41) and (42).

(b) The Spectrum and the Correlation Matrix

Since $F(t)$ is Gaussian, it is clear that $y(t)$ will also be a Gaussian random process with a spectrum:

$$G_y(f) = \frac{4D}{|-(2\pi f)^2 + 2\pi i\beta f + \omega_0^2|^2}, \quad (49)$$

[27] It should be emphasized perhaps again that from the physical point of view these assumptions (and, therefore, also (41) and (42)) are necessarily only approximations. The basic equation is always Boltzmann's integral equation. Only when in each collision the velocity of the particle can change very little, then the Boltzmann equation can be approximated by the diffusion Eq. (47). It is very instructive to compare the derivation above with the derivation of (47) in the well-known Rayleigh model (Scientific Papers Vol. 3, p. 473) for the Brownian motion of a heavy particle.

[28] We shall not discuss the distribution function for the displacement of the particle, since it follows from the velocity distribution (see, for instance, Doob, reference 17) and since it is also a special case of the distribution functions derived in Section 10.

from which follows according to (11a)

$$\langle y(t)y(t+\tau)\rangle_{Av}$$
$$=\frac{2D}{\pi}\int_0^\infty \frac{\cos\omega\tau}{(\omega_0^2-\omega^2)^2+\beta^2\omega^2}d\omega$$
$$=\frac{D}{\beta\omega_0^2}e^{-\beta\tau/2}\left(\cos\omega_1\tau+\frac{\beta}{2\omega_1}\sin\omega_1\tau\right), \quad (50a)$$

where $\omega_1^2=\omega_0^2-(\beta^2/4)$; the formula will always be written for the *underdamped* case; for the aperiodic case let $\omega_1\to 0$ and for the overdamped case put $\omega_1=i\omega'$.

Of course, $y(t)$ is *not* more a Markoff process. However, from the physical situation and also from the general theorem of Doob one must expect that $y(t)$ is the "projection" of the two-dimensional Gaussian Markoff process $[y(t), p(t)]$ where $p(t)=dy/dt$. The correlation function (50a) must be extended to the correlation matrix:

$$\begin{pmatrix}\langle y(t)y(t+\tau)\rangle_{Av} & \langle y(t)p(t+\tau)\rangle_{Av}\\ \langle p(t)y(t+\tau)\rangle_{Av} & \langle p(t)p(t+\tau)\rangle_{Av}\end{pmatrix},$$

and one finds easily that:

$$\langle p(t)y(t+\tau)\rangle_{Av}=2\pi\int_0^\infty dfG_y(f)f\sin 2\pi f\tau,$$
$$=\frac{2D}{\pi}\int_0^\infty \frac{\omega\sin\omega\tau}{(\omega_0^2-\omega^2)^2+\beta^2\omega^2}d\omega,$$
$$=+\frac{D}{\beta\omega_1}e^{-\beta\tau/2}\sin\omega_1\tau, \quad (50b)$$

and:

$$\langle p(t)p(t+\tau)\rangle_{Av}$$
$$=4\pi^2\int_0^\infty dfG_y(f)f^2\cos 2\pi f\tau,$$
$$=\frac{2D}{\pi}\int_0^\infty \frac{\omega^2\cos\omega\tau}{(\omega_0^2-\omega^2)^2+\beta^2\omega^2}d\omega,$$
$$=\frac{D}{\beta}e^{-\beta\tau/2}\left(\cos\omega_1\tau-\frac{\beta}{2\omega_1}\sin\omega_1\tau\right). \quad (50c)$$

The complete description of the process will now be given by $W_2(y_1p_1, y_2p_2, t)$, which is a four-dimensional Gaussian distribution. We will not write it down since for the discussion it is easier to consider the conditional probability $P_2(y_1p_1|y_2p_2, t)$. Consider first, however,

(c) The Fokker-Planck Equation

Replacing the Langevin Eq. (48) by the simultaneous equations:

$$dy/dt=p \quad\text{and}\quad dp/dt+(\beta p+\omega_0^2y)=F(t), \quad (48a)$$

one finds easily for the average values occurring in the two-dimensional Fokker-Planck equation (cf. Eq. (39a)):

$$A_1=\text{Lim}\frac{\langle\Delta y\rangle_{Av}}{\Delta t}=p;$$
$$A_2=\text{Lim}\frac{\langle\Delta p\rangle_{Av}}{\Delta t}=-(\beta p+\omega_0^2y);$$
$$B_{11}=\text{Lim}\frac{\langle\Delta y^2\rangle_{Av}}{\Delta t}=0;$$
$$B_{12}=\text{Lim}\frac{\langle\Delta y\Delta p\rangle_{Av}}{\Delta t}-0;$$
$$B_{22}=\text{Lim}\frac{\langle\Delta p^2\rangle_{Av}}{\Delta t}=2D,$$

so that one gets:[29]

$$\frac{\partial P}{\partial t}=-p\frac{\partial P}{\partial y}+\frac{\partial}{\partial p}[(\beta p+\omega_0^2y)P]+D\frac{\partial^2P}{\partial p^2}, \quad (51)$$

which has to be solved with the initial condition:

$$P(yp, 0)=\delta(y-y_0)\delta(p-p_0).$$

For the solution it is simpler to work with the independent variables:

$$z_1=p+ay;\quad z_2=p+by, \quad (52)$$

where:

$$a=\tfrac{1}{2}\beta+i\omega_1 \quad\text{and}\quad b=\tfrac{1}{2}\beta-i\omega_1.$$

Equation (51) is then transformed into the more symmetrical form:

$$\frac{\partial P}{\partial t}=b\frac{\partial}{\partial z_1}(z_1P)+a\frac{\partial}{\partial z_2}(z_2P)$$
$$+D\left(\frac{\partial}{\partial z_1}+\frac{\partial}{\partial z_2}\right)^2P, \quad (51a)$$

and this is a special case of the equation solved in Note IV of the appendix. One finds that the

[29] This is a special case of the *equation of Kramers*, Physica 7, 284 (1940). Kramers takes a general force $K(y)$ instead of the harmonic force $-\omega_0^2y$. His derivation is essentially the same as the one given above. One should emphasize perhaps, that with a general force $K(y)$ the process $[y(t), p(t)]$ is *still* Markoffian, but it is *not more* Gaussian, since the basic Langevin equation is then not *linear* anymore.

fundamental solution of (51a) is a two-dimensional Gaussian distribution in z_1 and z_2 with the average values:

$$\langle z_1\rangle_{Av}=z_{10}e^{-bt},\quad \langle z_2\rangle_{Av}=z_{20}e^{-at}, \tag{53}$$

and the variances:

$$\begin{aligned} \langle(z_1-\bar{z}_1)^2\rangle_{Av}&=\frac{D}{b}(1-e^{-2bt}),\\ \langle(z_2-\bar{z}_2)^2\rangle_{Av}&=\frac{D}{a}(1-e^{-2at}),\\ \langle(z_1-\bar{z}_1)(z_2-\bar{z}_2)\rangle_{Av}&=\frac{2D}{a+b}(1-e^{-(a+b)t}),\end{aligned} \tag{54}$$

where z_{10}, z_{20} are the initial values of z_1, z_2 corresponding to y_0 and p_0.

(d) Discussion

Since z_1 and z_2 are connected with p and y by the linear relations (52) it is clear that $P(p_0y_0|py, t)$ will also be a two-dimensional Gaussian distribution in p and y. One obtains from (53) and (54) for the average values and the variances the expressions:

$$\left.\begin{aligned} \langle p\rangle_{Av}&=\frac{p_0}{\omega_1}e^{-\frac{1}{2}\beta t}\left(\omega_1\cos\omega_1 t-\frac{\beta}{2}\sin\omega_1 t\right)\\ &\qquad-\frac{\omega_0^2}{\omega_1}y_0e^{-\frac{1}{2}\beta t}\sin\omega_1 t,\\ \langle y\rangle_{Av}&=\frac{p_0}{\omega_1}e^{-\frac{1}{2}\beta t}\sin\omega_1 t\\ &\qquad+\frac{y_0}{\omega_1}e^{-\frac{1}{2}\beta t}\left(\omega_1\cos\omega_1 t+\frac{\beta}{2}\sin\omega_1 t\right),\\ \langle(p-\bar{p})^2\rangle_{Av}&=\frac{D}{\beta}\left[1-\frac{1}{\omega_1^2}e^{-\beta t}\right.\\ &\qquad\times(\omega_1^2+\tfrac{1}{2}\beta^2\sin^2\omega_1 t\\ &\qquad\left.-\beta\omega_1\sin\omega_1 t\cos\omega_1 t)\right],\\ \langle\omega_0^2(y-\bar{y})^2\rangle_{Av}&=\frac{D}{\beta}\left[1-\frac{1}{\omega_1^2}e^{-\beta t}\right.\\ &\qquad\times(\omega_1^2+\tfrac{1}{2}\beta^2\sin^2\omega_1 t\\ &\qquad\left.+\beta\omega_1\sin\omega_1 t\cos\omega_1 t)\right],\\ \langle\omega_0(p-\bar{p})(y-\bar{y})\rangle_{Av}&=\frac{D\omega_0}{\omega_1^2}e^{-\beta t}\sin^2\omega_1 t.\end{aligned}\right\} \tag{55}$$

One has, of course,

$$\bar{p}=\frac{d\bar{y}}{dt},$$

$$\frac{d^2\bar{y}}{dt^2}+\beta\frac{d\bar{y}}{dt}+\omega_0^2\bar{y}=0.$$

The center of the Gaussian distribution moves, therefore, like the harmonic oscillator starting from the initial values p_0, y_0. In the (p, y) plane one gets (in the periodic case) for the orbit the well known spirals. For small t:

$$\begin{aligned}\langle(p-\bar{p})^2\rangle_{Av}&\cong 2Dt,\\ \langle(y-\bar{y})^2\rangle_{Av}&\cong\tfrac{2}{3}Dt^3,\\ \langle(p-\bar{p})(y-\bar{y})\rangle_{Av}&\cong Dt^2.\end{aligned}$$

One sees, therefore, that the initial two-dimensional δ function $\delta(p-p_0)\delta(y-y_0)$ will become first a narrow ellipse elongated in the p direction. The distribution ellipse will then turn and broaden out till at the time $t=\pi/\omega_1$ it has become again a circle. This process will repeat on a larger and larger scale with the period π/ω_1 (see Fig. 1). The center of the distribution will come nearer and nearer to the origin and finally of course the $P(p_0y_0|py, t)$ will become the Maxwell-Boltzmann distribution.

11. THE BROWNIAN MOTION OF A SYSTEM OF COUPLED HARMONIC OSCILLATORS

The generalization to more complicated systems does not involve anything new, so that we will only give an outline of the main results. We will use for a change the electrical language and we will consider, therefore, an arbitrary linear network of n meshes. The circuit equations are then:[30]

$$\sum_{j=1}^{n}\left(L_{ij}\frac{d^2y_j}{dt^2}+R_{ij}\frac{dy_j}{dt}+G_{ij}Y_j\right)=\sum_{j=1}^{n}E_{ij};\qquad i=1,2\cdots n. \tag{56}$$

The E_{ii} is the fluctuating thermal e.m.f. in that part of the resistance of the ith mesh which is

[30] For the precise definition of the matrices L_{ij}, R_{ij}, G_{ij} see, for instance, E. A. Guillemin, *Communication Networks*, Vol. I, Chap. IV. All these matrices are *symmetrical*. Note, however, that R_{ij} $(i\neq j)$ does *not* need to be positive. It is negative, when in the resistance common to the ith and jth mesh the positive directions chosen for the currents are opposite to each other. The y_j are the mesh-charges.

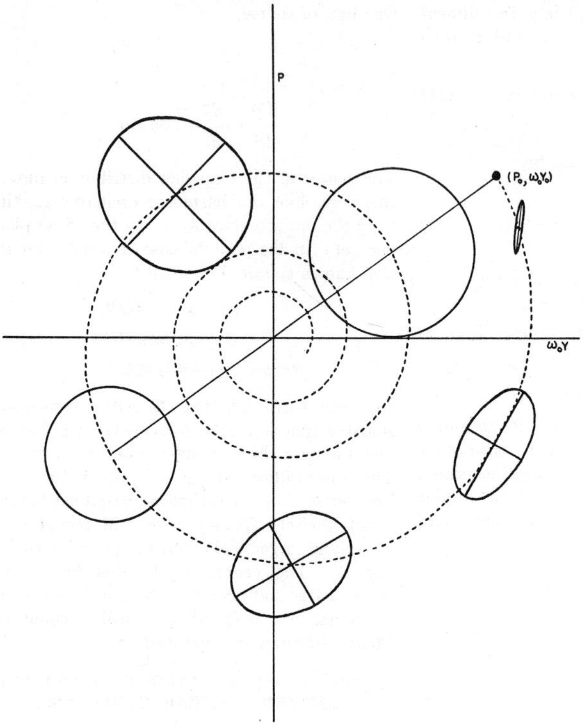

FIG. 1.

not in common with any other meshes. The E_{ij} ($i \neq j$) is the fluctuating thermal e.m.f. in the resistance R_{ij}; $E_{ij} = E_{ji}$ if in R_{ij} the positive directions chosen for the currents are in the same direction; if they are opposite to each other then $E_{ij} = -E_{ji}$. The E_{ij} are again supposed to be Gaussian random processes with a constant spectrum. We assume especially:

$$\left.\begin{aligned} &\langle E_{ij}\rangle_{\text{Av}} = 0; \\ &\langle E_{ij}(t_1)E_{ji}(t_2)\rangle_{\text{Av}} = 2R_{ij}kT\delta(t_2 - t_1); \\ &\langle E_{ij}(t_1)E_{ij}(t_2)\rangle_{\text{Av}} = 2|R_{ij}|kT\delta(t_2 - t_1); \\ &\langle E_{ij}E_{kl}\rangle_{\text{Av}} = 0. \end{aligned}\right\} \quad (57)$$

In addition one needs, of course, assumptions analogous to (42).

Since the Eqs. (56) are *linear* it is clear that each of the $y_j(t)$ will be a Gaussian random process. From the physical situation one must expect in addition that the $2n$ variables $[y_1(t)\cdots y_n(t),\ dy_1/dt\cdots dy_n/dt]$ will form a $2n$-dimensional Markoff process, governed by the Fokker-Planck equation:

$$\frac{\partial P}{\partial t} = -\sum_{i=1}^{2n} \frac{\partial}{\partial x_i}(A_i P) + \frac{1}{2}\sum_{i,\ j=1}^{2n} \frac{\partial^2}{\partial x_i \partial x_j}(D_{ij}P), \quad (58)$$

where $x_1\cdots x_{2n}$ denote the variables $y_1\cdots y_n$, $dy_1/dt\cdots dy_n/dt$. From (56) and (57) one finds further that:

$$A_i = \textstyle\sum_k a_{ik}x_k, \quad (59)$$

where the $2n$ by $2n$ matrix $\mathbf{a}$ is of the form:

$$\mathbf{a} = \begin{pmatrix} 0 & \mathbf{I} \\ -\mathbf{L}^{-1}\mathbf{G} & -\mathbf{L}^{-1}\mathbf{R} \end{pmatrix}. \quad (60)$$

Finally one gets for the $2n$ by $2n$ matrix $\mathbf{D}$ the constant matrix:

$$\mathbf{D}=\begin{pmatrix} 0 & 0 \\ 0 & 2kT\mathbf{L}^{-1}\mathbf{R}\mathbf{L}^{-1} \end{pmatrix}. \quad (61)$$

To find the fundamental solution of (58), it is best to make first a linear transformation of the x_i (analogous to (52) in §10):

$$z_i=\sum_j c_{ij}x_j \quad (62)$$

where the matrix $\mathbf{c}$ is the matrix which diagonaluzes $\mathbf{a}$, so that:

$$\sum_j c_{ij}a_{jl}=\lambda_i c_{il}. \quad (63)$$

The eigenvalues λ_i are of course the $2n$ roots of the equation:

$$Det(a_{ij}-\lambda\delta_{ij})=0. \quad (64)$$

One can easily show that this equation is identical with

$$Det(L_{ij}\lambda^2+R_{ij}\lambda+G_{ij})=0, \quad (64a)$$

and it is well known that for a linear passive network the roots of this equation must have a negative real part, and the same must hold therefore for the λ_i. The Fokker-Planck equation now becomes:

$$\frac{\partial P}{\partial t}=-\sum_i \lambda_i\frac{\partial}{\partial z_i}(z_iP)+\frac{1}{2}\sum_{i,j}\sigma_{ij}\frac{\partial^2 P}{\partial z_i\partial z_j}, \quad (65)$$

where:

$$\boldsymbol{\sigma}=\mathbf{c}\mathbf{D}\tilde{\mathbf{c}}. \quad (66)$$

The fundamental solution of (65) is derived in Note IV of the appendix. One gets a $2n$-dimensional Gaussian distribution with the average values:

$$\langle z_i\rangle_{\text{Av}}=z_{i0}\exp(\lambda_i t) \quad (67)$$

and the variances:

$$\mu_{ij}=\langle(z_i-\bar{z}_i)(z_j-\bar{z}_j)\rangle_{\text{Av}}$$

$$=-\frac{\sigma_{ij}}{\lambda_i+\lambda_j}[1-\exp(\lambda_i+\lambda_j)t], \quad (68)$$

where z_{i0} are the initial values of the z_i. Transforming back to the original variables x_i one gets of course again a $2n$-dimensional Gaussian distribution. Combining the variables x_i in a column matrix $\mathbf{x}$, one can write for the average values the matrix equation:

$$\bar{\mathbf{x}}=e^{\mathbf{a}t}\mathbf{x}_0. \quad (69)$$

This follows from (67) which can be written as

$$\bar{\mathbf{z}}=e^{\mathbf{\Lambda}t}\mathbf{z}_0,$$

where the diagonal matrix $\Lambda_{ij}=\lambda_i\delta_{ij}$. Now $\mathbf{z}=\mathbf{c}\mathbf{x}$ so that $\mathbf{x}=\mathbf{c}^{-1}\mathbf{z}$ and:

$$\bar{\mathbf{x}}=\mathbf{c}^{-1}\bar{\mathbf{z}}=\mathbf{c}^{-1}e^{\mathbf{\Lambda}t}\mathbf{z}_0=\sum_{n=0}^{\infty}\frac{t^n}{n!}\mathbf{c}^{-1}\mathbf{\Lambda}^n\mathbf{z}_0. \quad (70)$$

Equation (63) can be written in the matrix form:

$$\mathbf{c}\mathbf{a}\mathbf{c}^{-1}=\mathbf{\Lambda}, \quad (71)$$

from which follows that $\mathbf{\Lambda}^n=\mathbf{c}\mathbf{a}^n\mathbf{c}^{-1}$. Substituting this in (70) one obtains (69).

For the matrix of the variances:

$$b_{ij}=\langle(x_i-\bar{x}_i)(x_j-\bar{x}_j)\rangle_{\text{Av}},$$

one obtains from (68):

$$\mathbf{b}=\mathbf{c}^{-1}\boldsymbol{\mu}\tilde{\mathbf{c}}^{-1}. \quad (72)$$

Since the real parts of the λ_i are negative it is clear that for $t\rightarrow\infty$ all the average values $\bar{x}_i$ go to zero. The distribution function $P(x_{i0}|x_i,t)$ must become in the limit $t\rightarrow\infty$ the Maxwell-Boltzmann distribution law, which means that:

$$\lim_{t\rightarrow\infty}\mathbf{b}^{-1}=\begin{pmatrix} \frac{1}{kT}\mathbf{G} & 0 \\ 0 & \frac{1}{kT}\mathbf{L} \end{pmatrix}. \quad (73)$$

To show this from (72), one starts from Eq. (68) which in the limit $t\rightarrow\infty$ can be written in the form:

$$\mathbf{\Lambda}\boldsymbol{\mu}+\boldsymbol{\mu}\mathbf{\Lambda}=-\boldsymbol{\sigma}.$$

Substituting $\boldsymbol{\mu}=\mathbf{c}\mathbf{b}\tilde{\mathbf{c}}$ and using (66) and (71) one finds that $\mathbf{b}$ has to fulfill (always in the limit $t\rightarrow\infty$) the equation:

$$\mathbf{a}\mathbf{b}+\mathbf{b}\tilde{\mathbf{a}}=-\mathbf{D}. \quad (74)$$

This determines, of course, the matrix $\mathbf{b}$ uniquely. From (72) follows also that $\mathbf{b}$ is symmetric so that we can put:

$$\mathbf{b}=\begin{pmatrix} \mathbf{X}_1 & \mathbf{X}_2 \\ \mathbf{X}_2 & \mathbf{X}_3 \end{pmatrix},$$

where $\mathbf{X}_1$ and $\mathbf{X}_3$ are symmetric n by n matrices.

From (74) follows then:

$$\mathbf{X}_2=0,\quad \mathbf{LX}_3=\mathbf{GX}_1,\quad \mathbf{RX}_1\mathbf{G}+\mathbf{GX}_1\mathbf{R}=2kT\mathbf{R}.$$

By inspection one sees that these equations are fulfilled by:

$$\mathbf{X}_1=kT\mathbf{G}^{-1};\quad \mathbf{X}_3=kT\mathbf{L}^{-1};\quad \mathbf{X}_2=0, \tag{75}$$

and since the solution of (74) is unique this must also be the only solution. Equation (75) is, of course, equivalent with (73).

The same distribution function follows, of course, from the method of Rice. One has to start then from the correlation matrix, for which one finds from the circuit Eqs. (56):

$$\langle y_r(t)y_s(t+\tau)\rangle_{\mathrm{Av}}=\frac{kT}{\pi}\int_{-\infty}^{+\infty}d\omega e^{i\omega\tau}\times[\mathbf{Z}^{-1}(i\omega)\mathbf{R}\mathbf{Z}^{-1}(-i\omega)]_{rs}, \tag{76}$$

where $\mathbf{Z}(p)$ is the matrix:

$$\mathbf{Z}(p)=\mathbf{L}p^2+\mathbf{R}p+\mathbf{G}.$$

In order to see how in the method of Rice the Maxwell-Boltzmann distribution is reached, it is sufficient to prove the theorem of the equipartition of energy:

$$\left\langle y_r\frac{\partial V}{\partial y_r}\right\rangle_{\mathrm{Av}}=kT, \tag{77}$$

where V is the potential energy

$$V=\tfrac{1}{2}\sum_{r,s}G_{rs}y_ry_s.$$

From (76) one obtains:

$$\left\langle y_r\frac{\partial V}{\partial y_r}\right\rangle_{\mathrm{Av}}=\sum_s G_{rs}\langle y_ry_s\rangle_{\mathrm{Av}}$$

$$=\frac{kT}{\pi}\int_{-\infty}^{+\infty}d\omega[\mathbf{G}\mathbf{Z}^{-1}(i\omega)\mathbf{R}\mathbf{Z}^{-1}(-i\omega)]_{rr}. \tag{78}$$

To calculate the integral,[31] observe that:

$$\mathbf{Z}(i\omega)-\mathbf{Z}(-i\omega)=2i\omega\mathbf{R}.$$

Eliminating $\mathbf{R}$ in the integral (78) one finds:

$$\left\langle y_r\frac{\partial V}{\partial y_r}\right\rangle_{\mathrm{Av}}=-\frac{kT}{\pi i}\int_{-\infty}^{+\infty}\frac{d\omega}{\omega}[\mathbf{G}\mathbf{Z}^{-1}(i\omega)]_{rr}, \tag{79}$$

where one has to take the principal value of the integral. Since the determinant of $\mathbf{Z}(i\omega)$ has no zeros in the lower half of the complex ω-plane, it can be easily seen that the integral in (79) is $-\pi i$ times the residue of the integrand at $\omega=0$. Since $\mathbf{Z}^{-1}(0)=\mathbf{G}^{-1}$ it is clear that the residue is unity, so that one obtains the equipartition theorem (77).

12. SOME UNSOLVED PROBLEMS

Since we may have created the false impression that with the derivation of the fundamental probability distribution all problems in the theory of the Brownian motion have been solved, it may be useful to list a number of unsolved or partially solved problems.

(a) The Approach to the Barometric Distribution

It should be emphasized that *only* for harmonic forces one gets the simple theory of the Gaussian random process. For a constant force the problem becomes already much more complicated. For instance for a gravitational field (directed towards the negative x axis) the Kramers equation becomes:

$$\frac{\partial P}{\partial t}=-p\frac{\partial P}{\partial x}+g\frac{\partial P}{\partial p}+\frac{\partial}{\partial p}\left[\beta pP+D\frac{\partial P}{\partial p}\right]. \tag{80}$$

The trouble is now, that one needs a *reflecting* boundary, say at $x=0$ in order to prevent the particles from disappearing towards $x=-\infty$. We feel sure that this means the condition:

$$P(0,p,t)=P(0,-p,t). \tag{81}$$

We have been unable to find the solution of (80) (for $x\geq 0$, $-\infty<p<+\infty$) which fulfills the condition (81) and which for $t=0$ becomes $\delta(x-x_0)\delta(p-p_0)$. One can show[32] that for the stationary case the Kramers equation and the boundary condition (81) determines *uniquely* the barometric distribution:

$$P(x,p,\infty)=C\exp\left(-\frac{\beta}{2D}p^2-\frac{g\beta}{D}x\right).$$

(b) First Passage Time Problems

One may ask for the probability that the random variable y starting from the value $y=y_0$

[31] For this proof we are indebted to Dr. J. Schwinger.

[32] A proof was communicated to us by Mr. M. Dresden.

reaches the value $y=y_1$ for the first time in a time between t and $t+dt$. In the usual theory of the Brownian motion (which is based on the ordinary diffusion Eq. (33))[33] such first passage time problems have been considered and solved by Smoluchowski and others.[34] The method of Smoluchowski can also be used for a one-dimensional Gaussian Markoff process $y(t)$. One can show, for instance, that the probability distribution $W(y_0, t)dt$ of the first passage times to reach $y=0$ starting from y_0 is given by:

$$W(y_0, t)dt = y_0\left(\frac{2\beta}{\pi D}\right)^{\frac{1}{2}} \exp\left(-\frac{\beta y_0^2}{2D}z^2\right)dz, \quad (82)$$

where

$$z = e^{-\beta t}(1-e^{-2\beta t})^{-\frac{3}{2}}.$$

However, the generalization to processes $y(t)$ which are "projections" of Markoff processes seems quite complicated to us. For instance, we have *not* succeeded in finding, even for a free particle, the distribution of the one-sided first passage times if the damping is small, so that one has to use the exact Kramers equation. To extend the method of Smoluchowski it is necessary to introduce the idea of an *absorbing* boundary say at $x=0$. We feel sure that this means the condition:

$$P(0, p, t) = 0 \text{ for all } p > 0 \quad (83)$$

when the particle has started from $x_0 > 0$. However, to find solutions of the Kramers equation with the boundary condition (83) seems even more difficult than to find solutions with the boundary condition (81).

(c) The Recurrence Time Problem

One may ask for the probability that the random variable y starting from $y=a$ returns to the value $y=a$ for the first time in a time between t and $t+dt$.[35] Or in other words, what is the distribution of the time intervals between successive a-values of the random function $y(t)$. A formal solution of this problem has been given by Rice (II, Section 3.4, Eq. (11), p. 64). However, even for the simple case of the harmonic oscillator the actual discussion of the solution has not been achieved.

(d) The Distribution of the Average Value

One may ask for the probability distributions of the random variable:

$$z(t) = \int K(s-t)y(s)ds, \quad (84)$$

where $K(x)$ is a given function. A special case of (84) is the *average* of the random process $y(t)$ over a time interval of length T. Of course, when $y(t)$ is a Gaussian random process then also $z(t)$ will be Gaussian, and the problem is trivial. But for other types of processes $y(t)$ the problem is quite difficult. Rice (II, Section 3.9) has discussed some of the average values of $z(t)$. Recently M. Kac and A. Siegert have succeeded in finding the complete solution for the case that $y(t)$ is the sum of the squares of two independent Gaussian processes, which have the same probability distributions.

(e) The Distribution of the Absolute Maximum of a Random Function $y(t)$ in a Given Time Interval T

For Markoff processes one can show that this problem is equivalent with the first passage time problem, so that it is of the same degree of difficulty.

APPENDIX

Note I. Proof of Eq. (22)

One uses the integral representation of the Dirac δ function:

$$\delta(x-x') = \frac{1}{2\pi}\int_{-\infty}^{+\infty} dt \exp[it(x-x')],$$

[33] Kramers, reference 29, has shown that one obtains this equation (or the corresponding one if there is an outside force $K(y)$) from the Kramers Eq. (51) (with $-\omega_0^2 y$ replaced by $K(y)$) in the limit of strong damping.

[34] Cf. for instance, R. Furth, reference 6, and for the corresponding problems for random series M. Kac, Ann. Math. Stat. **16**, 62 (1945).

[35] For the definition of the mean recurrence time and the mean persistence time see the basic paper of Smoluchowski, Wien. Sitz. Ber. **124**, 339 (1915). Smoluchowski restricts himself mainly to discrete random series. Already in this case the question of the distribution of the recurrence times seems to be quite difficult. For the simple discrete random walk problem (example (a) of Section 5) the first passage time and the recurrence time problem can be solved exactly. But already for example (b) of Section 5 we have failed to find the solution.

which allows one to write:

$$P(y_1\cdots y_s)=\frac{1}{(2\pi)^{n/2+s}\sigma_1\sigma_2\cdots\sigma_n}$$

$$\times\int_{-\infty}^{+\infty}\cdots\int dx_1\cdots dx_n \exp\left[-\tfrac{1}{2}\sum_1^n (x_i^2/\sigma_i^2)\right]$$

$$\cdot\int_{-\infty}^{+\infty}\cdots\int dt_1\cdots dt_s \prod_{k=1}^{s}\exp\left[it_k(y_k-\sum_1^n a_{ki}x_i)\right].$$

Interchanging the integrations over the x_i with those over the t_k one can easily carry out the integrations over the x_i and one gets:

$$P(y_1\cdots y_s)=\frac{1}{(2\pi)^s}\int_{-\infty}^{+\infty}\cdots\int dt_1\cdots dt_s$$

$$\times\exp\left[i\sum_1^s y_k t_k-\tfrac{1}{2}\sum_{k,l=1}^{s} b_{kl}t_k t_l\right],\quad (85)$$

where the b_{kl} are given by (23). One sees, therefore, that $\exp(-\frac{1}{2}\sum b_{kl}t_kt_l)$ is the characteristic function of the probability distribution $P(y_1\cdots y_s)$. It is a standard result[36] that from (85) follows that P is an s-dimensional Gaussian distribution whose matrix is the inverse of the matrix b_{kl}, and this is just what is expressed by Eq. (22).

Note II. Proof of the General Theorem of Doob

Denote the $3n$ components of the vectors $\mathbf{y}(t_1)$, $\mathbf{y}(t_2)$, $\mathbf{y}(t_3)$ by $z_1z_2\cdots z_{3n}$. According to the general theorem (22) one then can write the third probability distribution in the form:

$$W_3(\mathbf{y}(t_1), \mathbf{y}(t_2), \mathbf{y}(t_3))=C_3\exp\left(-\tfrac{1}{2}\sum_{j,k=1}^{3n}\alpha_{jk}z_jz_k\right),$$

where the matrix α_{jk} is the inverse of the $3n$ by $3n$ matrix:

$$\mathbf{M}=\begin{pmatrix}\mathbf{R}(0) & \mathbf{R}(t_2-t_1) & \mathbf{R}(t_3-t_1)\\ \mathbf{R}(t_1-t_2) & \mathbf{R}(0) & \mathbf{R}(t_3-t_2)\\ \mathbf{R}(t_1-t_3) & \mathbf{R}(t_2-t_3) & \mathbf{R}(0)\end{pmatrix}.\quad (86)$$

[36] Comp. for instance H. Cramér, Random Variables and Probability Distributions (Cambridge Tracts No. 36, 1937), p. 110.

Note that $\mathbf{M}$ is symmetric because of (30). Analogously one has for the second probability distribution

$$W_2(\mathbf{y}(t_1), \mathbf{y}(t_2))=C_2\exp\left(-\tfrac{1}{2}\sum_{j,k=1}^{2n}\beta_{jk}z_jz_k\right),$$

where the $2n$ by $2n$ matrix β_{jk} is the inverse of

$$\begin{pmatrix}\mathbf{R}(0) & \mathbf{R}(t_2-t_1)\\ \mathbf{R}(t_1-t_2) & \mathbf{R}(0)\end{pmatrix}.$$

For a Markoff process $P(\mathbf{y}(t_1)\mathbf{y}(t_2)\,|\,\mathbf{y}(t_3))=W_3/W_2$ should be independent of $\mathbf{y}(t_1)$. This leads to the conditions:

$$\alpha_{jk}=0,\ \text{when } j=1,2\cdots n \text{ and } k=2n+1,\cdots 3n,\quad (87)$$

$$\alpha_{jk}=\beta_{jk},\ \text{when } j=1,2\cdots n \text{ and } k=1,2\cdots 2n.\quad (88)$$

In order to calculate the inverse of $\mathbf{M}$ we resort to the following formal trick. We treat $\mathbf{M}$ not as a numerical matrix but as a matrix whose elements are matrices. The rules of multiplication are the same but in taking inverses one must be careful because of possible non-commutativity. We want then a 3 by 3 matrix $\mathbf{X}$, whose elements are n by n matrices such that:

$$\mathbf{M}\times\begin{pmatrix}\mathbf{X}_{11} & \mathbf{X}_{12} & \mathbf{X}_{13}\\ \mathbf{X}_{21} & \mathbf{X}_{22} & \mathbf{X}_{23}\\ \mathbf{X}_{31} & \mathbf{X}_{32} & \mathbf{X}_{33}\end{pmatrix}=\begin{pmatrix}\mathbf{I} & \mathbf{O} & \mathbf{O}\\ \mathbf{O} & \mathbf{I} & \mathbf{O}\\ \mathbf{O} & \mathbf{O} & \mathbf{I}\end{pmatrix}.$$

Using (86) this leads for instance to the three matrix equations:

$$\mathbf{R}(0)\mathbf{X}_{13}+\mathbf{R}(t_2-t_1)\mathbf{X}_{23}+\mathbf{R}(t_3-t_1)\mathbf{X}_{33}=\mathbf{O},$$
$$\mathbf{R}(t_1-t_2)\mathbf{X}_{13}+\mathbf{R}(0)\mathbf{X}_{23}+\mathbf{R}(t_3-t_2)\mathbf{X}_{33}=\mathbf{O},$$
$$\mathbf{R}(t_1-t_3)\mathbf{X}_{13}+\mathbf{R}(t_2-t_3)\mathbf{X}_{23}+\mathbf{R}(0)\mathbf{X}_{33}=\mathbf{I}.$$

Condition (87) says that $\mathbf{X}_{13}=0$; assuming further that $\mathbf{R}(0)=\mathbf{I}$, which as mentioned in footnote 19 is no loss in generality, these equations become:

$$\mathbf{R}(t_2-t_1)\mathbf{X}_{23}+\mathbf{R}(t_3-t_1)\mathbf{X}_{33}=0,\quad (89a)$$

$$\mathbf{X}_{23}+\mathbf{R}(t_3-t_2)\mathbf{X}_{33}=0,\quad (89b)$$

$$\mathbf{R}(t_2-t_3)\mathbf{X}_{23}+\mathbf{X}_{33}=\mathbf{I}.\quad (89c)$$

$\mathbf{R}$ is a non-singular matrix. One can eliminate $\mathbf{X}_{23}$ from the last two equations and one gets:

$$\{\mathbf{R}^{-1}(t_2-t_3)-\mathbf{R}(t_3-t_2)\}\mathbf{X}_{33}=\mathbf{R}^{-1}(t_2-t_3).$$

The matrix in the curly brackets cannot be singular since $DetR^{-1}(t_2-t_3)\neq 0$, and therefore:

$$\mathbf{X}_{33}=\{\mathbf{R}^{-1}(t_2-t_3)-\mathbf{R}(t_3-t_2)\}^{-1}\mathbf{R}^{-1}(t_2-t_3),$$

and:

$$\mathbf{X}_{23}=-\mathbf{R}(t_3-t_2)\{\mathbf{R}^{-1}(t_2-t_3)-\mathbf{R}(t_3-t_2)\}^{-1}\mathbf{R}^{-1}(t_2-t_3).$$

Substituting in (89a) one gets as the condition on $\mathbf{R}$:

$$\mathbf{R}(t_3-t_1)=\mathbf{R}(t_2-t_1)\mathbf{R}(t_3-t_2), \tag{90}$$

which is Eq. (31). One must calculate also the other elements of $\mathbf{X}$ and in the same way the matrix β_{kl} in order to show that the condition (88) is now automatically satisfied so that (90) is also the *only* condition to be imposed on the correlation matrix $\mathbf{R}$.

Note III. Proof that the Properties (42) Imply that $F(t)$ is a Gaussian Process

From (42a) follows immediately that the average values of all the *odd* powers of the Fourier coefficient:

$$a_k=\frac{2}{T}\int_0^T dt\cos 2\pi f_k tF(t)$$

are zero. One gets further:

$$\begin{aligned}\langle a_k^2\rangle_{\text{Av}}&=\frac{4}{T^2}\int_0^T\!\!\int dt_1dt_2\cos 2\pi f_kt_1\times\cos 2\pi f_kt_2\langle F(t_1)F(t_2)\rangle_{\text{Av}}\\&=\frac{8D}{T^2}\int_0^T\!\!\int dt_1dt_2\cos 2\pi f_kt_1\times\cos 2\pi f_kt_2\delta(t_1-t_2)\\&=\frac{8D}{T^2}\int_0^T dt_1\cos^2 2\pi f_kt_1=\frac{4D}{T}.\end{aligned}\tag{91}$$

From (42b) follows then that the average values of the *even* powers of a_k are given by:

$$\langle a_k^{2n}\rangle_{\text{Av}}=1\cdot 3\cdot 5\cdots(2n-1)\langle a_k^2\rangle_{\text{Av}}^n \tag{92}$$

since the number of ways in which we can divide the $2n$ time points $t_1, t_2\cdots t_{2n}$ into n pairs is $1\cdot 3\cdot 5\cdots(2n-1)$. Equation (92) is characteristic for the Gaussian distribution. One can show[37] that still more explicitly by calculating the characteristic function of the distribution function $W(a_k)$, which is given by:

$$\begin{aligned}\langle\exp(i\xi a_k)\rangle_{\text{Av}}&=\sum_{n=0}^{\infty}\frac{(i\xi)^m}{m!}\langle a_k^m\rangle_{\text{Av}}\\&=\sum_{n=0}^{\infty}(-1)^n\frac{1}{n!}\left(\frac{2D}{T}\right)^n\xi^{2n}=\exp\left(-\frac{2D\xi^2}{T}\right)\end{aligned}$$

using (91) and (92). Therefore:

$$W(a_k)=\frac{1}{2\pi}\int_{-\infty}^{+\infty}d\xi\exp(-i\xi a_k)\cdot\exp\left(-\frac{2D\xi^2}{T}\right)=\left(\frac{T}{8\pi D}\right)^{\frac{1}{2}}\exp\left(-\frac{a_k^2T}{8D}\right).$$

The same distribution one finds for:

$$b_k=\frac{2}{T}\int_0^T dt\sin 2\pi f_ktF(t),$$

and it is also easy to show that the different a_k and b_l are independent of each other, so that the complete distribution function $W(a_1a_2\cdots;b_1b_2\cdots)$ will be given by (20) with $\sigma_k^2=4D/T$.

Note IV. The Fundamental Solution of Equation (65)

The problem is to find the solution of:

$$\frac{\partial P}{\partial t}=-\sum_i\lambda_i\frac{\partial}{\partial y_i}(y_iP)+\tfrac{1}{2}\sum_{ij}\sigma_{ij}\frac{\partial^2P}{\partial y_i\partial y_j}, \tag{93}$$

which for $t=0$ becomes:

$$P=\delta(y_1-y_{10})\delta(y_2-y_{20})\cdots\delta(y_n-y_{n0}). \tag{94}$$

Introduce instead of P its Fourier transform:

$$f(\xi_1\cdots\xi_n,t)=\int_{-\infty}^{+\infty}\!\!\cdots\int dy_1\cdots dy_nP(y_1\cdots y_n,t)\times\exp[-i\textstyle\sum_j\xi_jy_j]. \tag{95}$$

From (93) follows that f has to fulfill the linear, first-order partial differential equation:

$$\frac{\partial f}{\partial t}-\sum_i\lambda_i\xi_i\frac{\partial f}{\partial\xi_i}=-\tfrac{1}{2}f\sum_{ij}\sigma_{ij}\xi_i\xi_j. \tag{96}$$

[37] This was pointed out to us by Dr. A. Siegert.

The subsidiary equations are:

$$\frac{dt}{1}=-\frac{d\xi_1}{\lambda_1\xi_1}=-\frac{d\xi_2}{\lambda_2\xi_2}=\cdots=-\frac{d\xi_n}{\lambda_n\xi_n}=-\frac{df}{\frac{1}{2}f\sum\sigma_{ij}\xi_i\xi_j}.$$

These can easily be integrated; one finds that the general solution of (96) is given by:

$$f(\xi_1\cdots\xi_n,t)=\psi(\xi_1\exp(\lambda_1 t),\xi_2\exp(\lambda_2 t)\cdots\xi_n\exp(\lambda_n t))\times\exp\left[+\frac{1}{2}\sum_{ij}\sigma_{ij}\frac{\xi_i\xi_j}{\lambda_i+\lambda_j}\right],\quad(97)$$

where ψ is an arbitrary function. Now for $t=0$ one sees from (94) and (95) that:

$$f(\xi_1\cdots\xi_n,0)=\exp\left[-i\sum_j\xi_j y_{j0}\right].$$

Therefore the arbitrary function ψ must be:

$$\psi(\xi_1\cdots\xi_n,0)=\exp\left[-\frac{1}{2}\sum_{ij}\sigma_{ij}\frac{\xi_i\xi_j}{\lambda_i+\lambda_j}-i\sum_j\xi_j y_{j0}\right],$$

and one obtains for f:

$$f=\exp\left[-i\sum_j\xi_j y_{j0}\exp(\lambda_j t)+\frac{1}{2}\sum_{ij}\sigma_{ij}\frac{\xi_i\xi_j}{\lambda_i+\lambda_j}\{1-\exp[+(\lambda_i+\lambda_j)t]\}\right].\quad(98)$$

This is the Fourier transform of an n-dimensional Gaussian distribution with the average values

$$\langle y_i\rangle_{Av}=y_{i0}\exp(\lambda_i t),$$

and the variances:

$$\langle(y_i-\bar{y}_i)(y_j-\bar{y}_j)\rangle_{Av}=-\frac{\sigma_{ij}}{\lambda_i+\lambda_j}[1-\exp(\lambda_i+\lambda_j)t].$$

Mathematical Analysis of Random Noise

By S. O. RICE

INTRODUCTION

THIS paper deals with the mathematical analysis of noise obtained by passing random noise through physical devices. The random noise considered is that which arises from shot effect in vacuum tubes or from thermal agitation of electrons in resistors. Our main interest is in the statistical properties of such noise and we leave to one side many physical results of which Nyquist's law may be given as an example.[1]

About half of the work given here is believed to be new, the bulk of the new results appearing in Parts III and IV. In order to provide a suitable introduction to these results and also to bring out their relation to the work of others, this paper is written as an exposition of the subject indicated in the title.

When a broad band of random noise is applied to some physical device, such as an electrical network, the statistical properties of the output are often of interest. For example, when the noise is due to shot effect, its mean and standard deviations are given by Campbell's theorem (Part I) when the physical device is linear. Additional information of this sort is given by the (auto) correlation function which is a rough measure of the dependence of values of the output separated by a fixed time interval.

The paper consists of four main parts. The first part is concerned with shot effect. The shot effect is important not only in its own right but also because it is a typical source of noise. The Fourier series representation of a noise current, which is used extensively in the following parts, may be obtained from the relatively simple concepts inherent in the shot effect.

The second part is devoted principally to the fundamental result that the power spectrum of a noise current is the Fourier transform of its correlation function. This result is used again and again in Parts III and IV.

A rather thorough discussion of the statistics of random noise currents is given in Part III. Probability distributions associated with the maxima of the current and the maxima of its envelope are developed. Formulas for the expected number of zeros and maxima per second are given, and a start is made towards obtaining the probability distribution of the zeros.

When a noise voltage or a noise voltage plus a signal is applied to a non-

[1] An account of this field is given by E. B. Moullin, "Spontaneous Fluctuations of Voltage," Oxford (1938).

linear device, such as a square-law or linear rectifier, the output will also contain noise. The methods which are available for computing the amount of noise and its spectrál distribution are discussed in Part IV.

Acknowledgement

I wish to thank my friends for many helpful suggestions and discussions regarding the subject of this paper. Although it has been convenient to acknowledge some of this assistance in the text, I appreciate no less sincerely the considerable amount which is not mentioned. In particular, I am indebted to Miss Darville for computing the curves in Parts III and IV.

Summary of Results

Before proceeding to the main body of the paper, we shall state some of the principal results. It is hoped that this summary will give the casual reader an over-all view of the material covered and at the same time guide the reader who is interested in obtaining some particular item of information to those portions of the paper which may possibly contain it.

Part I—Shot Effect

Shot effect noise results from the superposition of a great number of disturbances which occur at random. A large class of noise generators produce noise in this way.

Suppose that the arrival of an electron at the anode of the vacuum tube at time $t = 0$ produces an effect $F(t)$ at some point in the output circuit. If the output circuit is such that the effects of the various electrons add linearly, the total effect at time t due to all the electrons is

$$I(t) = \sum_{k=-\infty}^{+\infty} F(t - t_k) \tag{1.2–1}$$

where the kth electron arrives at t_k and the series is assumed to converge. Although the terminology suggests that $I(t)$ is a current, and it will be spoken of as a noise current, it may be any quantity expressible in the form (1.2–1).

1. Campbell's theorem: The average value of $I(t)$ is

$$\overline{I(t)} = \nu \int_{-\infty}^{+\infty} F(t)\, dt \tag{1.2–2}$$

and the mean square value of the fluctuation about this average is

$$\text{ave.}\, [I(t) - \overline{I(t)}]^2 = \nu \int_{-\infty}^{+\infty} F^2(t)\, dt \tag{1.2–3}$$

where ν is the average number of electrons arriving per second at the anode. In this expression the electrons are supposed to arrive independently and at random. $\nu e^{-\nu t}\, dt$ is the probability that the length of the interval between two successive arrivals lies between t and $t + dt$.

2. Generalization of Campbell's theorem. Campbell's theorem gives information about the average value and the standard deviation of the probability distribution of $I(t)$. A generalization of the theorem gives information about the third and higher order moments. Let

$$I(t) = \sum_{-\infty}^{+\infty} a_k F(t - t_k) \tag{1.5–1}$$

where $F(t)$ and t_k are of the same nature as those in (1.2–1) and $\cdots a_1$, $a_2, \cdots a_k, \cdots$ are independent random variables all having the same distribution. Then the nth semi-invariant of the probability density $P(I)$ of $I = I(t)$ is

$$\lambda_n = \nu\, \overline{a^n} \int_{-\infty}^{+\infty} [F(t)]^n\, dt \tag{1.5–2}$$

The semi-invariants are defined as the coefficients in the expansion of the characteristic function $f(u)$:

$$\log_e f(u) = \sum_{n=1} \frac{\lambda_n}{n!} (iu)^n \tag{1.5–3}$$

where

$$f(u) = \text{ave. } e^{iIu} = \int_{-\infty}^{+\infty} P(I) e^{iIu}\, dI$$

The moments may be computed from the λ's.

3. As $\nu \to \infty$ the probability density $P(I)$ of the shot effect current approaches a normal law. The way it is approached is given by

$$P(I) \sim \sigma^{-1} \varphi^{(0)}(x) - \frac{\lambda_3 \sigma^{-4}}{3!} \varphi^{(3)}(x) + \left[\frac{\lambda_4 \sigma^{-5}}{4!} \varphi^{(4)}(x) + \frac{\lambda_3^2 \sigma^{-7}}{72} \varphi^{(6)}(x)\right] + \cdots \tag{1.6–3}$$

where the λ's are given by (1.5–2) and

$$\sigma^2 = \lambda_2 \qquad x = \frac{I - \overline{I}}{\sigma} \qquad \varphi^{(n)}(x) = \frac{1}{\sqrt{2\pi}} \frac{d^n}{dx^n} e^{-x^2/2}$$

Since the λ's are of the order of ν, σ is of the order of $\nu^{1/2}$ and the orders of σ^{-1}, $\lambda_3\sigma^{-4}$, $\lambda_4\sigma^{-5}$ and $\lambda_3^2\sigma^{-7}$ are $\nu^{-1/2}$, ν^{-1}, $\nu^{-3/2}$ and $\nu^{-3/2}$ respectively. A possible use of this result is to determine whether a noise due to random independent events occuring at the rate of ν per second may be regarded as "random noise" in the sense of this work.

4. When $I(t)$, as given by (1.5–1), is analyzed as a Fourier series over an interval of length T a set of Fourier coefficients is obtained. By taking many different intervals, all of length T, many sets of coefficients are obtained. If ν is sufficiently large these coefficients tend to be distributed normally and independently. A discussion of this is given in section 1.7.

Part II—Power Spectra and Correlation Functions

1. Suppose we have a curve, such as an oscillogram of a noise current, which extends from $t = 0$ to $t = \infty$. Let this curve be denoted by $I(t)$. The correlation function of $I(t)$ is $\psi(\tau)$ which is defined as

$$\psi(\tau) = \operatorname*{Limit}_{T\to\infty} \frac{1}{T} \int_0^T I(t)I(t + \tau)\, dt \tag{2.1–4}$$

where the limit is assumed to exist. This function is closely connected with another function, the power spectrum, $w(f)$, of $I(t)$. $I(t)$ may be regarded as composed of many sinusoidal components. If $I(t)$ were a noise current and if it were to flow through a resistance of one ohm the average power dissipated by those components whose frequencies lie between f and $f + df$ would be $w(f)\, df$.

The relation between $w(f)$ and $\psi(\tau)$ is

$$w(f) = 4 \int_0^\infty \psi(\tau) \cos 2\pi f\tau\, d\tau \tag{2.1–5}$$

$$\psi(\tau) = \int_0^\infty w(f) \cos 2\pi f\tau\, df \tag{2.1–6}$$

When $I(t)$ has no d.c. or periodic components,

$$w(f) = \operatorname*{Limit}_{T\to\infty} \frac{2\,|\,S(f)\,|^2}{T} \tag{2.1–3}$$

where

$$S(f) = \int_0^T I(t)e^{-2\pi i f t}\, dt.$$

The correlation function for

$$I(t) = A + C \cos (2\pi f_0 t - \varphi)$$

is

$$\psi(\tau) = A^2 + \frac{C^2}{2} \cos 2\pi f_0 \tau \tag{2.2–3}$$

These results are discussed in sections 2.1 to 2.4 inclusive.

2. So far we have supposed $I(t)$ to be some definite function for which a curve may be drawn. Now consider $I(t)$ to be given by a mathematical expression into which, besides t, a number of parameters enter. $w(f)$ and $\psi(\tau)$ are now obtained by averaging the integrals over the possible values of the parameters. This is discussed in section 2.5.

3. The correlation function for the shot effect current of (1.2–1) is

$$\psi(\tau) = \nu \int_{-\infty}^{+\infty} F(t)F(t+\tau)\, dt + \left[\nu \int_{-\infty}^{+\infty} F(t)\, dt\right]^2 \qquad (2.6\text{–}2)$$

The distributed portion of the power spectrum is

$$w_1(f) = 2\nu \mid s(f) \mid^2$$

where

$$s(f) = \int_{-\infty}^{+\infty} F(t) e^{-2\pi i f t}\, dt \qquad (2.6\text{–}5)$$

The complete power spectrum has in addition to $w_1(f)$ an impulse function representing the d.c. component $\overline{I}(t)$.

In the formulas above for the shot effect it was assumed that the expected number, ν, of electrons per second did not vary with time. A case in which ν does vary with time is briefly discussed near the end of Section 2.6.

4. Random telegraph signal. Let $I(t)$ be equal to either a or $-a$ so that it is of the form of a flat top wave, and let the lengths of the tops and bottoms be distributed independently and exponentially. The correlation function and power spectrum of I are

$$\psi(\tau) = a^2 e^{-2\mu|\tau|} \qquad (2.7\text{–}4)$$

$$w(f) = \frac{2a^2 \mu}{\pi^2 f^2 + \mu^2} \qquad (2.7\text{–}5)$$

where μ is the expected number of changes of sign per second.

Another type of random telegraph signal may be formed as follows: Divide the time scale into intervals of equal length h. In an interval selected at random the value of $I(t)$ is independent of the value in the other intervals and is equally likely to be $+a$ or $-a$. The correlation function of $I(t)$ is zero for $|\tau| > h$ and is

$$a^2\left(1 - \frac{|\tau|}{h}\right)$$

for $0 \leq |\tau| < h$ and the power spectrum is

$$w(f) = 2h\left(\frac{a \sin \pi f h}{\pi f h}\right)^2 \qquad (2\,7\text{-}9)$$

5. There are two representations of a random noise current which are especially useful. The first one is

$$I(t) = \sum_{n=1}^{N} (a_n \cos \omega_n t + b_n \sin \omega_n t) \tag{2.8-1}$$

where a_n and b_n are independent random variables which are distributed normally about zero with the standard deviation $\sqrt{w(f_n)\Delta f}$ and where

$$\omega_n = 2\pi f_n, \qquad f_n = n\Delta f$$

The second one is

$$I(t) = \sum_{n=1}^{N} c_n \cos (\omega_n t - \varphi_n) \tag{2.8-6}$$

where φ_n is a random phase angle distributed uniformly over the range $(0, 2\pi)$ and

$$c_n = [2w(f_n)\Delta f]^{1/2}$$

At an appropriate point in the analysis N and Δf are made to approach infinity and zero, respectively, in such a manner that the entire frequency band is covered by the summations (which then become integrations).

6. The normal distribution in several variables and the central limit theorem are discussed in sections 2.9 and 2.10.

Part III—Statistical Properties of Noise Current

1. The noise current is distributed normally. This has already been discussed in section 1.6 for the shot-effect. It is discussed again in section 3.1 using the concepts introduced in Part II, and the assumption, used throughout Part III, that the average value of the noise current $I(t)$ is zero. The probability that $I(t)$ lies between I and $I + dI$ is

$$\frac{dI}{\sqrt{2\pi\psi_0}} e^{-I^2/2\psi_0} \tag{3.1-3}$$

where ψ_0 is the value of the correlation function, $\psi(\tau)$, of $I(t)$ at $\tau = 0$

$$\psi_0 = \psi(0) = \int_0^\infty w(f)\, df, \tag{3.1-2}$$

$w(f)$ being the power spectrum of $I(t)$. ψ_0 is the mean square value of $I(t)$, i.e., the r.m.s. value of $I(t)$ is $\psi_0^{1/2}$.

The characteristic function (ch. f.) of this distribution is

$$\text{ave. } e^{iuI(t)} = \exp - \frac{\psi_0}{2} u^2 \tag{3.1-6}$$

2. The probability that $I(t)$ lies between I_1 and $I_1 + dI$, and $I(t + \tau)$ lies between I_2 and $I_2 + dI_2$ when t is chosen at random is

$$[\psi_0^2 - \psi_\tau^2]^{-1/2} \frac{dI_1\, dI_2}{2\pi} \exp\left[\frac{-\psi_0 I_1^2 - \psi_0 I_2^2 + 2\psi_\tau I_1 I_2}{2(\psi_0^2 - \psi_\tau^2)}\right] \quad (3.2\text{-}4)$$

where ψ_τ is the correlation function $\psi(\tau)$ of $I(t)$:

$$\psi(\tau) = \int_0^\infty w(f) \cos 2\pi f\tau \, df \quad (3.2\text{-}3)$$

The ch. f. for this distribution is

$$\text{ave. } e^{iuI(t)+ivI(t+\tau)} = \exp\left[-\frac{\psi_0}{2}(u^2 + v^2) - \psi_\tau uv\right] \quad (3.2\text{-}7)$$

3. The expected number of zeros per second of $I(t)$ is

$$\frac{1}{\pi}\left[-\frac{\psi''(0)}{\psi(0)}\right]^{1/2} = 2\left[\frac{\int_0^\infty f^2 w(f)\, df}{\int_0^\infty w(f)\, df}\right]^{1/2} \quad (3.3\text{-}11)$$

assuming convergence of the integrals. The primes denote differentiation with respect to τ:

$$\psi''(\tau) = \frac{d^2}{d\tau^2}\psi(\tau).$$

For an ideal band-pass filter whose pass band extends from f_a to f_b the expected number of zeros per second is

$$2\left[\frac{1}{3}\frac{f_b^3 - f_a^3}{f_b - f_a}\right]^{1/2} \quad (3.3\text{-}12)$$

When f_a is zero this becomes $1.155\, f_b$ and when f_a is very nearly equal to f_b it approaches $f_b + f_a$.

4. The problem of determining the distribution function for the length of the interval between two successive zeros of $I(t)$ seems to be quite difficult. In section 3.4 some related results are given which lead, in some circumstances, to approximations to the distribution. For example, for an ideal narrow band-pass filter the probability that the distance between two successive zeros lies between τ and $\tau + d\tau$ is approximately

$$\frac{d\tau}{2} \frac{a}{[1 + a^2(\tau - \tau_1)^2]^{3/2}}$$

where

$$a = \sqrt{3}\,\frac{(f_b + f_a)^2}{f_b - f_a}, \qquad \tau_1 = \frac{1}{f_b + f_a}$$

f_b and f_a being the upper and lower cut-off frequencies.

5. In section 3.5 several multiple integrals which occur in the work of Part III are discussed.

6. The distribution of the maxima of $I(t)$ is discussed in section 3.6. The expected number of maxima per second is

$$\frac{1}{2\pi}\left[-\frac{\psi_0^{(4)}}{\psi_0''}\right]^{1/2} = \left[\frac{\int_0^\infty f^4 w(f)\,df}{\int_0^\infty f^2 w(f)\,df}\right]^{1/2} \tag{3.6-6}$$

For a band-pass filter the expected number of maxima per second is

$$\left[\frac{3}{5}\,\frac{f_b^5 - f_a^5}{f_b^3 - f_a^3}\right]^{1/2} \tag{3.6-7}$$

For a low-pass filter where $f_a = 0$ this number is $0.775 f_b$.

The expected number of maxima per second lying above the line $I(t) = I_1$ is approximately, when I_1 is large,

$$e^{-I_1^2/2\psi_0} \times \tfrac{1}{2}[\text{the expected number of zeros of } I \text{ per second}] \tag{3.6-11}$$

where ψ_0 is the mean square value of $I(t)$.

For a low-pass filter the probability that a maximum chosen at random from the universe of maxima lies between I and $I + dI$ is approximately, when I is large,

$$\frac{\sqrt{5}}{3}\, y e^{-y^2/2}\,\frac{dI}{\psi_0^{1/2}} \tag{3.6-9}$$

where

$$y = \frac{I}{\psi_0^{1/2}}$$

7. When we pass noise through a relatively narrow band-pass filter one of the most noticeable features of an oscillogram of the output current is its fluctuating envelope. In sections 3.7 and 3.8 some statistical properties of this envelope, denoted by R or $R(t)$, are derived.

The probability that the envelope lies between R and $R + dR$ is

$$\frac{R}{\psi_0}\, e^{-R^2/2\psi_0}\, dR \tag{3.7-10}$$

where ψ_0 is the mean square value of $I(t)$. The probability that $R(t)$ lies between R_1 and $R_1 + dR_1$ and at the same time $R(t + \tau)$ lies between R_2 and $R_2 + dR_2$ when t is chosen at random is obtained by multiplying (3.7–13) by $dR_1\, dR_2$. For an ideal band-pass filter, the expected number of maxima of the envelope in one second is

$$.64110(f_b - f_a) \tag{3.8–15}$$

When R is large, say $y > 2.5$ where

$$y = \frac{R}{\psi_0^{1/2}}, \qquad \psi_0^{1/2} = \text{r.m.s. value of } I(t),$$

the probability that a maximum of the envelope, selected at random from the universe of such maxima, lies between R and $R + dR$ is approximately

$$1.13(y^2 - 1)e^{-y^2/2} \frac{dR}{\psi_0^{1/2}}$$

A curve for the corresponding probability density is shown for the range $0 \leq y \leq 4$. Curves which compare the distribution function of the maxima of R with other distribution functions of the same type are also given.

8. In section 3.9 some information is given regarding the statistical behavior of the random variable:

$$E = \int_{t_1}^{t_1+T} I^2(t)\, dt \tag{3.9–1}$$

where t_1 is chosen at random and $I(t)$ is a noise current with the power spectrum $w(f)$ and the correlation function $\psi(\tau)$. The average value m_T of E is $T\psi_0$ and its standard deviation σ_T is given by (3.9–9). For a relatively narrow band-pass filter

$$\frac{\sigma_T}{m_T} \sim \frac{1}{\sqrt{T(f_b - f_a)}}$$

when $T(f_b - f_a) \gg 1$. This follows from equation (3.9–10). An expression which is believed to approximate the distribution of E is given by (3.9–20).

9. In section 3.10 the distribution of a noise current plus one or more sinusoidal currents is discussed. For example, if I consists of two sine waves plus noise:

$$I = P \cos pt + Q \cos qt + I_N, \tag{3.10–20}$$

where p and q are incommensurable and the r.m.s. value of the noise current I_N is $\psi_0^{1/2}$, the probability density of the envelope R is

$$R \int_0^\infty rJ_0(Rr)J_0(Pr)J_0(Qr)e^{-\psi_0 r^2/2}\, dr \tag{3.10–21}$$

where $J_0(\ )$ is a Bessel function.

Curves showing the probability density and distribution function of R, when $Q = 0$, for various ratios of P/r.m.s. I_N are given.

10. In section 3.11 it is pointed out that the representations (2.8–1) and (2.8–6) of the noise current as the sum of a great number of sinusoidal components are not the only ones which may be used in deriving the results given in the preceding sections of Part III. The shot effect representation

$$I(t) = \sum_{-\infty}^{+\infty} F(t - t_k)$$

studied in Part I may also be used.

Part IV—Noise Through Non-Linear Devices

1. Suppose that the power spectrum of the voltage V applied to the square-law device

$$I = \alpha V^2 \tag{4.1–1}$$

is confined to a relatively narrow band. The total low-frequency output current $I_{t\ell}$ may be expressed as the sum

$$I_{t\ell} = I_{dc} + I_{\ell f} \tag{4.1–2}$$

where I_{dc} is the d.c. component and $I_{\ell f}$ is the variable component. When none of the low-frequency band is eliminated (by audio frequency filters)

$$I_{t\ell} = \frac{\alpha R^2}{2} \tag{4.1–6}$$

where R is the envelope of V. If V is of the form

$$V = V_N + P \cos pt + Q \cos qt, \tag{4.1–4}$$

where V_N is a noise voltage whose mean square value is ψ_0, then

$$I_{dc} = \alpha\left(\psi_0 + \frac{P^2}{2} + \frac{Q^2}{2}\right)$$

$$\overline{I_{\ell f}^2} = \alpha^2\left[\psi_0^2 + P^2\psi_0 + Q^2\psi_0 + \frac{P^2Q^2}{2}\right] \tag{4.1–16}$$

2. If instead of a square-law device we have a linear rectifier,

$$I = \begin{cases} 0 & V < 0 \\ \alpha V, & V > 0 \end{cases} \tag{4.2–1}$$

the total low-frequency output is

$$I_{t\ell} = \frac{\alpha R}{\pi} \tag{4.2–2}$$

When V is a sine wave plus noise, $V_N + P\cos pt$,

$$I_{dc} = \alpha\left(\frac{\psi_0}{2\pi}\right)^{1/2} {}_1F_1(-\tfrac{1}{2}; 1; -x) \qquad (4.2\text{–}3)$$

$$\overline{I_{t\ell}^2} = \frac{\alpha^2}{\pi^2}(P^2 + 2\psi_0) \qquad (4.2\text{–}6)$$

where ${}_1F_1$ is a hypergeometric function and

$$x = \frac{P^2}{2\psi_0} = \frac{\text{Ave. sine wave power}}{\text{Ave. noise power}} \qquad (4.2\text{–}4)$$

When x is large

$$\overline{I_{\ell f}^2} \sim \frac{\alpha^2\psi_0}{\pi^2}\left[1 - \frac{1}{4x}\cdots\right] \qquad (4.2\text{–}7)$$

If V consists of two sine waves plus noise, I_{dc} consists of a hypergeometric function of two variables. The equations running from (4.2–9) to (4.2–15) are concerned with this case. About the only simple equation is

$$\overline{I_{t\ell}^2} = \frac{\alpha^2}{\pi^2}[2\psi_0 + P^2 + Q^2] \qquad (4.2\text{–}14)$$

3. The expressions (4.1–6) and (4.2–2) for $I_{t\ell}$ in terms of the envelope R of V, namely

$$\frac{\alpha R^2}{2} \quad \text{and} \quad \frac{\alpha R}{\pi},$$

are special cases of a more general result

$$I_{t\ell} = A_0(R) = \frac{1}{2\pi}\int_C F(iu)J_0(uR)\,du. \qquad (4.3\text{–}11)$$

In this expression $J_0(uR)$ is a Bessel function. The path of integration C and the function $F(iu)$ are chosen so that the relation between I and V may be expressed as

$$I = \frac{1}{2\pi}\int_C F(iu)e^{iVu}\,du. \qquad (4A\text{–}1)$$

A table giving $F(iu)$ and C for a number of common non-linear devices is shown in Appendix 4A.

If this relation is used to study the biased linear rectifier.

$$I = \begin{cases} 0, & V < B \\ V - B, & V > B \end{cases}$$

for the case in which V is $V_N + P \cos pt$, we find

$$
\begin{aligned}
I_{dc} &\sim -\frac{B}{2} + \frac{P}{\pi} + \frac{B^2 + \psi_0}{2\pi P} \\
\overline{I_{\ell f}^2} &\sim \frac{P^2 - B^2}{\pi^2 P^2}\, \psi_0
\end{aligned}
\qquad (4.3\text{–}17)
$$

when $P \gg |B|$, $P^2 \gg \psi_0$ where ψ_0 is the mean square value of V_N.

4. When V is confined to a relatively narrow band and there are no audio-frequency filters, the probability density and all the associated statistical properties of $I_{t\ell}$ may be obtained by expressing $I_{t\ell}$ as a function of the envelope R of V and then using the probability density of R. When V is $V_N + P \cos pt + Q \cos qt$ this probability density is given by the integral, (3.10–21) (which is the integral containing three Bessel functions stated in the above summary of Part III). When V consists of three sine waves plus noise there are four J_0's in the integrand, and so on. Expressions for $\overline{R^n}$ when R has the above distribution are given by equations (3.10–25) and (3.10–27).

When audio-frequency filters remove part of the low-frequency band the statistical properties, except the mean square value, of the resulting current are hard to compute. In section 4.3 it is shown that as the output band is chosen narrower and narrower, the statistical properties of the output current approach those of a random noise current.

5. The sections in Part IV from 4.4 onward are concerned with the problem: Given a non-linear device and an input voltage consisting of noise alone or of a signal plus noise. What is the power spectrum of the output? A survey of the methods available for the solution of this problem is given in section 4.4.

6. When a noise voltage V_N with the power spectrum $w(f)$ is applied to the square-law device

$$I = \alpha V^2 \qquad (4.1\text{–}1)$$

the power spectrum of the output current I is, when $f \neq 0$,

$$W(f) = \alpha^2 \int_{-\infty}^{+\infty} w(x) w(f - x)\, dx \qquad (4.5\text{–}5)$$

where $w(-x)$ is defined to equal $w(x)$. The power spectrum of I when V is either $P \cos pt + V_N$ or

$$Q(1 + k \cos pt) \cos qt + V_N$$

is considered in the portion of section 4.5 containing equations (4.5–10) to (4.5–17).

7. A method discovered independently by Van Vleck and North shows that the correlation function $\Psi(\tau)$ of the output current for an unbiased linear rectifier is

$$\Psi(\tau) = \frac{\psi_\tau}{4} + \frac{\psi_0}{2\pi}\,{}_2F_1\left[-\tfrac{1}{2}, -\tfrac{1}{2}; \tfrac{1}{2}; \frac{\psi_\tau^2}{\psi_0^2}\right] \qquad (4.7\text{–}6)$$

where the input voltage is V_N. The correlation function $\psi(\tau)$ of V_N is denoted by ψ_τ and the mean square value of V_N is ψ_0. The power spectrum $W(f)$ of I may be obtained from

$$W(f) = 4\int_0^\infty \Psi(\tau)\cos 2\pi f\tau\, d\tau \qquad (4.6\text{–}1)$$

by expanding the hypergeometric function and integrating termwise using

$$G_n(f) = \int_0^\infty \psi_\tau^n \cos 2\pi f\tau\, d\tau. \qquad (4C\text{–}1)$$

Appendix 4C is devoted to the problem of evaluating the integral for $G_n(f)$.

8. Another method of obtaining the correlation function $\psi(\tau)$ of I, termed the "characteristic function method," is explained in section 4.8. It is illustrated in section 4.9 where formulas for $\Psi(\tau)$ and $W(f)$ are developed when the voltage $P\cos pt + V_N$ is applied to a general non-linear device.

9. Several miscellaneous results are given in section 4.10. The characteristic function method is used to obtain the correlation function for a square-law device. The general formulas of section 4.9 are applied to the case of a νth law rectifier when the input noise spectrum has a normal law distribution. Some remarks are also made concerning the audio-frequency output of a linear rectifier when the input voltage V is

$$Q(1 + r\cos pt)\cos qt + V_N.$$

10. A discussion of the hypergeometric function ${}_1F_1(a; c; x)$, which often occurs in problems concerning a sine wave plus noise, is given in Appendix 4B.

PART I

THE SHOT EFFECT

The shot effect in vacuum tubes is a typical example of noise. It is due to fluctuations in the intensity of the stream of electrons flowing from the cathode to the anode. Here we analyze a simplified form of the shot effect.

1.1 The Probability of Exactly K Electrons Arriving at the Anode in Time T

The fluctuations in the electron stream are supposed to be random. We shall treat this randomness as follows. We count the number of electrons flowing in a long interval of time T measured in seconds. Suppose there are K_1. Repeating this counting process for many intervals all of length T gives a set of numbers $K_2, K_3 \cdots K_M$ where M is the total number of intervals. The average number ν, of electrons per second is defined as

$$\nu = \lim_{M \to \infty} \frac{K_1 + K_2 + \cdots + K_M}{MT} \tag{1.1-1}$$

where we assume that this limit exists. As M is increased with T being held fixed some of the K's will have the same value. In fact, as M increases the number of K's having any particular value will tend to increase. This of course is based on the assumption that the electron stream is a steady flow upon which random fluctuations are superposed. The probability of getting K electrons in a given trial is defined as

$$p(K) = \lim_{M \to \infty} \frac{\text{Number of trials giving exactly } K \text{ electrons}}{M} \tag{1.1-2}$$

Of course $p(K)$ also depends upon T. We assume that the randomness of the electron stream is such than the probability that an electron will arrive at the anode in the interval $(t, t + \Delta t)$ is $\nu\Delta t$ where Δt is such that $\nu\Delta t \ll 1$, and that this probability is independent of what has happened before time t or will happen after time $t + \Delta t$.

This assumption is sufficient to determine the expression for $p(K)$ which is

$$p(K) = \frac{(\nu T)^K}{K!} e^{-\nu T} \tag{1.1-3}$$

This is the "law of small probabilities" given by Poisson. One method of derivation sometimes used can be readily illustrated for the case $K = 0$. Thus, divide the interval, $(0, T)$ into M intervals each of length $\Delta t = \dfrac{T}{M}$. Δt is taken so small that $\nu\Delta t$ is much less than unity. (This is the "small probability" that an electron will arrive in the interval Δt). The probability that an electron will not arrive in the first sub-interval is $(1 - \nu\Delta t)$. The probability that one will not arrive in either the first or the second sub-interval is $(1 - \nu\Delta t)^2$. The probability that an electron will not arrive in any of the M intervals is $(1 - \nu\Delta t)^M$. Replacing M by $T/\Delta t$ and letting $\Delta t \to 0$ gives

$$p(0) = e^{-\nu T}$$

The expressions for $p(1)$, $p(2)$, $\cdots$ $p(K)$ may be derived in a somewhat similar fashion.

1.2 Statement of Campbell's Theorem

Suppose that the arrival of an electron at the anode at time $t = 0$ produces an effect $F(t)$ at some point in the output circuit. If the output circuit is such that the effects of the various electrons add linearly, the total effect at time t due to all the electrons is

$$I(t) = \sum_{k=-\infty}^{+\infty} F(t - t_k) \tag{1.2-1}$$

where the kth electron arrives at t_k and the series is assumed to converge.

Campbell's theorem[2] states that the average value of $I(t)$ is

$$\overline{I(t)} = \nu \int_{-\infty}^{+\infty} F(t)\,dt \tag{1.2-2}$$

and the mean square value of the fluctuation about this average is

$$\overline{(I(t) - \overline{I(t)})^2} = \nu \int_{-\infty}^{+\infty} F^2(t)\,dt \tag{1.2-3}$$

where ν is the average number of electrons arriving per second.

The statement of the theorem is not precise until we define what we mean by "average". From the form of the equations the reader might be tempted to think of a time average; e.g. the value

$$\underset{T\to\infty}{\text{Lim}} \frac{1}{T} \int_0^T I(t)\,dt \tag{1.2-4}$$

However, in the proof of the theorem the average is generally taken over a great many intervals of length T with t held constant. The process is somewhat similar to that employed in (1.1) and in order to make it clear we take the case of $\overline{I(t)}$ for illustration. We observe $I(t)$ for many, say M, intervals each of length T where T is large in comparison with the interval over which the effect $F(t)$ of the arrival of a single electron is appreciable. Let ${}_nI(t')$ be the value of $I(t)$, t' seconds after the beginning of the nth interval. t' is equal to t plus a constant depending upon the beginning time of the interval. We put the subscript in front because we wish to reserve the usual place for another subscript later on. The value of $\overline{I(t')}$ is then defined as

$$\overline{I(t')} = \underset{M\to\infty}{\text{Limit}} \frac{1}{M} [{}_1I(t') + {}_2I(t') + \cdots + {}_MI(t')] \tag{1.2-5}$$

and this limit is assumed to exist. The mean square value of the fluctuation of $I(t')$ is defined in much the same way.

[2] *Proc. Camb. Phil. Soc.* 15 (1909), 117–136, 310–328. Our proof is similar to one given by J. M. Whittaker, *Proc. Camb. Phil. Soc.* 33 (1937), 451–458.

Actually, as the equations (1.2–2) and (1.2–3) of Campbell's theorem show, these averages and all the similar averages encountered later turn out to be independent of the time. When this is true and when the M intervals in (1.2–5) are taken consecutively the time average (1.2–4) and the average (1.2–5) become the same. To show this we multiply both sides of (1.2–5) by dt' and integrate from 0 to T:

$$\begin{aligned} \overline{I(t')} &= \underset{M\to\infty}{\text{Limit}} \frac{1}{MT} \sum_{m=1}^{M} \int_0^T {}_mI(t')\, dt' \\ &= \underset{M\to\infty}{\text{Limit}} \frac{1}{MT} \int_0^{MT} I(t)\, dt \end{aligned} \tag{1.2–6}$$

and this is the same as the time average (1.2–4) if the latter limit exists.

1.3 Proof of Campbell's Theorem

Consider the case in which exactly K electrons arrive at the anode in an interval of length T. Before the interval starts, we think of these K electrons as fated to arrive in the interval $(0, T)$ but any particular electron is just as likely to arrive at one time as any other time. We shall number these fated electrons from one to K for purposes of identification but it is to be emphasized that the numbering has nothing to do with the order of arrival. Thus, if t_k be the time of arrival of electron number k, the probability that t_k lies in the interval $(t, t + dt)$ is dt/T.

We take T to be very large compared with the range of values of t for which $F(t)$ is appreciably different from zero. In physical applications such a range usually exists and we shall call it Δ even though it is not very definite. Then, when exactly K electrons arrive in the interval $(0, T)$ the effect is approximately

$$I_K(t) = \sum_{k=1}^{K} F(t - t_k) \tag{1.3–1}$$

the degree of approximation being very good over all of the interval except within Δ of the end points.

Suppose we examine a large number M of intervals of length T. The number having exactly K arrivals will be, to a first approximation $M\,p(K)$ where $p(K)$ is given by (1.1–3). For a fixed value of t and for each interval having K arrivals, $I_K(t)$ will have a definite value. As $M \to \infty$, the average value of the $I_K(t)$'s, obtained by averaging over the intervals, is

$$\begin{aligned} \overline{I_K(t)} &= \int_0^T \frac{dt_1}{T} \cdots \int_0^T \frac{dt_K}{T} \sum_{k=1}^{K} F(t - t_k) \\ &= \sum_{k=1}^{K} \int_0^T \frac{dt_k}{T} F(t - t_k) \end{aligned} \tag{1.3–2}$$

and if $\Delta < t < T - \Delta$, we have effectively

$$\overline{I_K(t)} = \frac{K}{T} \int_{-\infty}^{+\infty} F(t)\, dt \tag{1.3-3}$$

If we now average $I(t)$ over all of the M intervals instead of only over those having K arrivals, we get, as $M \to \infty$,

$$\begin{aligned} \overline{I(t)} &= \sum_{K=0}^{\infty} p(K)\overline{I_K(t)} \\ &= \sum_{K=0}^{\infty} \frac{K}{T} \frac{(\nu T)^K}{K!} e^{-\nu T} \int_{-\infty}^{+\infty} F(t)\, dt \\ &= \nu \int_{-\infty}^{+\infty} F(t)\, dt \end{aligned} \tag{1.3-4}$$

and this proves the first part of the theorem. We have used this rather elaborate proof to prove the relatively simple (1.3–4) in order to illustrate a method which may be used to prove more complicated results. Of course, (1.3–4) could be established by noting that the integral is the average value of the effect produced by one arrival, the average being taken over one second, and that ν is the average number of arrivals per second.

In order to prove the second part, (1.2–3) of Campbell's theorem we first compute $\overline{I^2(t)}$ and use

$$\begin{aligned} \overline{(I(t) - \overline{I(t)})^2} &= \overline{I^2(t)} - 2\,\overline{I(t)\overline{I(t)}} + \overline{I(t)}^2 \\ &= \overline{I^2(t)} - \overline{I(t)}^2 \end{aligned} \tag{1.3-5}$$

From the definition (1.3–1) of $I_K(t)$,

$$I_K^2(t) = \sum_{k=1}^{K} \sum_{m=1}^{K} F(t - t_k)F(t - t_m)$$

Averaging this over all values of $t_1, t_2, \cdots t_K$ with t held fixed as in (1.3–2),

$$\overline{I_K^2(t)} = \sum_{k=1}^{K} \sum_{m=1}^{K} \int_0^T \frac{dt_1}{T} \cdots \int_0^T \frac{dt_K}{T} F(t - t_k)F(t - t_m)$$

The multiple integral has two different values. If $k = m$ its value is

$$\int_0^T F^2(t - t_k) \frac{dt_k}{T}$$

and if $k \neq m$ its value is

$$\int_0^T F(t - t_k) \frac{dt_k}{T} \int_0^T F(t - t_m) \frac{dt_m}{T}$$

Counting up the number of terms in the double sum shows that there are K of them having the first value and $K^2 - K$ having the second value. Hence, if $\Delta < t < T - \Delta$ we have

$$\overline{I_K^2(t)} = \frac{K}{T} \int_{\infty}^{+\infty} F^2(t)\, dt + \frac{K(K-1)}{T^2} \left[\int_{\infty}^{+\infty} F(t)\, dt \right]^2$$

Averaging over all the intervals instead of only those having K arrivals gives

$$\overline{I^2(t)} = \sum_{K=0}^{\infty} p(K)\, \overline{I_K^2(t)}$$

$$= \nu \int_{\infty}^{+\infty} F^2(t)\, dt + \overline{I(t)}^2$$

where the summation with respect to K is performed as in (1.3–4), and after summation the value (1.3–4) for $\overline{I(t)}$ is used. Comparison with (1.3–5) establishes the second part of Campbell's theorem.

1.4 The Distribution of $I(t)$

When certain conditions are satisfied the proportion of time which $I(t)$ spends in the range I, $I + dI$ is $P(I)dI$ where, as $\nu \to \infty$, the probability density $P(I)$ approaches

$$\frac{1}{\sigma_I \sqrt{2\pi}} e^{-(I-\bar{I})^2/2\sigma_I^2} \tag{1.4–1}$$

where $\bar{I}$ is the average of $I(t)$ given by (1.2–2) and the square of the standard deviation σ_I, i.e. the variance of $I(t)$, is given by (1.2–3). This normal distribution is the one which would be expected by virtue of the "central limit theorem" in probability. This states that, under suitable conditions, the distribution of the sum of a large number of random variables tends toward a normal distribution whose variance is the sum of the variances of the individual variables. Similarly the average of the normal distribution is the sum of the averages of the individual variables.

So far, we have been speaking of the limiting form of the probability density $P(I)$. It is possible to write down an explicit expression for $P(I)$, which, however, is quite involved. From this expression the limiting form may be obtained. We now obtain this expression. In line with the discussion given of Campbell's theorem, we seek the probability density $P(I)$ of the values of $I(t)$ observed at t seconds from the beginning of each of a large number, M, of intervals, each of length T.

Probability that $I(t)$ lies in range $(I, I + dI)$

$$= \sum_{K=0}^{\infty} \text{(Probability of exactly } K \text{ arrivals)} \times$$

$$\text{(Probability that if there are exactly } K \text{ arrivals, } I_K(t) \text{ lies in } (I, I + dI)).$$

Denoting the last probability in the summation by $P_K(I)dI$, using notation introduced earlier, and cancelling out the factor dI gives

$$P(I) = \sum_{K=0}^{\infty} p(K)P_K(I) \tag{1.4-2}$$

We shall compute $P_K(I)$ by the method of "characteristic functions"[3] from the definition

$$I_K(t) = \sum_{k=1}^{K} F(t - t_k) \tag{1.3-1}$$

of $I_K(t)$. The method will be used in its simplest form: the probability that the sum

$$x_1 + x_2 + \cdots + x_K$$

of K independent random variables lies between X and $X + dX$ is

$$dX \frac{1}{2\pi} \int_{-\infty}^{+\infty} e^{-iXu} \prod_{k=1}^{K} (\text{average value of } e^{ix_k u})\, du \tag{1.4-3}$$

The average value of $e^{ix_k u}$, i.e., the characteristic function of the distribution of x_k, is obtained by averaging over the values of x_k. Although this is the simplest form of the method it is also the least general in that the integral does not converge for some important cases. The distribution which gives a probability of $\frac{1}{2}$ that $x_k = -1$ and $\frac{1}{2}$ that $x_k = +1$ is an example of such a case. However, we may still use (1.4-3) formally in such cases by employing the relation

$$\int_{-\infty}^{+\infty} e^{-iau}\, du = 2\pi\delta(a) \tag{1.4-4}$$

where $\delta(a)$ is zero except at $a = 0$ where it is infinite and its integral from $a = -\epsilon$ to $a = +\epsilon$ is unity where $\epsilon > 0$.

When we identify x_k with $F(t - t_k)$ we see that the average value of $e^{ix_k u}$ is

$$\frac{1}{T} \int_0^T \exp\,[iuF(t - t_k)]\, dt_k$$

[3] The essentials of this method are due to Laplace. A few remarks on its history are given by E. C. Molina, *Bull. Amer. Math. Soc.*, 36 (1930), pp. 369–392. An account of the method may be found in any one of several texts on probability theory. We mention "Random Variables and Probability Distributions," by H. Cramér, Camb. Tract in Math. and Math. Phys. No. 36 (1937), Chap. IV. Also "Introduction to Mathematical Probability," by J. V. Uspensky, McGraw-Hill (1937), pages 240, 264, and 271–278.

All of the K characteristic functions are the same and hence, from (1.4–3), $P_K(I)dI$ is

$$dI\,\frac{1}{2\pi}\int_{-\infty}^{+\infty} e^{-iIu}\left(\frac{1}{T}\int_0^T \exp\,[iuF(t-\tau)]\,d\tau\right)^K du$$

Although in deriving this relation we have taken $K > 0$, it also holds for $K = 0$ (provided we use (1.4–4)). In this case $P_0(I) = \delta(I)$, because $I = 0$ when no electrons arrive.

Inserting our expression for $P_K(I)$ and the expression (1.1–3) for $p(K)$ in (1.4–2) and performing the summation gives

$$P(I) = \frac{1}{2\pi}\int_{-\infty}^{+\infty} \exp\left(-iIu - \nu T + \nu\int_0^T \exp\,[iuF(t-\tau)]\,d\tau\right)du \qquad (1.4\text{–}5)$$

The first exponential may be simplified somewhat. Using

$$\nu T = \nu\int_0^T d\tau$$

permits us to write

$$-\nu T + \nu\int_0^T \exp\,[iuF(t-\tau)]\,d\tau = \nu\int_0^T (\exp\,[iuF(t-\tau)] - 1)\,d\tau$$

Suppose that $\Delta < t < T - \Delta$ where Δ is the range discussed in connection with equation (1.3–1). Taking $|F(t-\tau)| = 0$ for $|t-\tau| > \Delta$ then enables us to write the last expression as

$$\nu\int_{-\infty}^{+\infty} [e^{iuF(t)} - 1]\,dt \qquad (1.4\text{–}6)$$

Placing this in (1.4–5) yields the required expression for $P(I)$:

$$P(I) = \frac{1}{2\pi}\int_{-\infty}^{+\infty} \exp\left(-iIu + \nu\int_{-\infty}^{+\infty} [e^{iuF(t)} - 1]\,dt\right)du \qquad (1.4\text{–}7)$$

An idea of the conditions under which the normal law (1.4–1) is approached may be obtained from (1.4–7) by expanding (1.4–6) in powers of u and determining when the terms involving u^3 and higher powers of u may be neglected. This is taken up for a slightly more general form of current in section 1.6.

1.5 Extension of Campbell's Theorem

In section 1.2 we have stated Campbell's theorem. Here we shall give an extension of it. In place of the expression (1.2–1) for the $I(t)$ of the shot effect we shall deal with the current

$$I(t) = \sum_{k=-\infty}^{+\infty} a_k F(t - t_k) \tag{1.5–1}$$

where $F(t)$ is the same sort of function as before and where $\cdots a_1, a_2, \cdots a_k, \cdots$ are independent random variables all having the same distribution. It is assumed that all of the moments $\overline{a^n}$ exist, and that the events occur at random

The extension states that the nth semi-invariant of the probability density $P(I)$ of I, where I is given by (1.5–1), is

$$\lambda_n = \nu \overline{a^n} \int_{-\infty}^{+\infty} [F(t)]^n \, dt \tag{1.5–2}$$

where ν is the expected number of events per second. The semi-invariants of a distribution are defined as the coefficients in the expansion

$$\log_e (\text{ave. } e^{iIu}) = \sum_{n=1}^{N} \frac{\lambda_n}{n!} (iu)^n + o(u^N) \tag{1.5–3}$$

i.e. as the coefficients in the expansion of the logarithm of the characteristic function. The λ's are related to the moments of the distribution. Thus if $m_1, m_2, \cdots$ denote the first, second $\cdots$ moments about zero we have

$$\text{ave. } e^{iIu} = 1 + \sum_{n=1}^{N} \frac{m_n}{n!} (iu)^n + o(u^N)$$

By combining this relation with the one defining the λ's it may be shown that

$$\begin{aligned} \bar{I} &= m_1 = \lambda_1 \\ \overline{I^2} &= m_2 = \lambda_2 + \lambda_1 m_1 \\ \overline{I^3} &= m_3 = \lambda_3 + 2\lambda_2 m_1 + \lambda_1 m_2 \\ &\cdots\cdots\cdots \end{aligned}$$

It follows that $\lambda_1 = \bar{I}$ and $\lambda_2 = \text{ave. } (I - \bar{I})^2$. Hence (1.5–2) yields the original statement of Campbell's theorem when we set n equal to one and two and also take all the a's to be unity.

The extension follows almost at once from the generalization of expression (1.4–7) for the probability density $P(I)$. By proceeding as in section 1.4 and identifying x_k with $a_k F(t - t_k)$ we see that

$$\text{ave. } e^{ix_k u} = \frac{1}{T} \int_{-\infty}^{+\infty} q(a) \int_0^T \exp [iuaF(t - t_k)] \, dt_k$$

where $q(a)$ is the probability density function for the a's. It turns out that the probability density $P(I)$ of I as defined by (1.5–1) is

$$P(I) = \frac{1}{2\pi} \int_{-\infty}^{+\infty} \exp\left(-iIu + \nu \int_{-\infty}^{+\infty} q(a)\, da \int_{-\infty}^{+\infty} [e^{iuaF(t)} - 1]\, dt\right) du \qquad (1.5\text{–}4)$$

The logarithm of the characteristic function of $P(I)$ is, from (1.5–4),

$$\nu \int_{-\infty}^{+\infty} q(a)\, da \int_{-\infty}^{+\infty} [e^{iuaF(t)} - 1]\, dt = \sum_{n=1}^{\infty} \frac{(iu)^n}{n!} \nu \int_{-\infty}^{+\infty} q(a)\, da\, a^n \int_{-\infty}^{+\infty} F^n(t)\, dt$$

Comparison with the series (1.5–3) defining the semi-invariants gives the extension of Campbell's theorem stated by (1.5–2).

Other extensions of Campbell's theorem may be made. For example, suppose in the expression (1.5–1) for $I(t)$ that $t_1, t_2, \cdots t_k, \cdots$ while still random variables, are no longer necessarily distributed according to the laws assumed above. Suppose now that the probability density $p(x)$ is given where x is the interval between two successive events:

$$t_2 = t_1 + x_1 \qquad (1.5\text{–}5)$$

$$t_3 = t_2 + x_2 = t_1 + x_1 + x_2$$

and so on. For the case treated above

$$p(x) = \nu e^{-\nu x}. \qquad (1.5\text{–}6)$$

We assume that the expected number of events per second is still ν.

Also we take the special, but important, case for which

$$F(t) = 0, \qquad t < 0 \qquad (1.5\text{–}7)$$

$$F(t) = e^{-\alpha t}, \qquad t > 0.$$

For a very long interval extending from $t = t_1$ to $t = T + t_1$ inside of which there are exactly K events we have, if t is not near the ends of the interval,

$$I(t) = a_1F(t - t_1) + a_2F(t - t_1 - x_1) + \cdots + a_{K+1}F(t - t_1 - x_1 \cdots - x_K)$$

$$= a_1F(t') + a_2F(t' - x_1) + \cdots + a_{K+1}F(t' - x_1 - \cdots - x_K)$$

$$I^2(t) = a_1^2F^2(t') + a_2^2F^2(t' - x_1) + \cdots + a_{K+1}^2F^2(t' - x_1 \cdots - x_K)$$
$$+ 2a_1a_2F(t')F(t' - x_1) + \cdots + 2a_1a_{K+1}F(t')F(t' - x_1 \cdots - x_K)$$
$$+ 2a_2a_3F(t' - x_1)F(t' - x_1 - x_2) + \cdots + \cdots$$

where $t' = t - t_1$. If we integrate $I^2(t)$ over the entire interval $0 < t' < T$ and drop the primes we get approximately

$$\int_0^T I^2(t)dt = (a_1^2 + \cdots + a_{K+1}^2)\varphi(0)$$
$$+ 2a_1a_2\varphi(x_1) + 2a_1a_3\varphi(x_1 + x_2) + \cdots + 2a_1a_{K+1}\varphi(x_1 + \cdots + x_K)$$
$$+ 2a_2a_3\varphi(x_2) + \cdots + \cdots + 2a_Ka_{K+1}\varphi(x_K)$$

where

$$\varphi(x) = \int_{\infty}^{+\infty} F(t)F(t - x)\, dt$$

When we divide both sides by T and consider K and T to be very large,

$$\frac{K}{T}\frac{a_1^2 + \cdots + a_{K+1}^2}{K}\varphi(0) \approx \nu\overline{a^2}\varphi(0)$$

$$\frac{1}{T}[a_1 a_2 \varphi(x_1) + a_2 a_3 \varphi(x_2) + \cdots + a_K a_{K+1} \varphi(x_K)] = \frac{K}{T} \text{ average } a_k a_{k+1} \varphi(x_k)$$
$$\approx \nu\bar{a}^2 \int_0^\infty \varphi(x)p(x)\, dx$$

$$\frac{1}{T}[a_1 a_3 \varphi(x_1 + x_2) + \cdots] = \frac{K-1}{T} \text{ ave. } a_k a_{k+3} \varphi(x_k + x_{k+1})$$
$$\approx \nu\bar{a}^2 \int_0^\infty dx_1 \int_0^\infty dx_2\, p(x_1)p(x_2)\varphi(x_1 + x_2)$$

Consequently

$$\overline{I^2(t)} = \lim_{T\to\infty} \frac{1}{T}\int_0^T I^2(t)\, dt$$
$$= \nu\overline{a^2}\varphi(0) + 2\nu\bar{a}^2\left[\int_0^\infty p(x)\varphi(x)\, dx\right.$$
$$\left. + \int_0^\infty dx_1 \int_0^\infty dx_2\, p(x_1)p(x_2)\varphi(x_1 + x_2) + \cdots\right]$$

For our special exponential form (1.5–7) for $F(t)$,

$$\varphi(x) = \frac{e^{-\alpha x}}{2\alpha}$$

and the multiple integrals occurring in the expression for $\overline{I^2(t)}$ may be written in terms of powers of

$$q = \int_0^\infty p(x)e^{-\alpha x}\,dx \tag{1.5-8}$$

Thus

$$2\alpha\overline{I^2(t)} = \nu\overline{a^2} + 2\bar{a}^2\nu\frac{q}{1-q}$$

and since

$$\bar{I}(t) = \nu\bar{a}\int_{-\infty}^{+\infty} F(t)\,dt = \nu\bar{a}/\alpha$$

we have

$$\overline{I^2(t)} - \overline{I(t)}^2 = \frac{\nu\overline{a^2}}{2\alpha} + \left(\frac{\nu\bar{a}}{\alpha}\right)^2\left[\frac{\alpha q}{\nu(1-q)} - 1\right] \tag{1.5-9}$$

Equations (1.5–8) and (1.5–9) give us an extension of Campbell's theorem subject to the restrictions discussed in connection with equations (1.5–5) and (1.5–7). Other generalizations have been made[4] but we shall leave the subject here. The reader may find it interesting to verify that (1.5–9) gives the correct answer when $p(x)$ is given by (1.5–6), and also to investigate the case when the events are spaced equally.

1.6 Approach of Distribution of I to a Normal Law

In section 1.5 we saw that the probability density $P(I)$ of the noise current I may be expressed formally as

$$P(I) = \frac{1}{2\pi}\int_{-\infty}^{+\infty} \exp\left[-iIu + \sum_{n=1}^{\infty} (iu)^n\lambda_n/n!\right] du \tag{1.6-1}$$

where λ_n is the nth semi-invariant given by (1.5–2). By setting

$$\lambda_2 = \sigma^2$$

$$x = \frac{I - \lambda_1}{\sigma} = \frac{I - \bar{I}}{\sigma} \tag{1.6-2}$$

[4] See E. N. Rowland, *Proc. Camb. Phil. Soc.* 32 (1936), 580–597. He extends the theorem to the case where there are two functions instead of a single one, which we here denote by $I(t)$. According to a review in the Zentralblatt für Math., 19, p. 224, Khintchine in the Bull. Acad. Sci. URSS, sér. Math. Nr. 3 (1938), 313–322, has continued and made precise the earlier work of Rowland.

expanding

$$\exp \sum_{n=3}^{\infty} (iu)^n \lambda_n / n!$$

as a power series in u, integrating termwise using

$$\frac{1}{2\pi} \int_{-\infty}^{+\infty} (iu\sigma)^n \exp\left[-iu\sigma x - \frac{u^2\sigma^2}{2}\right] du = (-)^n \sigma^{-1} \varphi^{(n)}(x),$$

$$\varphi^{(n)}(x) = \frac{1}{\sqrt{2\pi}} \frac{d^n}{dx^n} e^{-x^2/2}$$

and finally collecting terms according to their order in powers of $\nu^{-1/2}$, gives

$$P(I) \sim \sigma^{-1} \varphi^{(0)}(x) - \frac{\lambda_3 \sigma^{-4}}{3!} \varphi^{(3)}(x) + \left[\frac{\lambda_4 \sigma^{-5}}{4!} \varphi^{(4)}(x) + \frac{\lambda_3^2 \sigma^{-7}}{72} \varphi^{(6)}(x)\right] + \cdots \tag{1.6-3}$$

The first term is $0(\nu^{-1/2})$, the second term is $0(\nu^{-1})$, and the term within brackets is $0(\nu^{-3/2})$. This is Edgeworth's series.[5] The first term gives the normal distribution and the remaining terms show how this distribution is approached as $\nu \to \infty$.

1.7 The Fourier Components of $I(t)$

In some analytical work noise current is represented as

$$I(t) = \frac{a_0}{2} + \sum_{n=1}^{N} \left(a_n \cos \frac{2\pi nt}{T} + b_n \sin \frac{2\pi nt}{T}\right) \tag{1.7-1}$$

where at a suitable place in the work T and N are allowed to become infinite. The coefficients a_n and b_n, $1 \leq n \leq N$, are regarded as independent random variables distributed about zero according to a normal law.

It appears that the association of (1.7–1) with a sequence of disturbances occurring at random goes back many years. Rayleigh[6] and Gouy suggested that black-body radiation and white light might both be regarded as sequences of irregularly distributed impulses.

Einstein[7] and von Laue have discussed the normal distribution of the coefficients in (1.7–1) when it is used to represent black-body radiation, this radiation being the resultant produced by a great many independent os-

[5] See, for example, pp. 86–87, in "Random Variables and Probability Distributions" by H. Cramér, *Cambridge Tract* No. 36 (1937).

[6] *Phil. Mag.* Ser. 5, Vol. 27 (1889) pp. 460–469.

[7] A. Einstein and L. Hopf, *Ann. d. Physik* 33 (1910) pp. 1095–1115.
M. V. Laue, *Ann. d. Physik* 47 (1915) pp. 853–878.
A. Einstein, *Ann. d. Physik* 47 (1915) pp. 879–885.
M. V. Laue, *Ann. d. Physik* 48 (1915) pp. 668–680.
I am indebted to Prof. Goudsmit for these references.

cillators. Some argument arose as to whether the coefficients in (1.7–1) were statistically independent or not. It was finally decided that they are independent.

The shot effect current has been represented in this way by Schottky.[8] The Fourier series representation has been discussed by H. Nyquist[9] and also by Goudsmit and Weiss. Remarks made by A. Schuster[10] are equivalent to the statement that a_n and b_n are distributed normally.

In view of this wealth of information on the subject it may appear superfluous to say anything about it. However, for the sake of completeness, we shall outline the thoughts which lead to (1.7–1).

In line with our usual approach to the shot effect, we suppose that exactly K electrons arrive during the interval $(0, T)$, so that the noise current for the interval is

$$I_K(t) = \sum_{k=1}^{K} F(t - t_k) \tag{1.7–2}$$

The coefficients in the Fourier series expansion of $I_K(t)$ over the interval $(0, T)$ are a_{nK} and b_{nK} where

$$\begin{aligned} a_{nK} - ib_{nK} &= \frac{2}{T} \sum_{k=1}^{K} \int_0^T F(t - t_k) \exp\left[-i\,\frac{2\pi nt}{T}\right] dt \\ &\approx \frac{2}{T} \sum_{k=1}^{K} \int_{-\infty}^{+\infty} F(t) \exp\left[-i\,\frac{2\pi n}{T}\,(t + t_k)\right] dt \\ &= R_n e^{-i\varphi_n} \sum_{k=1}^{K} e^{-in\theta_k} \end{aligned} \tag{1.7–3}$$

In this expression

$$\begin{aligned} \theta_k &= \frac{2\pi t_k}{T} \\ R_n e^{-i\varphi_n} &= C_n - iS_n = \frac{2}{T} \int_{-\infty}^{+\infty} F(t) e^{-i2\pi nt/T}\, dt \end{aligned} \tag{1.7–4}$$

In the earlier sections the arrival times $t_1, t_2, \cdots t_K$ were regarded as K independent random variable each distributed uniformly over the interval (O, T). Hence the θ_k's may be regarded as random variables distributed uniformly over the interval 0 to 2π.

Incidentally, it will be noted that in (1.7–3) there occurs the sum of K randomly oriented unit vectors. When K becomes very large, as it does

[8] *Ann. d. Physik*, 57 (1918) pp. 541–567.

[9] Unpublished Memorandum, "Fluctuations in Vacuum Tube Noise and the Like," March 17, 1932.

[10] Investigation of Hidden Periodicities, Terrestrial Magnetism, 3 (1898), pp. 13–41. See especially propositions 1 and 2 on page 26 of Schuster's paper.

when $\nu \to \infty$, it is known that the real and imaginary parts of this sum are random variables, which tend to become independent and normally distributed about zero. This suggests the manner in which the normal distribution of the coefficients arises. Averaging over the θ_k's in (1.7–3) gives when $n > 0$

$$\bar{a}_{nK} = \bar{b}_{nK} = 0 \tag{1.7–5}$$

Some further algebra gives

$$\begin{aligned} \overline{a_{nK}^2} &= \overline{b_{nK}^2} = \frac{K}{2} R_n^2 \\ \overline{a_{nK}\, b_{nK}} &= \overline{a_{nK}\, a_{mK}} = \overline{b_{nK}\, b_{mK}} = 0 \end{aligned} \tag{1.7–6}$$

where $n \neq m$ and $n, m > 0$.

So far, we have been considering the case of exactly K arrivals in our interval of length T. Now we pass to the general case of any number of arrivals by making use of formulas analogous to

$$\overline{a_n^2} = \sum_{K=0}^{\infty} p(K)\overline{a_{nK}^2} \tag{1.7–7}$$

as has been done in section 1.3. Thus, for $n > 0$,

$$\begin{aligned} \bar{a}_n &= \bar{b}_n = 0 \\ \overline{a_n^2} &= \overline{b_n^2} = \frac{\nu T}{2} R_n^2 = \sigma_n^2 \\ \overline{a_n\, b_n} &= \overline{a_n\, a_m} = \overline{b_n\, b_m} = 0, \qquad n \neq m \end{aligned} \tag{1.7–8}$$

In the second line we have used σ_n to denote the standard deviation of a_n and b_n. We may put the expression for σ_n^2 in a somewhat different form by writing

$$f_n = \frac{n}{T} = n\Delta f, \qquad \Delta f = \frac{1}{T} \tag{1.7–9}$$

where f_n is the frequency of the nth component. Using (1.7-4),

$$\sigma_n^2 = 2\nu\Delta f \left| \int_{-\infty}^{+\infty} F(t)e^{-i2\pi f_n t}\, dt \right|^2 \tag{1.7–10}$$

Thus, σ_n^2 is proportional to ν/T.

The probability density function $P(a_1, \cdots a_N, b_1, \cdots b_N)$ for the $2N$ coefficients, $a_1, \cdots a_N, b_1, \cdots b_N$ may be derived in much the same fashion as was the probability density of the noise current in section 1.4. Here N

is arbitrary but fixed. The expression analogous to (1.4–5) is the $2N$ fold integral

$$P(a_1, \cdots, b_N) = (2\pi)^{-2N} \int_{-\infty}^{+\infty} du_1 \cdots \int_{-\infty}^{+\infty} dv_N \qquad (1.7\text{–}11)$$
$$\exp[-i(a_1 u_1 + \cdots + b_N v_N) - \nu T + \nu T E]$$

where

$$E = \frac{1}{2\pi} \int_0^{2\pi} d\theta \exp\left[i \sum_{n=1}^{N} (u_n C_n + v_n S_n) \cos n\theta + (v_n C_n - u_n S_n) \sin n\theta \right] \qquad (1.7\text{–}12)$$

in which $C_n - iS_n$ is defined as the Fourier transform (1.7–4) of $F(t)$.

The next step is to show that (1.7–11) approaches a normal law in $2N$ dimensions as $\nu \to \infty$. This appears to be quite involved. It will be noted that the integrand in the integral defining E is composed of N factors of the form

$$\exp[i\rho_n \cos(n\theta - \psi_n)]$$
$$= J_0(\rho_n) + 2i \cos(n\theta - \psi_n) J_1(\rho_n) - 2\cos(2n\theta - 2\psi_n) J_2(\rho_n) + \cdots$$

where

$$\rho_n^2 = (u_n^2 + v_n^2)(C_n^2 + S_n^2) = \frac{2}{\nu T} \sigma_n^2 (u_n^2 + v_n^2).$$

As ν becomes large, it turns out that the integral (1.7–11) for the probability density obtains most of its contributions from small values of u and v. By substituting the product of the Bessel function series in the integral for E and integrating we find

$$E = \prod_{n=1}^{N} J_0(\rho_n) + A + B + C$$

where A is the sum of products such as

$$-2i \cos(\psi_{k+\ell} - \psi_k - \psi_\ell) J_1(\rho_k) J_1(\rho_\ell) J_1(\rho_{k+\ell}) \text{ times } N - 3\ J_0\text{'s}$$

in which $0 < k \leq l$ and $2 \leq k + l \leq N$. Similarly B is the sum of products of the form

$$-2i \cos(\psi_{2k} - 2\psi_k) J_1(\rho_{2k}) J_2(\rho_k) \text{ times } N - 2\ J_0\text{'s}$$

C consists of terms which give fourth and higher powers in u and v. There are roughly $N^2/4$ terms of form A and $N/2$ terms of form B.

Expanding the Bessel functions, neglecting all powers above the third and proceeding as in section 1.4, will give us the normal distribution plus the first correction term. It is rather a messy affair. An idea of how it looks may be obtained by taking the special case in which $F(t)$ is an even function of t and neglecting terms of type B. Then

$$P(a_1, \cdots a_N, b_1, \cdots b_N) = (1 + \eta) \prod_{n=1}^{N} \frac{e^{-(x_n^2+y_n^2)/2}}{2\pi\sigma_n^2} \quad (1.7\text{–}12)$$

where

$$x_n = \frac{a_n}{\sigma_n}, \qquad y_n = \frac{b_n}{\sigma_n}$$

$$\eta = (2\nu T)^{-1/2} \sum_{k,\ell} [x_{k+\ell}(x_k x_\ell - y_k y_\ell) + 2\, y_{k+\ell} y_k y_\ell] \quad (1.7\text{–}13)$$

and the summation extends over $2 \leq k + l \leq N$ with $k \leq l$.

It is seen that if T and N are held constant, the correction term η approaches zero as ν becomes very large. A very rough idea of the magnitude of η may be obtained by assuming that unity is a representative value of the x's and y's. Further assuming that there are N^2 terms in the summation each one of which may be positive or negative suggests that magnitude of the sum is of the order of N. Hence we might expect to find that η is of the order of $N(2\nu T)^{-1/2}$.

PART II

POWER SPECTRA AND CORRELATION FUNCTIONS

2.0 Introduction

A theory for analyzing functions of time, t, which do not die down and which remain finite as t approaches infinity has gradually been developed over the last sixty years. A few words of its history together with an extensive bibliography are given by N. Wiener in his paper on "Generalized Harmonic Analysis".[11] G. Gouy, Lord Rayleigh and A. Schuster were led to study this problem in their investigations of such things as white light and noise. Schuster[12] invented the "periodogram" method of analysis which has as its object the discovery of any periodicities hidden in a continuous curve representing meteorological or economic data.

[11] *Acta Math.*, Vol. 55, pp. 117–258 (1930). See also "Harmonic Analysis of Irregular Motion," *Jour. Math. and Phys.* 5 (1926) pp. 99–189.

[12] The periodogram was first introduced by Schuster in reference 10 cited in Section 1.7. He later modified its definition in the Trans. Camb. Phil. Soc. 18 (1900), pp. 107–135, and still later redefined it in "The Periodogram and its Optical Analogy," Proc. Roy. Soc., London, Ser. A, 77 (1906) pp. 136–140. In its final form the periodogram is equivalent to $\frac{1}{2}w(f)$, where $w(f)$ is the power spectrum defined in Section 2.1, plotted as a function of the period $T = (2\pi f)^{-1}$.

The correlation function, which turns out to be a very useful tool, apparently was introduced by G. I. Taylor.[13] Recently it has been used by quite a few writers[14] in the mathematical theory of turbulence.

In section 2.1 the power spectrum and correlation function of a specific function, such as one given by a curve extending to $t = \infty$, are defined by equations (2.1–3) and (2.1–4) respectively. That they are related by the Fourier inversion formulae (2.1–5) and (2.1–6) is merely stated; the discussion of the method of proof being delayed until sections 2.3 and 2.4. In section 2.3 a discussion based on Fourier series is given and in section 2.4 a parallel treatment starting with Parseval's integral theorem is set forth. The results as given in section 2.1 have to be supplemented when the function being analyzed contains a d.c. or periodic components. This is taken up in section 2.2.

The first four sections deal with the analysis of a specific function of t. However, most of the applications are made to functions which behave as though they are more or less random in character. In the mathematical analysis this randomness is introduced by assuming the function of t to be also a function of suitable parameters, and then letting these parameters be random variables. This question is taken up in section 2.5. In section 2.6 the results of 2.5 are applied to determine the average power spectrum and the average correlation function of the shot effect current. The same thing is done in 2.7 for a flat top wave, the tops (and bottoms) being of random length. The case in which the intervals are of equal length but the sign of the wave is random is also discussed in 2.7. The representation of the noise current as a trigonometrical series with random variable coefficients is taken up in 2.8. The last two sections 2.9 and 2.10 are devoted to probability theory. The normal law and the central limit theorem, respectively, are discussed.

2.1 Some Results of Generalized Harmonic Analysis

We shall first state the results which we need, and then show that they are plausible by methods which are heuristic rather than rigorous. Suppose that $I(t)$ is one of the functions mentioned above. We may think of it as being specified by a curve extending from $t = -\infty$ to $t = \infty$. $I(t)$ may be regarded as composed of a great number of sinusoidal components whose frequencies range from 0 to $+\infty$. It does not necessarily have to be a noise current, but if we think of it as such, then, in flowing through a resistance of one ohm it will dissipate a certain average amount of power, say p watts.

[13] Diffusion by Continuous Movements, *Proc. Lond. Math. Soc.*, Ser. 2, 20, pp. 196–212 (1920).

[14] See the text "Modern Developments in Fluid Dynamics" edited by S. Goldstein, Oxford (1938).

That portion of ρ arising from the components having frequencies between f and $f + df$ will be denoted by $w(f)df$, and consequently

$$\rho = \int_0^\infty w(f)df \tag{2.1-1}$$

Since $w(f)$ is the spectrum of the average power we shall call it the "power spectrum" of $I(t)$. It has the dimensions of energy and on this account is frequently called the "energy-frequency spectrum" of $I(t)$. A mathematical formulation of this discussion leads to a clear cut definition of $w(f)$.

Let $\Phi(t)$ be a function of t, which is zero outside the interval $0 \leq t \leq T$ and is equal to $I(t)$ inside the interval. Its spectrum $S(f)$ is given by

$$S(f) = \int_0^T I(t)e^{-2\pi ift}\, dt \tag{2.1-2}$$

The spectrum of the power, $w(f)$, is defined as

$$w(f) = \underset{T\to\infty}{\text{Limit}} \frac{2|S(f)|^2}{T} \tag{2.1-3}$$

where we consider only values of $f > 0$ and assume that this limit exists. This is substantially the definition of $w(f)$ given by J. R. Carson[15] and is useful when $I(t)$ has no periodic terms and no d.c. component. In the latter case (2.1–3) must either be supplemented by additional definitions or else a somewhat different method of approach used. These questions will be discussed in section 2.2.

The correlation function $\psi(\tau)$ of $I(t)$ is defined by the limit

$$\psi(\tau) = \underset{T\to\infty}{\text{Limit}} \frac{1}{T}\int_0^T I(t)I(t+\tau)\, dt \tag{2.1-4}$$

which is assumed to exist. $\psi(\tau)$ is closely related to the correlation coefficients used in statistical theory to measure the correlation of two random variables. In the present case the value of $I(t)$ at time t is one variable and its value at a different time $t + \tau$ is the other variable.

The spectrum of the power $w(f)$ and the correlation function $\psi(\tau)$ are related by the equations

$$w(f) = 4\int_0^\infty \psi(\tau) \cos 2\pi f\tau\, d\tau \tag{2.1-5}$$

$$\psi(\tau) = \int_0^\infty w(f) \cos 2\pi f\tau\, df \tag{2.1-6}$$

[15] "The Statistical Energy-Frequency Spectrum of Random Disturbances," *B.S.T.J.*, Vol. 10, pp. 374–381 (1931).

It is seen that $\psi(\tau)$ is an even function of τ and that

$$\psi(0) = \rho \tag{2.1-7}$$

When either $\psi(\tau)$ or $w(f)$ is known the other may be obtained provided the corresponding integral converges.

2.2 Power Spectrum for D.C. and Periodic Components

As mentioned in section 2.1, when $I(t)$ has a d.c. or a periodic component the limit in the definition (2.1–3) for $w(f)$ does not exist for f equal to zero or to the frequency of the periodic component. Perhaps the most satisfactory method of overcoming this difficulty, from the mathematical point of view, is to deal with the integral of the power spectrum.[16]

$$\int_0^f w(g)\, dg \tag{2.2-1}$$

instead of with $w(f)$ itself.

The definition (2.1–4) for $\psi(\tau)$ still holds. If, for example,

$$I(t) = A + C \cos (2\pi f_0 t - \varphi) \tag{2.2-2}$$

$\psi(\tau)$ as given by (2.1–4) is

$$\psi(\tau) = A^2 + \frac{C^2}{2} \cos 2\pi f_0 \tau \tag{2.2-3}$$

The inversion formulas (2.1–5) and (2.1–6) give

$$\begin{aligned} \int_0^f w(g)\, dg &= \frac{2}{\pi} \int_0^\infty \psi(\tau) \frac{\sin 2\pi f \tau}{\tau}\, d\tau \\ \psi(\tau) &= \int_0^\infty \cos 2\pi f \tau\, d\left[\int_0^f w(g)\, dg\right] \end{aligned} \tag{2.2-4}$$

[16] This is done by Wiener,[11] loc. cit., and by G. W. Kenrick, "The Analysis of Irregular Motions with Applications to the Energy Frequency Spectrum of Static and of Telegraph Signals," *Phil. Mag.*, Ser. 7, Vol. 7, pp. 176–196 (Jan. 1929). Kenrick appears to be one of the first to apply, to noise problems, the correlation function method of computing the power spectrum (one of his problems is discussed in Sec. 2.7). He bases his work on results due to Wiener. Khintchine, in "Korrelationstheorie der stationären stochastischen Prozesse," *Math. Annalen*, 109 (1934), pp. 604–615, proves the following theorem: A necessary and sufficient condition that a function $R(t)$ may be the correlation function of a continuous, stationary, stochastic process is that $R(t)$ may be expressed as

$$R(t) = \int_{-\infty}^{+\infty} \cos tx\, dF(x)$$

where $F(x)$ is a certain distribution function. This expression for $R(t)$ is essentially the second of equations (2.2–4). Khintchine's work has been extended by H. Cramér, "On the theory of stationary random processes," *Ann. of Math.*, Ser. 2, Vol. 41 (1940), pp. 215–230. However, Khintchine and Cramér appear to be interested primarily in questions of existence, representation, etc., and do not stress the concept of the power spectrum.

where the last integral is to be regarded as a Stieltjes' integral. When the expression (2.2–3) for $\psi(\tau)$ is placed in the first formula of (2.2–4) we get

$$\int_0^f w(g)\,dg = \begin{cases} A^2 & \text{when} \quad 0 < f < f_0 \\ A^2 + \dfrac{C^2}{2}, & \text{“} \quad f > f_0 \end{cases} \tag{2.2–5}$$

When this expression is used in the second formula of (2.2–4), the increments of the differential are seen to be A^2 at $f = 0$ and $C^2/2$ at $f = f_0$. The resulting expression for $\psi(\tau)$ agrees with the original one.

Here we desire to use a less rigorous, but more convenient, method of dealing with periodic components. By examining the integral of $w(f)$ as given by (2.2–5) we are led to write

$$w(f) = 2A^2\delta(f) + \frac{C^2}{2}\delta(f - f_0) \tag{2.2–6}$$

where $\delta(x)$ is an even unit impulse function so that if $\epsilon > 0$

$$\int_0^\epsilon \delta(x)\,dx = \frac{1}{2}\int_{-\epsilon}^\epsilon \delta(x)\,dx = \tfrac{1}{2} \tag{2.2–7}$$

and $\delta(x) = 0$ except at $x = 0$, where it is infinite. This enables us to use the simpler inversion formulas of section 2.1. The second of these, (2.1–6), is immediately seen to give the correct expression for $\psi(\tau)$. The first one, (2.1–5), gives the correct expression for $w(f)$ provided we interpret the integrals as follows:

$$\begin{aligned} \int_0^\infty \cos 2\pi f\tau\,d\tau &= \tfrac{1}{2}\delta(f) \\ \int_0^\infty \cos 2\pi f_0\tau \cos 2\pi f\tau\,d\tau &= \tfrac{1}{4}\delta(f - f_0) \end{aligned} \tag{2.2–8}$$

It is not hard to show that these are in agreement with the fundamental interpretation

$$\int_{-\infty}^{+\infty} e^{-i2\pi ft}\,dt = \int_{-\infty}^{+\infty} e^{i2\pi ft}\,dt = \delta(f) \tag{2.2–9}$$

which in its turn follows from a formal application of the Fourier integral formula and

$$\int_{-\infty}^{+\infty} \delta(f)e^{i2\pi ft}\,df = \int_{-\infty}^{+\infty} \delta(f)e^{-i2\pi ft}\,df = 1 \tag{2.2–10}$$

We must remember that $f_0 > 0$ and $f \geq 0$ in (2.2–8) so that $\delta(f + f_0) = 0$ for $f \geq 0$.

The definition (2.1–3) for $w(f)$ gives the continuous part of the power spectrum. In order to get the part due to the d.c. and periodic components, which is exemplified by the expression (2.2–6) for $w(f)$ involving the δ functions, we must supplement (2.1–3) by adding terms of the type

$$\begin{aligned} 2A^2\delta(f) + \frac{C^2}{2}\delta(f-f_0) &= \left[\operatorname*{Limit}_{T\to\infty} \frac{2|S(0)|^2}{T^2}\right]\delta(f) \\ &\quad + \left[\operatorname*{Limit}_{T\to\infty} \frac{2|S(f_0)|^2}{T^2}\right]\delta(f-f_0) \end{aligned} \tag{2.2–11}$$

The correctness of this expression may be verified by calculating $S(f)$ for the $I(t)$ of this section given by (2.2–2), and actually carrying out the limiting process.

2.3 Discussion of Results of Section One—Fourier Series

The fact that the spectrum of power $w(f)$ and the correlation function $\psi(\tau)$ are related by Fourier inversion formulas is closely connected with Parseval's theorems for Fourier series and integrals. In this section we shall not use Parseval's theorems explicitly. We start with Fourier's series and use the concept of each component dissipating its share of energy independently of the behavior of the other components.

Let that portion of $I(t)$ which lies in the interval $0 \le t < T$ be expanded in the Fourier series

$$I(t) = \frac{a_0}{2} + \sum_{n=1}^{\infty}\left(a_n \cos\frac{2\pi nt}{T} + b_n \sin\frac{2\pi nt}{T}\right) \tag{2.3–1}$$

where

$$\begin{aligned} a_n &= \frac{2}{T}\int_0^T I(t)\cos\frac{2\pi nt}{T}\,dt \\ b_n &= \frac{2}{T}\int_0^T I(t)\sin\frac{2\pi nt}{T}\,dt \end{aligned} \tag{2.3–2}$$

Then for the interval $-\tau \le t < T - \tau$,

$$I(t+\tau) = \frac{a_0}{2} + \sum_{n=1}^{\infty}\left(a_n \cos\frac{2\pi n(t+\tau)}{T} + b_n \sin\frac{2\pi n(t+\tau)}{T}\right) \tag{2.3–3}$$

Multiplying the series for $I(t)$ and $I(t+\tau)$ together and integrating with respect to t gives, after some reduction,

$$\begin{aligned} &\frac{1}{T}\int_0^T I(t)I(t+\tau)\,dt \\ &= \frac{a_0^2}{4} + \sum_{n=1}^{\infty}\frac{1}{2}(a_n^2 + b_n^2)\cos\frac{2\pi n}{T}\tau + O\left(\frac{\tau I^2}{T}\right) \end{aligned} \tag{2.3–4}$$

where the last term represents correction terms which must be added because the series (2.3–3) does not represent $I(t + \tau)$ in the interval $(T - \tau, T)$ when $\tau > 0$, or in the interval $(0, -\tau)$ if $\tau < 0$.

If $I(t)$ were a current and if it were to flow through one ohm for the interval $(0, T)$, each component would dissipate a certain average amount of power. The average power dissipated by the component of frequency $f_n = n/T$ cycles per second would be, from the Fourier series and elementary principles,

$$\begin{aligned} &\frac{1}{2}(a_n^2 + b_n^2) \text{ watts}, \qquad && n \neq 0 \\ &\frac{a_0^2}{4} \qquad \text{watts}, && n = 0 \end{aligned} \tag{2.3–5}$$

The band width associated with the nth component is the difference in frequency between the $n + 1$ th and nth components:

$$f_{n+1} - f_n = \frac{n+1}{T} - \frac{n}{T} = \frac{1}{T}\, cps$$

Hence if the average power in the band $f, f + df$ is defined as $w(f)df$, the average power in the band $f_{n+1} - f_n$ is

$$w(f_n)(f_{n+1} - f_n) = w\left(\frac{n}{T}\right)\frac{1}{T}$$

and, from (2.3–5), this is given by

$$\begin{aligned} &w\left(\frac{n}{T}\right)\frac{1}{T} = \frac{1}{2}(a_n^2 + b_n^2), \qquad && n \neq 0 \\ &w(0)\,\frac{1}{T} = \frac{a_0^2}{4}, && n = 0 \end{aligned} \tag{2.3–6}$$

When the coefficients in (2.3–4) are replaced by their expressions in terms of $w(f)$ we get

$$\begin{aligned} \frac{1}{T}\int_0^T I(t)I(t + \tau)\, dt + O\left(\frac{\tau I^2}{T}\right) &= \frac{1}{T}\sum_{n=0}^{\infty} w\left(\frac{n}{T}\right)\cos\frac{2\pi n\tau}{T} \\ &= \int_0^{\infty} w\left(\frac{n}{T}\right)\cos\frac{2\pi n\tau}{T}\frac{dn}{T} \\ &= \int_0^{\infty} w(f)\cos 2\pi f\tau\, df \end{aligned} \tag{2.3–7}$$

where we have assumed T so large and $w(f)$ of such a nature that the summation may be replaced by integration.

If I remains finite, then as $T \to \infty$ with τ held fixed, the correction term on the left becomes negligibly small and we have, upon using the definitions (2.1–4) for the correlation function $\psi(\tau)$, the second of the fundamental inversion formulas (2.1–6). The first inversion formula may be obtained from this at once by using Fourier's double integral for $w(f)$.

Incidentally, the relation (2.3–6) between $w(f)$ and the coefficients a_n and b_n is in agreement with the definition (2.1–3) for $w(f)$ as a limit involving $|S(f)|^2$. From the expressions (2.3–2) for a_n and b_n, the spectrum $S(f_n)$ given by (2.1–2) is

$$S(f_n) = \frac{T}{2}(a_n - ib_n)$$

Then, from (2.1–3) $w(f_n)$ is given by the limit, as $T \to \infty$, of

$$\begin{aligned} \frac{2}{T}|S(f_n)|^2 &= \frac{2}{T}\cdot\frac{T^2}{4}(a_n^2 + b_n^2) \\ &= \frac{T}{2}(a_n^2 + b_n^2) \end{aligned}$$

and this is the expression for $w\left(\frac{n}{T}\right)$ given by (2.3–6).

2.4 Discussion of Results of Section One—Parseval's Theorem

The use of Parseval's theorem[17] enables us to derive the results of section 2.1 more directly than the method of the preceding section. This theorem states that

$$\int_{-\infty}^{+\infty} F_1(f)F_2(f)\,df = \int_{-\infty}^{+\infty} G_1(t)G_2(-t)\,dt \tag{2.4-1}$$

where F_1, G_1 and F_2, G_2 are Fourier mates related by

$$\begin{aligned} F(f) &= \int_{-\infty}^{+\infty} G(t)e^{-i2\pi ft}\,dt \\ G(t) &= \int_{-\infty}^{+\infty} F(f)e^{i2\pi ft}\,df \end{aligned} \tag{2.4-2}$$

It may be proved in a formal manner by replacing the F_1 on the left of (2.4–1) by its expression as an integral involving $G_1(t)$. Interchanging the

[17] E. C. Titchmarsh, Introduction to the Theory of Fourier Integrals, Oxford (1937).

order of integration and using the second of (2.4–2) to replace F_2 by G_2 gives the right hand side.

We now set $G_1(t)$ and $G_2(t)$ equal to zero except for intervals of length T. These intervals and the corresponding values of G_1 and G_2 are

$$\begin{aligned} G_1(t) &= I(t), & 0 < t < T \\ G_2(t) &= I(-t + \tau), & \tau - T < t < \tau \end{aligned} \tag{2.4–3}$$

From (2.4-3) it follows that $F_1(f)$ is the spectrum $S(f)$ of $I(t)$ given by equation (2.1–2). Since $I(t)$ is real it follows from the first of equations (2.4–2) that

$$S(-f) = S^*(f), \tag{2.4–4}$$

where the star denotes conjugate complex, and hence that $|S(f)|^2$ is an even function of f.

The first of equations (2.4–2) also gives

$$\begin{aligned} F_2(f) &= \int_{\tau-T}^{\tau} I(-t+\tau)e^{-i2\pi ft}\,dt \\ &= \int_0^T I(t)e^{i2\pi f(t-\tau)}\,dt \\ &= S^*(f)e^{-i2\pi f\tau} \end{aligned} \tag{2.4–5}$$

When these G's and F's are placed in (2.4–1) we obtain

$$\int_{-\infty}^{+\infty} |S(f)|^2 e^{-2\pi f\tau i}\,df = \int_0^{T-\tau} I(t)I(t+\tau)\,dt \tag{2.4–6}$$

where we have made use of the fact that $G_2(-t)$ is zero except in the interval $-\tau < t < T - \tau$ and have assumed $\tau > 0$. If $\tau < 0$ the limits of integration on the right would be $-\tau$ and T.

Since $|S(f)|^2$ is an even function of f we may write (2.4–6) as

$$\frac{1}{T}\int_0^T I(t)I(t+\tau)\,dt + O\left(\frac{\tau I^2}{T}\right) = \int_0^\infty \frac{2|S(f)|^2}{T}\cos 2\pi f\tau\,df \tag{2.4–7}$$

If we now define the correlation function $\psi(\tau)$ as the limit, as $T \rightarrow \infty$, of the left hand side and define $w(f)$ as the function

$$w(f) = \operatorname*{Limit}_{T\rightarrow\infty} \frac{2|S(f)|^2}{T}, \qquad f > 0 \tag{2.1–3}$$

we obtain the second, (2.1–6), of the fundamental inversion formulas. As before, the first may be obtained from Fourier's integral theorem.

In order to obtain the interpretation of $w(f)df$ as the average power dissipated in one ohm by those components of $I(t)$ which lie in the band f, $f + df$, we set $\tau = 0$ in (2.4–7):

$$\underset{T\to\infty}{\text{Limit}} \frac{1}{T} \int_0^T I^2(t)\, dt = \int_0^\infty w(f)\, df \tag{2.4–8}$$

The expression on the left is certainly the total average power which would be dissipated in one ohm and the right hand side represents a summation over all frequencies extending from 0 to ∞. It is natural therefore to interpret $w(f)df$ as the power due to the components in $f, f + df$.

The preceding sections have dealt with the power spectrum $w(f)$ and correlation function $\psi(\tau)$ of a very general type of function. It will be noted that a knowledge of $w(f)$ does not enable us to determine the original function. In obtaining $w(f)$, as may be seen from the definition (2.1–3) or from (2.3–6), the information carried by the phase angles of the various components of $I(t)$ has been dropped out. In fact, as we may see from the Fourier series representation (2.3–1) of $I(t)$ and from (2.3–6), it is possible to obtain an infinite number of different functions all of which have the same $w(f)$, and hence the same $\psi(\tau)$. All we have to do is to assign different sets of values to the phase angles of the various components, thereby keeping $a_n^2 + b_n^2$ constant.

2.5 Harmonic Analysis for Random Functions

In many applications of the theory discussed in the foregoing sections $I(t)$ is a function of t which has a certain amount of randomness associated with it. For example $I(t)$ may be a curve representing the price of wheat over a long period of years, a component of air velocity behind a grid placed in a wind tunnel, or, of primary interest here, a noise current.

In some mathematical work this randomness is introduced by considering $I(t)$ to involve a number of parameters, and then taking the parameters to be random variables. Thus, in the shot effect the arrival times $t_1, t_2, \cdots t_K$ of the electrons were taken to be the parameters and each was assumed to be uniformly distributed over an interval $(0, T)$.

For any particular set of values of the parameters, $I(t)$ has a definite power spectrum $w(f)$ and correlation function $\psi(\tau)$. However, now the principal interest is not in these particular functions, but in functions which give the average values of $w(f)$ and $\psi(\tau)$ for fixed f and τ. These functions are obtained by averaging $w(f)$ and $\psi(\tau)$ over the ranges of the parameters, using, of course, the distribution functions of the parameters.

By averaging both sides of the appropriate equations in sections 2.1 and

2.2 it is seen that our fundamental inversion formulae (2.1–5) and (2.1–6) are unchanged. Thus,

$$\bar{w}(f) = 4\int_0^\infty \bar{\psi}(\tau)\cos 2\pi f\tau\, d\tau \tag{2.5–1}$$

$$\bar{\psi}(\tau) = \int_0^\infty \bar{w}(f)\cos 2\pi f\tau\, df \tag{2.5–2}$$

where the bars indicate averages taken over the parameters with f or τ held constant.

The definitions of $\bar{w}$ and $\bar{\psi}$ appearing in these equations are likewise obtained from (2.1–3) and (2.1–4)

$$\bar{w}(f) = \operatorname*{Limit}_{T\to\infty} \frac{2\overline{|S(f)|^2}}{T} \tag{2.5–3}$$

and

$$\bar{\psi}(\tau) = \operatorname*{Limit}_{T\to\infty} \frac{1}{T}\int_0^T \overline{I(t)I(t+\tau)}\, dt \tag{2.5–4}$$

The values of t and τ are held fixed while averaging over the parameters. In (2.5–3) $S(f)$ is regarded as a function of the parameters obtained from $I(t)$ by

$$S(f) = \int_0^T I(t)e^{-2\pi i f t}\, dt \tag{2.1–2}$$

Similar expressions may be obtained for the average power spectrum for d.c. and periodic components. All we need to do is to average the expression (2.2–11)

Sometimes the average value of the product $I(t)I(t + \tau)$ in the definition (2.5–4) of $\bar{\psi}(\tau)$ is independent of the time T. This enables us to perform the integration at once and obtain

$$\bar{\psi}(\tau) = \overline{I(t)I(t+\tau)} \tag{2.5–5}$$

This introduces a considerable simplification and it appears that the simplest method of computing $\bar{w}(f)$ for an $I(t)$ of this sort is first to compute $\bar{\psi}(\tau)$, and then use the inversion formula (2.5–1).

2.6 First Example—The Shot Effect

We first compute the average on the right of (2.5–5). By using the nethod of averaging employed many times in part I, we have

$$\overline{I(t)I(t+\tau)} = \sum_{K=0}^{\infty} p(K)\, \overline{I_K(t)I_K(t+\tau)} \tag{2.6–1}$$

where $p(K)$ is the probability of exactly K electrons arriving in the interval $(0, T)$,

$$p(K) = \frac{(\nu T)^K}{K!} e^{-\nu T} \tag{1.1–3}$$

and

$$I_K(t) = \sum_{k=1}^{K} F(t - t_k) \tag{1.3–1}$$

Multiplying $I_K(t)$ and $I_K(t + \tau)$ together and averaging $t_1, t_2, \cdots t_K$ over their ranges gives

$$\overline{I_K(t)I_K(t+\tau)} = \sum_{k=1}^{K} \sum_{m=1}^{K} \int_0^T \frac{dt_1}{T} \cdots \int_0^T \frac{dt_K}{T} F(t - t_k)F(t + \tau - t_m)$$

This is similar to the expression for $\overline{I_K^2(t)}$ which was used in section 1.3 to prove Campbell's theorem and may be treated in much the same way. Thus, if t and $t + \tau$ lie between Δ and $T - \Delta$, the expression above becomes

$$\frac{K}{T} \int_{-\infty}^{+\infty} F(t)F(t+\tau)\, dt + \frac{K(K-1)}{T^2} \left[\int_{-\infty}^{+\infty} F(t)\, dt \right]^2$$

When this is placed in (2.6–1) and the summation performed we obtain an expression independent of T. Consequently we may use (2.5–5) and get

$$\bar{\psi}(\tau) = \nu \int_{-\infty}^{+\infty} F(t)F(t+\tau)\, dt + \overline{I(t)}^2 \tag{2.6–2}$$

where we have used the expression for the average current

$$\overline{I(t)} = \nu \int_{-\infty}^{+\infty} F(t)\, dt \tag{1.3–4}$$

In order to compute $\bar{w}(f)$ from $\bar{\psi}(\tau)$ it is convenient to make use of the fact that $\psi(\tau)$ is always an even function of τ and hence (2.5–1) may also be written as

$$\bar{w}(f) = 2 \int_{-\infty}^{+\infty} \bar{\psi}(\tau) \cos 2\pi f\tau\, d\tau \tag{2.6–3}$$

Then

$$\bar{w}(f) = 2\nu \int_{-\infty}^{+\infty} dt\, F(t) \int_{-\infty}^{+\infty} d\tau\, F(t+\tau) \cos 2\pi f\tau$$
$$+ 2 \int_{-\infty}^{+\infty} \overline{I(t)}^2 \cos 2\pi f\tau\, d\tau$$

$$= 2\nu \text{ Real Part of } \int_{-\infty}^{+\infty} dt\, F(t)e^{-2\pi i f t} \int_{-\infty}^{+\infty} dt'\, F(t')e^{2\pi i f t'}$$

$$+ 2\overline{I(t)}^2 \int_{-\infty}^{+\infty} e^{i2\pi f \tau}\, d\tau$$

$$= 2\nu\,|\,s(f)\,|^2 + 2\overline{I(t)}^2\delta(f) \tag{2.6–4}$$

In going from the first equation to the second we have written $t' = t + \tau$ and have considered $\cos 2\pi f\tau$ to be the real part of the corresponding exponential. In going from the second equation to the third we have set

$$s(f) = \int_{-\infty}^{+\infty} F(t)e^{-2\pi i f t}\, dt \tag{2.6–5}$$

and have used

$$\int_{-\infty}^{+\infty} e^{i2\pi f t}\, dt = \delta(f) \tag{2.2–9}$$

The term in $\bar{w}(f)$ involving $\delta(f)$ represents the average power which would be dissipated by the d.c. component of $I(t)$ in flowing through one ohm. It is in agreement with the concept that the average power in the band $0 \leq f < \epsilon$, $\epsilon > 0$ but very small, is

$$\begin{aligned} \int_0^\epsilon \bar{w}(f)\, df &= 2\overline{I(t)}^2 \int_0^\epsilon \delta(f)\, df \\ &= \overline{I(t)}^2 \end{aligned} \tag{2.6–6}$$

The expression (2.6–4) for $\bar{w}(f)$ may also be obtained from the definition (2.5–3) for $\bar{w}(f)$plus the additional term due to the d.c. component obtained by averaging the expressions (2.2–11). We leave this as an exercise for the reader. He will find it interesting to study the steps in Carson's[15] paper leading up to equation (8). Carson's $R(\omega)$ is related to our $\bar{w}(f)$ by

$$\bar{w}(f) = 2\pi R(\omega)$$

and his $f(i\omega)$ is equal to our $s(f)$.

Integrating both sides of (2.6–4) with respect to f from 0 to ∞ and using

$$\overline{I^2} = \int_0^\infty \bar{w}(f)\, df$$

gives the result

$$\overline{I^2} - \bar{I}^2 = 2\nu \int_0^{+\infty} |\, s(f)\,|^2\, df \tag{2.6–7}$$

[15] Loc. cit.

This may be obtained immediately from Campbell's theorem by applying Parseval's theorem.

As an example of the use of these formulas we derive the power spectrum of the voltage across a resistance R when a current consisting of a great number of very short pulses per second flows through R. Let $F(t - t_k)$ be the voltage produced by the pulse occurring at time t_k. Then

$$F(t) = R\varphi(t)$$

where $\varphi(t)$ is the current in the pulse. We confine our interest to relatively low frequencies such that we may make the approximation

$$s(f) = \int_{-\infty}^{+\infty} R\varphi(t)e^{-2\pi i f t}\,dt$$

$$\approx R\int_{-\infty}^{+\infty} \varphi(t)\,dt = Rq$$

where q is the charge carried through the resistance by one pulse. From (2.6–4) it follows that for these low frequencies the continuous portion of the power spectrum for the voltage is constant and equal to

$$\bar{w}(f) = 2\nu R^2q^2 = 2\bar{I}R^2q \qquad (2.6\text{–}8)$$

where $\bar{I} = \nu q$ is the average current flowing through R. This result is often used in connection with the shot effect in diodes.

In the study of the shot effect it was assumed that the probability of an event (electron arriving at the anode) happening in dt was νdt where ν is the expected number of events per second. This probability is independent of the time t. Sometimes we wish to introduce dependency on time.[18] As an example, consider a long interval extending from 0 to T. Let the probability of an event happening in $t, t + dt$ be $\bar{K}p(t)dt$ where $\bar{K}$ is the average number of events during T and $p(t)$ is a given function of t such that

$$\int_0^T p(t)\,dt = 1$$

For the shot effect $p(t) = 1/T$.

What is the probability that exactly K events happen in T? As in the case of the shot effect, section 1.1, we may divide $(0, T)$ into N intervals each of length Δt so that $N\Delta t = T$. The probability of no event happening in the first Δt is

$$1 - \bar{K}p\left(\frac{\Delta t}{2}\right)\Delta t$$

[18] A careful discussion of this subject is given by Hurwitz and Kac in "Statistical Analysis of Certain Types of Random Functions." I understand that this paper will soon appear in the Annals of Math. Statistics.

The product of N such probabilities is, as $N \to \infty$, $\Delta t \to 0$;

$$\exp\left[-\bar{K}\int_0^T p(t)\,dt\right] = e^{-\bar{K}}$$

This is the probability that exactly 0 events happen in T. In the same way we are led to the expression

$$\frac{\bar{K}^K}{K!}e^{-\bar{K}} \tag{2.6-9}$$

for the probability that exactly K events happen in T.

When we consider many intervals $(0, T)$ we obtain many values of K and also many values of I measured t seconds from the beginning of each interval. These values of I define the distribution of I at time t. By proceeding as in section 1.4 we find that the probability density of I is

$$P(I, t) = \frac{1}{2\pi}\int_{-\infty}^{+\infty} du \exp\left[-iuI + \bar{K}\int_0^T p(x)(e^{iuF(t-x)} - 1)\,dx\right]$$

The corresponding average and variance are

$$\bar{I} = \bar{K}\int_0^T p(x)F(t-x)\,dx$$

$$\overline{(I - \bar{I})^2} = \bar{K}\int_0^T p(x)F^2(t-x)\,dx \tag{2.6-10}$$

If $S(f)$ is given by (2.1–2) and $s(f)$ by (2.6–5) (assuming the duration of $F(t)$ short in comparison with T) the average value of $|S(f)|^2$ may be obtained by putting (1.3-1) in (2.1–2) to get

$$S_K(f) = s(f)\sum_1^K e^{-2\pi i f t_k}$$

Expressing $S_K(f)\,S_K^*(f)$, where the star denotes conjugate complex, as a double sum and averaging over the t_k's, using $p(t)$, and then averaging over the K's gives

$$\overline{|S(f)|^2} = \bar{K}|s(f)|^2\left[1 + \bar{K}\left|\int_0^T p(x)e^{-2\pi i f x}\,dx\right|^2\right] \tag{2.6-11}$$

This may be used to compute the power spectrum from (2.5–3) provided $p(x)$ is not periodic. If $p(x)$ is periodic then the method of section 2.2 should be used at the harmonic frequencies. If the fluctuations of $p(t)$ are slow in comparison with the fluctuations of $F(t)$ the second term within the brackets of (2.6–11) may generally be neglected since there are no values of

f which make both it and $s(f)$ large at the same time. On the other hand. if both $p(t)$ and $F(t)$ fluctuate at about the same rate this term must be considered.

2.7 Second Example—Random Telegraph Signal[16]

Let $I(t)$ be equal to either a or $-a$ so that it is of the form of a flat top wave. Let the intervals between changes of sign, i.e. the lengths of the tops and bottoms, be distributed exponentially. We are led to this distribution by assuming that, if on the average there are μ changes of sign per second, the probability of a change of sign in $t, t + dt$ is μdt and is independent of what happens outside the interval $t, t + dt$. From the same sort of reasoning as employed in section 1.1 for the shot effect we see that the probability of obtaining exactly K changes of sign in the interval $(0, T)$ is

$$p(K) = \frac{(\mu T)^K}{K!} e^{-\mu T} \tag{2.7–1}$$

We consider the average value of the product $I(t)I(t + \tau)$. This product is a^2 if the two I's are of the same sign and is $-a^2$ if they are of opposite sign. In the first case there are an even number, including zero, of changes of sign in the interval $(t, t + \tau)$, and in the second case there are an odd number of changes of sign. Thus

$$\begin{aligned} &\text{Average value of } I(t)I(t+\tau) && (2.7\text{–}2)\\ &= a^2 \times \text{probability of an even number of changes of sign in } t, t+\tau \\ &- a^2 \times \text{probability of an odd number of changes of sign in } t, t+\tau \end{aligned}$$

The length of the interval under consideration is $|t + \tau - t| = |\tau|$ seconds. Since, by assumption, the probability of a change of sign in an elementary interval of length Δt is independent of what happens outside that interval, it follows that the same is true of any interval irrespective of when it starts. Hence the probabilities in (2.7–2) are independent of t and may be obtained from (2.7–1) by setting $T = |\tau|$. Then (2.7–2) becomes, assuming $\tau > 0$ for the moment,

$$\begin{aligned} \overline{I(t)I(t+\tau)} &= a^2[p(0) + p(2) + p(4) + \cdots] \\ &\qquad - a^2[p(1) + p(3) + p(5) + \cdots] \\ &= a^2 e^{-\mu\tau}\left[1 - \frac{\mu\tau}{1!} + \frac{(\mu\tau)^2}{2!} - \cdots\right] \\ &= a^2 e^{-2\mu\tau} \end{aligned} \tag{2.7–3}$$

[16] Kenrick, cited in Section 2.2.

From (2.5–5), this gives the correlation function for $I(t)$

$$\bar{\psi}(\tau) = a^2 e^{-2\mu|\tau|} \tag{2.7-4}$$

The corresponding power spectrum is, from (2.5–1),

$$\begin{aligned} \bar{w}(f) &= 4a^2 \int_0^\infty e^{-2\mu\tau} \cos 2\pi f\tau \, d\tau \\ &= \frac{2a^2 \mu}{\pi^2 f^2 + \mu^2} \end{aligned} \tag{2.7-5}$$

Correlation functions and power spectra of this type occur quite frequently. In particular, they are of use in the study of turbulence in hydrodynamics. We may also obtain them from our shot effect expressions if we disregard the d.c. component. All we have to do is to assume that the effect $F(t)$ of an electron arriving at the anode at time $t = 0$ is zero for $t < 0$, and that $F(t)$ decays exponentially with time after jumping to its maximum value at $t = 0$. This may be verified by substituting the value

$$F(t) = 2a\sqrt{\frac{\mu}{\nu}}\, e^{-2\mu t}, \qquad t > 0 \tag{2.7-6}$$

for $F(t)$ in the expressions (2.6–2) and (2.6–4) (after using 2.6–5) for the correlation function and energy spectrum of the shot effect.

The power spectrum of the current flowing through an inductance and a resistance in series in response to a very wide band thermal noise voltage is also of the form (2.7–5).

Incidentally, this gives us an example of two quite different $I(t)$'s, one a flat top wave and the other a shot effect current, which have the same correlation functions and power spectra, aside from the d.c. component.

There is another type of random telegraph signal which is interesting to analyze. The time scale is divided into intervals of equal length h. In an interval selected at random the value of $I(t)$ is independent of the values in the other intervals, and is equally likely to be $+a$ or $-a$. We could construct such a wave by flipping a penny. If heads turned up we would set $I(t) = a$ in $0 < t < h$. If tails were obtained we would set $I(t) = -a$ in this interval. Flipping again would give either $+a$ or $-a$ for the second interval $h < t < 2h$, and so on. This gives us one wave. A great many waves may be constructed in this way and we denote averages over these waves, with t held constant, by bars.

We ask for the average value of $I(t)I(t + \tau)$, assuming $\tau > 0$. First we note that if $\tau > h$ the currents correspond to different intervals for all values of t. Since the values in these intervals are independent we have

$$\overline{I(t)I(t+\tau)} = \overline{I(t)}\ \overline{I(t+\tau)} = 0$$

for all values of t when $\tau > h$.

To obtain the average when $\tau < h$ we consider t to lie in the first interval $0 < t < h$. Since all the intervals are the same from a statistical point of view we lose no generality in doing this. If $t + \tau < h$, i.e., $t < h - \tau$, both currents lie in the first interval and

$$\overline{I(t)I(t+\tau)} = a^2$$

If $t > h - \tau$ the current $I(t + \tau)$ corresponds to the second interval and hence the average value is zero.

We now return to (2.5–4). The integral there extends from 0 to T. When $\tau > h$, the integrand is zero and hence

$$\bar{\psi}(\tau) = 0, \qquad \tau > h \tag{2.7–7}$$

When $\tau < h$, our investigation of the interval $0 < t < h$ enables us to write down the portion of the integral extending from 0 to h:

$$\begin{aligned} \int_0^h I(t)I(t+\tau)\,dt &= \int_0^{h-\tau} a^2\,dt + \int_{h-\tau}^h 0\,dt \\ &= a^2(h-\tau) \end{aligned}$$

Over the interval of integration $(0, T)$ we have T/h such intervals each contributing the same amount. Hence, from (2.5–4),

$$\begin{aligned} \bar{\psi}(\tau) &= \operatorname*{Limit}_{T\to\infty} \frac{a^2}{T}\cdot\frac{T}{h}\,(h-\tau) \\ &= a^2\left(1 - \frac{\tau}{h}\right), \qquad 0 \leq \tau < h \end{aligned} \tag{2.7–8}$$

The power spectrum of this type of telegraph wave is thus

$$\begin{aligned} \bar{w}(f) &= 4a^2 \int_0^h \left(1 - \frac{\tau}{h}\right) \cos 2\pi f\tau\,d\tau \\ &= 2h\left(\frac{a \sin \pi f h}{\pi f h}\right)^2 \end{aligned} \tag{2.7–9}$$

This is seen to have the same general behavior as $\bar{w}(f)$ for the first type of telegraph signal given by (2.7–5), when we relate the average number, μ, of changes of sign per second to the interval length h by $\mu h = 1$.

2.8 Representation of Noise Current

In section 1.7 the Fourier series representation of the shot effect current was discussed. This suggests the representation*

$$I(t) = \sum_{n=1}^{N} (a_n \cos \omega_n t + b_n \sin \omega_n t) \tag{2.8–1}$$

where

$$\omega_n = 2\pi f_n , \qquad f_n = n\Delta f \tag{2.8–2}$$

a_n and b_n are taken to be independent random variables which are distributed normally about zero with the standard deviation $\sqrt{w(f_n)\Delta f}$. $w(f)$ is the power spectrum of the noise current, i.e., $w(f)\, df$ is the average power which would be dissipated by those components of $I(t)$ which lie in the frequency range $f, f + df$ if they were to flow through a resistance of one ohm.

The expression for the standard deviation of a_n and b_n is obtained when we notice that Δf is the width of the frequency band associated with the nth component. Hence $w(f_n)\Delta f$ is the average power which would be dissipated if the current

$$a_n \cos \omega_n t + b_n \sin \omega_n t$$

were to flow through a resistance of one ohm, this average being taken over all possible values of a_n and b_n. Thus

$$w(f_n)\Delta f = \overline{a_n^2} \cos^2 \omega_n t + \overline{2a_n b_n} \cos \omega_n t \sin \omega_n t + \overline{b_n^2} \sin^2 \omega_n t = \overline{a_n^2} = \overline{b_n^2} \tag{2.8–3}$$

The last two steps follow from the independence of a_n and b_n and the identity of their distributions. It will be observed that $w(f)$, as used with the representation (2.8–1), is the same sort of average as was denoted in section 2.5 by $\bar{w}(f)$. However, $w(f)$ is often given to us in order to specify the spectrum of a given noise.

For example, suppose we are interested in the output of a certain filter when a source of thermal noise is applied to the input. Let $A(f)$ be the absolute value of the ratio of the output current to the input current when a steady sinusoidal voltage of frequency f is applied to the input. Then

$$w(f) = cA^2(f) \tag{2.8–4}$$

* As mentioned in section 1.7 this sort of representation was used by Einstein and Hopf for radiation. Shottky (1918) used (2.8–1), apparently without explicitly taking the coefficients to be normally distributed. Nyquist (1932) derived the normal distribution from the shot effect.

If W is the average power dissipated in one ohm by $I(t)$,

$$
\begin{aligned}
W &= \operatorname*{Limit}_{T\to\infty} \frac{1}{T} \int_0^T I^2(t)\, dt = \int_0^\infty w(f)\, df \\
&= c \int_0^\infty A^2(f)\, df
\end{aligned}
\tag{2.8–5}
$$

which is an equation to determine c when W and $A(f)$ are known.

In using the representation (2.8–1) to investigate the statistical properties of $I(t)$ we first find the corresponding statistical properties of the summation on the right when the a's and b's are regarded as random variables distributed as mentioned above and t is regarded as fixed. In general, the time t disappears in this procedure just as it did in (2.8–3). We then let $N \to \infty$ and $\Delta f \to 0$ so that the summations may be replaced by integrations. Finally, the frequency range is extended to cover all frequencies from 0 to ∞.

The usual way of looking at the representation (2.8–1) is to suppose that we have an oscillogram of $I(t)$ extending from $t = 0$ to $t = \infty$. This oscillogram may be cut up into strips of length T. A Fourier analysis of $I(t)$ for each strip will give a set of coefficients. These coefficients will vary from strip to strip. Our representation ($T\Delta f = 1$) assumes that this variation is governed by a normal distribution. Our process for finding statistical properties by regarding the a's and b's as random variables while t is kept fixed corresponds to examining the noise current at a great many instants. Corresponding to each strip there is an instant, and this instant occurs at t (this is the t in (2.8–1)) seconds from the beginning of the strip. This is somewhat like examining the noise current at a great number of instants selected at random.

Although (2.8–1) is the representation which is suggested by the shot effect and similar phenomena, it is not the only representation, nor is it always the most convenient. Another representation which leads to the same results when the limits are taken is[19]

$$
I(t) = \sum_{n=1}^{N} c_n \cos(\omega_n t - \varphi_n) \tag{2.8–6}
$$

where $\varphi_1, \varphi_2, \cdots \varphi_N$ are angles distributed at random over the range $(0, 2\pi)$ and

$$
c_n = [2w(f_n)\Delta f]^{1/2}, \qquad \omega_n = 2\pi f_n, \qquad f_n = n\Delta f \tag{2.8-7}
$$

[19] This representation has often been used by W. R. Bennett in unpublished memoranda written in the 1930's.

In this representation $I(t)$ is regarded as the sum of a number of sinusoidal components with fixed amplitudes but random phase angles.

That the two different representations (2.8–1) and (2.8–6) of $I(t)$ lead to the same statistical properties is a consequence of the fact that they are always used in such a way that the "central limit theorem*" may be used in both cases.

This theorem states that under certain general conditions, the distribution of the sum of N random vectors approaches a normal law (it may be normal in several dimensions**) as $N \to \infty$. In fact from this theorem it appears that a representation such as

$$I(t) = \sum_{n=1}^{N} (a_n \cos \omega_n t + b_n \sin \omega_n t) \tag{2.8–8}$$

where a_n and b_n are independent random variables which take only the values $\pm [w(f_n)\Delta f]^{1/2}$, the probability of each value being $\frac{1}{2}$, will lead in the limit to the same statistical properties of $I(t)$ as do (2.8–1) and (2.8–6).

2.9 The Normal Distribution in Several Variables[20]

Consider a random vector r in K dimensions. The distribution of this vector may be specified by stating the distribution of the K components, $x_1, x_2, \cdots x_K$, of r. r is said to be normally distributed when the probability density function of the x's is of the form

$$(2\pi)^{-K/2} \mid M \mid^{-1/2} \exp [-\tfrac{1}{2}x'M^{-1}x] \tag{2.9–1}$$

where the exponent is a quadratic form in the x's. The square matrix M is composed of the second moments of the x's.

$$M = \begin{bmatrix} \mu_{11} & \mu_{12} & \cdots & \mu_{1K} \\ \cdot & & & \cdot \\ \mu_{1K} & & \cdots & \mu_{KK} \end{bmatrix} \tag{2.9–2}$$

where the second moments are defined by

$$\mu_{11} = \overline{x_1^2}, \qquad \mu_{12} = \overline{x_1 x_2}, \qquad \text{etc.} \tag{2.9–3}$$

$\mid M \mid$ represents the determinant of M and x' is the row matrix

$$x' = [x_1, x_2, \cdots x_K] \tag{2.9–4}$$

x is the column matrix obtained by transposing x'.

* See section 2.10.
** See section 2.9.
[20] H. Cramér, "Random Variables and Probability Distributions," Chap. X., Cambridge Tract No. 36 (1937).

The exponent in the expression (2.9–1) for the probability density may be written out by using

$$x'M^{-1}x = \sum_{r=1}^{K}\sum_{s=1}^{K} \frac{M_{rs}}{|M|} x_r x_s \tag{2.9–5}$$

where M_{rs} is the cofactor of μ_{rs} in $|M|$.

Sometimes there are linear relations between the x's so that the random vector r is restricted to a space of less than K dimensions. In this case the appropriate form for the density function may be obtained by considering a sequence of K-dimensional distributions which approach the one being investigated.

If r_1 and r_2 are two normally distributed random vectors their sum $r_1 + r_2$ is also normally distributed. It follows that the sum of any number of normally distributed random vectors is normally distributed.

The characteristic function of the normal distribution is

$$\text{ave. } e^{iz_1x_1+iz_2x_2+\cdots+iz_Kx_K} = \exp\left[-\frac{1}{2}\sum_{r=1}^{K}\sum_{s=1}^{K} \mu_{rs} z_r z_s\right] \tag{2.9–6}$$

2.10 Central Limit Theorem

The central limit theorem in probability states that the distribution of the sum of N independent random vectors $r_1 + r_2 + \cdots + r_N$ approaches a normal law as $N \rightarrow \infty$ when the distributions of $r_1, r_2, \cdots r_N$ satisfy certain general conditions.[7]

As an example we take the case in which $r_1, r_2, \cdots$ are two-dimensional vectors[21], the components of r_n being x_n and y_n. Without loss of generality we assume that

$$\bar{x}_n = 0, \qquad \bar{y}_n = 0.$$

The components of the resultant vector are

$$\begin{aligned} X &= x_1 + x_2 + \cdots + x_N \\ Y &= y_1 + y_2 + \cdots + y_N \end{aligned} \tag{2.10–1}$$

and, since $r_1, r_2, \cdots$ are independent vectors, the second moments of the resultant are

$$\begin{aligned} \mu_{11} &= \overline{X^2} = \overline{x_1^2} + \overline{x_2^2} + \cdots + \overline{x_N^2} \\ \mu_{22} &= \overline{Y^2} = \overline{y_1^2} + \overline{y_2^2} + \cdots + \overline{y_N^2} \\ \mu_{12} &= \overline{XY} = \overline{x_1y_1} + \overline{x_2y_2} + \cdots + \overline{x_Ny_N} \end{aligned} \tag{2.10–2}$$

[7] Incidentally, von Laue (see references in section 1.7) used this theorem in discussing the normal distribution of the coefficients in a Fourier series used to represent black-body radiation. He ascribed it to Markoff.

[21] This case is discussed by J. V. Uspensky, "Introduction to Mathematical Probability", McGraw-Hill (1937) Chap. XV.

Apparently there are several types of conditions which are sufficient to ensure that the distribution of the resultant approaches a normal law. One sufficient condition is that[21]

$$\mu_{11}^{-3/2} \sum_{n=1}^{N} \overline{|x_n|^3} \to 0$$
$$\mu_{22}^{-3/2} \sum_{n=1}^{N} \overline{|y_n|^3} \to 0 \qquad (2.10\text{-}3)$$

The central limit theorem tells us that the distribution of the random vector (X, Y) approaches a normal law as $N \to \infty$. The second moments of this distribution are given by (2.10–2). When we know the second moments of a normal distribution we may write down the probability density function at once. Thus from section 2.9

$$M = \begin{bmatrix} \mu_{11} & \mu_{12} \\ \mu_{12} & \mu_{22} \end{bmatrix}, \qquad M^{-1} = |M|^{-1} \begin{bmatrix} \mu_{22} & -\mu_{12} \\ -\mu_{12} & \mu_{11} \end{bmatrix}$$

$$|M| = \mu_{11}\mu_{22} - \mu_{12}^2$$

$$x' = [X, Y]$$

$$x'M^{-1}x = |M|^{-1}(\mu_{22}X^2 - 2\mu_{12}XY + \mu_{11}Y^2)$$

The probability density is therefore

$$\frac{(\mu_{11}\mu_{22} - \mu_{12}^2)^{-1/2}}{2\pi} \exp\left[\frac{-\mu_{22}X^2 - \mu_{11}Y^2 + 2\mu_{12}XY}{2(\mu_{11}\mu_{22} - \mu_{12}^2)}\right] \qquad (2.10\text{-}3)$$

Incidentally, the second moments are related to the standard deviations σ_1, σ_2 of X, Y and to the correlation coefficient τ of X and Y by

$$\mu_{11} = \sigma_1^2, \qquad \mu_{22} = \sigma_2^2, \qquad \mu_{12} = \tau\sigma_1\sigma_2 \qquad (2.10\text{-}4)$$

and the probability density takes the standard form

$$\frac{(1 - \tau^2)^{-1/2}}{2\pi\sigma_1\sigma_2} \exp\left[-\frac{1}{2(1 - \tau^2)}\left(\frac{X^2}{\sigma_1^2} - 2\tau\frac{XY}{\sigma_1\sigma_2} + \frac{Y^2}{\sigma_2^2}\right)\right] \qquad (2.10\text{-}5)$$

[21] This is used by Uspensky, loc. cit. Another condition analogous to the Lindeberg condition is given by Cramer,[20] loc. cit.

PART III

STATISTICAL PROPERTIES OF RANDOM NOISE CURRENTS

3.0 Introduction

In this section we use the representations of the noise currents given in section 2.8 to derive some statistical properties of $I(t)$. The first six sections are concerned with the probability distribution of $I(t)$ and of its zeros and maxima. Sections 3.7 and 3.8 are concerned with the statistical properties of the envelope of $I(t)$. Fluctuations of integrals involving $I^2(t)$ are discussed in section 3.9. The probability distribution of a sine wave plus a noise current is given in 3.10 and in 3.11 an alternative method of deriving the results of Part III is mentioned. Prof. Uhlenbeck has pointed out that much of the material in this Part is closely connected with the theory of Markoff processes. Also S. Chandrasekhar has written a review of a class of physical problems which is related, in a general way, to the present subject.[22]

3.1 The Distribution of the Noise Current[23]

In section 1.4 it has been shown that the distribution of a shot effect current approaches a normal law as the expected number of events per second, ν, increases without limit.

In line with the spirit of this Part, Part III, we shall use the representation

$$I(t) = \sum_{n=1}^{N} (a_n \cos \omega_n t + b_n \sin \omega_n t) \qquad (2.8\text{-}1)$$

to show that $I(t)$ is distributed according to a normal law. This is obtained at once when the procedure outlined in section 2.8 is followed. Since a_n and b_n are distributed normally, so are $a_n \cos \omega_n t$ and $b_n \sin \omega_n t$ when t is regarded as fixed. $I(t)$ is thus the sum of $2N$ independent normal variates and consequently is itself distributed normally.

[22] Stochastic Problems in Physics and Astronomy, *Rev. of Mod. Phys.*, Vol. 15, pp. 1–89 (1943).

[23] An interesting discussion of this subject by V. D. Landon and K. A. Norton is given in the *I.R.E. Proc.*, 30 (Sept. 1942) pp. 425–429.

The average value of $I(t)$ as given by (2.8–1) is zero since $\bar{a}_n = \bar{b}_n = 0$:

$$\bar{I}(t) = 0 \tag{3.1–1}$$

The mean square value of $I(t)$ is

$$\begin{aligned} \overline{I^2}(t) &= \sum_{n=1}^{N} (\overline{a_n^2} \cos^2 \omega_n t + \overline{b_n^2} \sin^2 \omega_n t) \\ &= \sum_{n=1}^{N} w(f_n)\Delta f \\ &\rightarrow \int_0^\infty w(f)\, df = \psi(0) \equiv \psi_0 \end{aligned} \tag{3.1–2}$$

In writing down (3.1–2) we have made use of the fact that all the a's and b's are independent and consequently the average of any cross product is zero. We have also made use of

$$\overline{a_n^2} = \overline{b_n^2} = w(f_n)\Delta f, \qquad f_n = n\Delta f, \qquad \omega_n = 2\pi f_n$$

which were given in 2.8. $\psi(\tau)$ is the correlation function of $I(t)$ and is related to $w(f)$ by

$$\psi_\tau \equiv \psi(\tau) = \int_0^\infty w(f) \cos 2\pi f\tau \, df \tag{2.1–6}$$

as is explained in section 2.1. In this part we shall write the argument of $\psi(\tau)$ as a subscript in order to save space.

Since we know that $I(t)$ is normal and since we also know that its average is zero and its mean square value is ψ_0, we may write down its probability density function at once. Thus, the probability of $I(t)$ being in the range $I, I + dI$ is

$$\frac{dI}{\sqrt{2\pi\psi_0}} e^{-I^2/2\psi_0} \tag{3.1–3}$$

This is the probability of finding the current between I and $I + dI$ at a time selected at random. Another way of saying the same thing is to state that (3.1–3) is the fraction of time the current spends in the range $I, I + dI$.

In many cases it is more convenient to use the representation (2.8–6)

$$I(t) = \sum_{n=1}^{N} c_n \cos (\omega_n t - \varphi_n), \qquad c_n^2 = 2w(f_n)\Delta f \tag{2.8–6}$$

in which $\varphi_1, \cdots \varphi_n$ are independent random phase angles. In order to deduce the normal distribution from this representation we first observe that (2.8–6) expresses $I(t)$ as the sum of a large number of independent random variables

$$I(t) = x_1 + x_2 + \cdots + x_N$$

$$x_n = c_n \cos (\omega_n t - \varphi_n)$$

and hence that as $N \to \infty$ $I(t)$ becomes distributed according to a normal law. In order to make the limiting process definite we first choose N and Δf such that $N\Delta f = F$ where

$$\int_F^\infty w(f)\, df < \epsilon \int_0^\infty w(f)\, df$$

where ϵ is some arbitrarily chosen small positive quantity. We now let $N \to \infty$ and $\Delta f \to 0$ in such a way that $N\Delta f$ remains equal to F. Then

$$\begin{aligned} A = \overline{x_1^2} + \overline{x_2^2} + \cdots + \overline{x_N^2} &= \sum_1^N 2w(f_n)\Delta f\, \overline{\cos^2 (\omega_n t - \varphi_n)} \\ &= \sum_1^N w(f_n)\Delta f \to \int_0^F w(f)\, df \end{aligned} \tag{3.1-4}$$

$$\begin{aligned} B = \overline{|x_1|^3} + \cdots + \overline{|x_N|^3} &= \sum_1^N (2w(f_n)\Delta f)^{3/2}\, \overline{|\cos (\omega_n t - \varphi_n)|^3} \\ &< 4(\Delta f)^{1/2} \int_0^F [w(f)]^{3/2}\, df \end{aligned}$$

where the bars denote averages with respect to the φ's, t being held constant. If we assume that the integrals are proper, the ratio $BA^{-3/2} \to 0$ as $N \to \infty$, and consequently the central limit theorem* may be used if $w(f) = 0$ for $f > F$. Since we may make F as large as we please by choosing ϵ small enough, we may cover as large a frequency range as we wish. For this reason we write ∞ in place of F.

Now that the central limit theorem has told us that the distribution of $I(t)$, as given by (2.8–6), approaches a normal law, there remains only the problem of finding the average and the standard deviation:

$$\overline{I(t)} = \sum_1^N c_n \overline{\cos (\omega_n t - \varphi_n)} = 0$$

$$\begin{aligned} \overline{I^2(t)} &= \sum_1^N c_n^2\, \overline{\cos^2 (\omega_n t - \varphi_n)} \\ &\to \int_0^\infty w(f)\, df = \psi_0 \end{aligned} \tag{3.1-5}$$

* Section 2.10.

This gives the probability density (3.1–3). Hence the two representations lead to the same result in this case. Evidently, they will continue to lead to identical results as long as the central limit theorem may be used. In the future use of the representation (2.8–6) we shall merely assume that the central limit theorem may be applied to show that a normal distribution is approached. We shall omit the work corresponding to equations (3.1–4).

The characteristic function for the distribution of $I(t)$ is

$$\text{ave. } e^{iuI(t)} = \exp - \frac{\psi_0}{2} u^2 \qquad (3.1\text{–}6)$$

3.2 The Distribution of $I\ (t)$ and $I\ (t + \tau)$

We require the two dimensional distribution in which the first variable is the noise current $I(t)$ and the second variable is its value $I(t + \tau)$ at some later time τ. It turns out that this distribution is normal[24], as we might expect from the analogy with section 3.1. The second moments of this distribution are

$$\begin{aligned} \mu_{11} &= \overline{I^2(t)} = \psi_0 = \int_0^\infty w(f)\, df \\ \mu_{22} &= \psi_0 \\ \mu_{12} &= \overline{I(t)I(t + \tau)} \\ &= \psi_\tau \end{aligned} \qquad (3.2\text{–}1)$$

The expression for μ_{12} is in line with our definition (2.1–4) for the correlation function:

$$\psi_\tau \equiv \psi(\tau) = \underset{T\to\infty}{\text{Limit}} \frac{1}{T} \int_0^T I(t)I(t + \tau)\, dt \qquad (2.1\text{–}4)$$

In order to get the distribution from the representation (2.8–6) we write

$$I_1 = I(t) = \sum_1^N c_n \cos (\omega_n t - \varphi_n)$$

$$I_2 = I(t + \tau) = \sum_1^N c_n \cos (\omega_n t - \varphi_n + \omega_n \tau)$$

[24] It seems that the first person to obtain this distribution in connection with noise was H. Thiede, *Elec. Nachr. Tek.* 13 (1936), 84–95.

From the central limit theorem for two dimensions it follows that I_1 and I_2 are distributed normally. As in (3.1)

$$\mu_{11} = \overline{I_1^2} = \sum_1^N c_n^2 \cdot \tfrac{1}{2} \to \int_0^\infty w(f)\, df = \psi_0$$

$$\mu_{22} = \overline{I_2^2} = \overline{I_1^2} = \psi_0 \qquad (3.2\text{–}2)$$

$$\mu_{12} = \overline{I_1 I_2} = \sum_1^N c_n^2 \text{ ave. } \{\cos(\omega_n t - \varphi_n) \cos(\omega_n t - \varphi_n + \omega_n \tau)\}$$

Now the quantity within the parenthesis is

$$\cos^2(\omega_n t - \varphi_n) \cos \omega_n \tau - \cos(\omega_n t - \varphi_n) \sin(\omega_n t - \varphi_n) \sin \omega_n \tau$$

and when we take the average with respect to φ_n the second term drops out, giving

$$\mu_{12} = \sum_1^N c_n^2 \cdot \tfrac{1}{2} \cos \omega_n \tau \to \int_0^\infty w(f) \cos 2\pi f \tau\, df = \psi_\tau \qquad (3.2\text{–}3)$$

where we have used $\omega_n = 2\pi f_n$ and the relation (2.1–6) between $w(f)$ and $\psi(\tau)$.

The probability density function for I_1 and I_2 may be stated. From the discussion of the normal law in 2.9 it is

$$\frac{[\psi_0^2 - \psi_\tau^2]^{-1/2}}{2\pi} \exp\left[\frac{-\psi_0 I_1^2 - \psi_0 I_2^2 + 2\psi_\tau I_1 I_2}{2(\psi_0^2 - \psi_\tau^2)}\right] \qquad (3.2\text{–}4)$$

For a band pass filter whose range extends from f_a to f_b we have

$$\begin{aligned} \psi_\tau &= \int_{f_a}^{f_b} w_0 \cos 2\pi f \tau\, df \\ &= w_0 \frac{\sin \omega_b \tau - \sin \omega_a \tau}{2\pi\tau} \qquad (3.2\text{–}5) \\ &= \frac{w_0}{\pi\tau} \sin \pi\tau(f_b - f_a) \cos \pi\tau(f_b + f_a) \\ \psi_0 &= w_0(f_b - f_a) \end{aligned}$$

where w_0 is the constant value of $w(f)$ in the pass band and

$$\begin{aligned} \omega_b &= 2\pi f_b \qquad (3.2\text{–}6) \\ \omega_a &= 2\pi f_a \end{aligned}$$

According to our formula (3.2–4), I_1 and I_2 are independent when ψ_τ is zero. For the τ's which make ψ_τ zero, a knowledge of I_1 does not add to our knowledge of I_2. For example, suppose we have a narrow filter. Then

$$\psi_\tau = 0 \text{ when } \tau = [2(f_b + f_a)]^{-1}$$

$$\psi_\tau \text{ is nearly } -\psi_0 \text{ when } \tau = [f_b + f_a]^{-1}$$

For the first value of τ, all we know is that I_2 is distributed about zero with $\overline{I_2^2} = \psi_0$. For the second value of τ I_2 is likely to be near $-I_1$. This is in line with the idea that the noise current through a narrow filter behaves like a sine wave of frequency $\frac{1}{2}(f_b + f_a)$ (and, incidentally, whose amplitude fluctuates with an irregular frequency of the order of $\frac{1}{2}(f_b - f_a)$). The first value of τ corresponds to a quarter-period of such a wave and the second value to a half-period. By drawing a sine wave and looking at points separated by quarter and half periods, the reader will see how the ideas agree.

The characteristic function for the distribution of I_1 and I_2 is

$$\text{ave. } e^{iuI_1 + ivI_2} = \exp\left[-\frac{\psi_0}{2}(u^2 + v^2) - \psi_\tau uv\right] \tag{3.2-7}$$

The three dimensional distribution in which

$$I_1 = I(t)$$
$$I_2 = I(t + \tau_1)$$
$$I_3 = I(t + \tau_1 + \tau_2)$$

where τ_1 and τ_2 are given and t is chosen at random is, as we might expect, normal in three dimensions. The moments, from which the distribution may be obtained by the method of Section 2.9, are

$$\mu_{11} = \mu_{22} = \mu_{33} = \psi_0$$
$$\mu_{12} = \psi_{\tau_1}$$
$$\mu_{23} = \psi_{\tau_2}$$
$$\mu_{13} = \psi(\tau_1 + \tau_2) = \psi_{\tau_1+\tau_2}$$

The characteristic function for I_1, I_2, I_3 is

$$\text{ave. } e^{iz_1I_1 + iz_2I_2 + iz_3I_3}$$
$$= \exp\left[-\frac{\psi_0}{2}(z_1^2 + z_2^2 + z_3^2) - \mu_{12}z_1z_2 - \mu_{23}z_2z_3 - \mu_{13}z_1z_3\right] \tag{3.2-8}$$

3.3 Expected Number of Zeros per Second

We shall use the following result. Let y be given by

$$y = F(a_1, a_2, \cdots a_N; x), \tag{3.3-1}$$

and let the a's be random variables. For a given set of a's, this equation gives a curve of y versus x. Since the a's are random variables we shall call this curve a random curve. Let us select a short interval x_1, $x_1 + dx$, and then draw a batch of a's. The probability that the curve obtained by putting these a's in (3.3–1) will have a zero in x_1, $x_1 + dx$ is

$$dx \int_{-\infty}^{+\infty} |\eta| \, p(0, \eta; x_1) \, d\eta \tag{3.3–2}$$

and the expected number of zeros in the interval (x_1, x_2) is

$$\int_{x_1}^{x_2} dx \int_{-\infty}^{+\infty} |\eta| \, p(0, \eta; x) \, d\eta \tag{3.3–3}$$

In these expressions $p(\xi, \eta; x)$ is the probability density function for the variables

$$\begin{aligned} \xi &= F(a_1, \cdots a_N; x) \\ \eta &= \frac{\partial F}{\partial x} \end{aligned} \tag{3.3–4}$$

Since the a's are random variables so are ξ and η, and their distribution will contain x as a parameter. This is indicated by the notation $p(\xi, \eta; x)$.

These results may be proved in much the same manner as are similar results for the distribution of the maxima of a random curve. This method of proof suffers from the restriction that the a's are required to be bounded.[25] Results equivalent to (3.3–2) and (3.3–3) have been obtained independently by M. Kac.[26] His method of proof has the advantage of not requiring the a's to be bounded.

Here we shall sketch the derivation of a closely related result: The probability that y will pass through zero in x_1, $x_1 + dx$ with positive slope is

$$dx \int_0^{\infty} \eta p(0, \eta; x_1) \, d\eta \tag{3.3–5}$$

We choose dx so small that the portions of all but a negligible fraction of the possible random curves lying in the strip $(x_1, x_1 + dx)$ may be regarded as straight lines. If $y = \xi$ at x_1 and passes through zero for $x_1 < x < x_1 + dx$, its intercept on $y = 0$ is $x_1 - \frac{\xi}{\eta}$ where η is the slope. Thus ξ and η must be of opposite sign and

$$x_1 < x_1 - \frac{\xi}{\eta} < x_1 + dx$$

[25] S. O. Rice, *Amer. Jour. Math.* Vol. 61, pp. 409–416 (1939). However, L. A. MacColl has pointed out to me that a set of sufficient conditions for (3.3–5) to hold is: (a) $p(\xi, \eta; x)$ is continuous with respect to (ξ, η) throughout the $\xi\eta$-plane; and (b) that the integral

$$\int_0^{\infty} p(a\eta, \eta; x_1) \, d\eta$$

converges uniformly with respect to a in some interval $-a_1 \leq a \leq a_2$, where a_1 and a_2 are positive. These conditions are satisfied in all the applications we shall make use of (3.3–5).

[26] M. Kac, *Bull. Amer. Math. Soc.* Vol. 49, pp. 314–320 (1943).

According to the statement of our problem, we are interested only in positive values of η, and we therefore write our inequality as

$$-\eta\, dx < \xi < 0$$

For a given random curve i.e. for a given set of a's ξ and η have the values given by

$$\xi = F(a_1, \cdots a_N ; x_1)$$

$$\eta = \left[\frac{\partial F}{\partial x}\right]_{x=x_1}$$

If these values of ξ and η satisfy our inequality, the curve goes through zero in x_1, $x_1 + dx$. The probability of this happening is[27]

$$\int_0^\infty d\eta \int_{-\eta\, dx}^0 d\xi\, p(\xi, \eta; x_1) = \int_0^\infty [0 - (-\eta\, dx)]p(0, \eta; x_1)\, d\eta$$

where we have made use of the fact that dx is so very small that ξ is effectively zero. The last expression is the same as (3.3–5).

In the same way it may be shown that the probability of y passing through zero in x_1, $x_1 + dx$ with a negative slope is

$$-dx \int_{-\infty}^0 \eta p(0, \eta; x_1)\, d\eta \tag{3.3–6}$$

Expression (3.3–2) is obtained by adding (3.3–5) and (3.3–6).

We are now ready to apply our formulas. We let t, $I(t)$ and φ_n play the roles of x, y, and a_n, respectively, and use

$$I(t) = \sum_{n=1}^N c_n \cos(\omega_n t - \varphi_n), \qquad c_n^2 = 2w(f)\Delta f \tag{2.8–6}$$

[27] MacColl has remarked that the step from the double integral on the left hand side of this equation to the final result (3.3–5) may be made as follows:
It is easily seen that the probability density we are seeking is

$$\left[\frac{d}{d(\Delta x)} \int_0^\infty d\eta \int_{-\eta\Delta x}^0 p(\xi, \eta; x)\, d\xi\right]_{\Delta x=0}$$

Proceeding formally, without regard to conditions validating the analytical operations (for such conditions see the footnote on page 52), we have

$$\frac{d}{d\Delta x} \int_0^\infty d\eta \int_{-\eta\Delta x}^0 p(\xi, \eta; x)\, d\xi = \int_0^\infty \eta p(-\eta\Delta x, \eta; x)\, d\eta$$

and hence the required probability density is

$$\int_0^\infty \eta p(0, \eta; x)\, d\eta$$

The first step is to find the probability density function of the two random variables

$$\xi = \sum_{n=1}^{N} c_n \cos(\omega_n t_1 - \varphi_n)$$
$$\eta = I'(t_1) = -\sum_{n=1}^{N} c_n \omega_n \sin(\omega_n t_1 - \varphi_n) \tag{3.3-7}$$

where the prime denotes differentiation with respect t. From section 2.10

$$\mu_{11} = \overline{\xi^2} = \psi_0$$

$$\mu_{22} = \overline{\eta^2} = \sum_{n=1}^{N} \overline{c_n^2 \omega_n^2 \sin^2(\omega_n t_1 - \varphi_n)}$$
$$= \sum_{n=1}^{N} (2\pi f_n)^2 w(f_n)\Delta f$$
$$\to 4\pi^2 \int_0^\infty f^2 w(f)\, df = -\psi_0''$$

$$\mu_{12} = \overline{\xi\eta} = -\sum_{n=1}^{N} \overline{c_n^2 \omega_n \cos(\omega_n t_1 - \varphi_n)\sin(\omega_n t_1 - \varphi_n)}$$
$$= 0$$

The expression for μ_{22} arises from (2.1–6) by differentiation. In this expression ψ_0'' denotes the second derivative of $\psi(\tau)$ with respect to τ at $\tau = 0$:

$$\psi''(\tau) = -4\pi^2 \int_0^\infty f^2 w(f) \cos 2\pi f\tau\, df \tag{3.3-8}$$

Hence the probability density is

$$p(\xi, \eta; t) = \frac{[-\psi_0\psi_0'']^{-1/2}}{2\pi} \exp\left[-\frac{\xi^2}{2\psi_0} + \frac{\eta^2}{2\psi_0''}\right] \tag{3.3-9}$$

where ψ_0'' is negative. It will be observed that the expression on the right is independent of t. Hence the probability of having a zero in t_1, $t_1 + dt$,

$$dt \int_{-\infty}^{+\infty} |\eta| \frac{[-\psi_0\psi_0'']^{-1/2}}{2\pi} e^{\eta^2/2\psi_0''}\, d\eta = \frac{dt}{\pi}\left[-\frac{\psi''(0)}{\psi(0)}\right]^{1/2} \tag{3.3-10}$$

which follows from (3.3–3), is independent of t.

The expected number of zeros per second, which may be obtained from (3.3–3) by integrating (3.3–10) over an interval of one second, is

$$\frac{1}{\pi}\left[-\frac{\psi''(0)}{\psi(0)}\right]^{1/2} = 2\left[\frac{\int_0^\infty f^2 w(f)\, df}{\int_0^\infty w(f)\, df}\right]^{1/2} \tag{3.3-11}$$

For an ideal band pass filter whose pass band extends from f_a to f_b the expected number of zeros per second is

$$2\left[\frac{1}{3}\frac{f_b^3 - f_a^3}{f_b - f_a}\right]^{1/2} \tag{3.3–12}$$

When f_a is zero this becomes 1.155 f_b and when f_a is very nearly equal to f_b it approaches $f_b + f_a$.

In a recent paper M. Kac[28] has given a result which, after a slight generalization, leads to

$$e^{-I^2/2\psi_0}\frac{1}{2\pi}\left[-\frac{\psi_0''}{\psi_0}\right]^{1/2} dt \tag{3.3–13}$$

for the probability that the noise current will pass through the value I with positive slope during the interval $t, t + dt$. The expected number of such passages per second is

$$e^{-I^2/2\psi_0} \times [\tfrac{1}{2} \text{ the expected number of zeros per second}] \tag{3.3–14}$$

The expression (3.3–13) may also be derived from analogue of (3.3–5) obtained by replacing the zero in $p(0, \eta; x_1)$ by y.

In some cases the integral

$$\psi_0'' = -4\pi^2 \int_0^\infty f^2 w(f)\, df$$

does not converge.

An example occurs when we apply a broad band noise voltage to a resistance and condenser in series. The power spectrum of the voltage across the condenser is of the form

$$w(f) = \frac{1}{f^2 + a^2} \tag{3.3–15}$$

Although ψ_0'' is infinite, ψ_0 is finite and equal to $\pi/2a$. A straightforward substitution in our formula (3.3–11) gives infinity as the expected number of zeros per second.

Some light is thrown on this breakdown of our formula when we consider a noise current consisting of two bands of noise. One band is confined to relatively low frequencies, and its power spectrum will be denoted by $w_1(f)$. The other band is very narrow and is centered at the relatively high frequency f_2. The complete power spectrum of our noise is then

$$w(f) = w_1(f) + A^2\delta(f - f_2)$$

[28] On the Distribution of Values of Trigonometric Sums with Linearly Independent Frequencies, *Amer. Jour. Math.*, Vol. LXV, pp 609–615, (1943).

where the unit impulse function δ is used to represent the very narrow band. The power spectrum of the narrow band is approximately the same as that of the wave $A\sqrt{2} \cos 2\pi f_2 t$.

The integrals occurring in our formula are

$$\int_0^\infty w(f)\, df = \int_0^\infty w_1(f)\, df + A^2$$

$$= W + A^2$$

$$\int_0^\infty w(f) f^2\, df = \int_0^\infty f^2 w_1(f)\, df + A^2 f_2^2$$

$$= U + A^2 f_2^2$$

We suppose that A and f_2 are such that

$$W \gg A^2$$

$$U \ll A^2 f_2^2 .$$

Then our formula (3.3–11) gives us the expected number of zeros

$$2 \frac{A f_2}{W^{1/2}}$$

We may give a qualitative explanation of this formula if we regard our noise current as composed of a small component

$$I_2 = 2^{1/2} A \cos 2\pi f_2 t$$

due to the narrow band superposed on a large, slowly varying component due to the lower band. Since the r.m.s. value of the second component is $W^{1/2}$ we may assign it a representative frequency f_1 and write it approximately as

$$I_1 = (2\mathrm{W})^{1/2} \cos 2\pi f_1 t$$

The zeros of the noise current are clustered around the zeros of the second wave. Near such a zero

$$I_1 = \pm(2W)^{1/2} 2\pi f_1 \Delta t$$

where Δt is the distance from the zero. The oscillations of I_1 produce zeros when $| I_1 |$ is less than the amplitude of I_2 or when

$$A > W^{1/2} 2\pi f_1 | \Delta t |$$

and the interval over which zeros are produced is given by

$$2\Delta t = \frac{A W^{-1/2}}{\pi f_1}$$

The number of zeros is this multiplied by $2f_2$. Since there are $2f_1$ such intervals per second the number of zeros per second is

$$\frac{4}{\pi} AW^{-1/2} f_2$$

This differs from the result given by our formula by a factor of $2/\pi$. This discrepancy is due to our representing the two bands by the sine waves I_1 and I_2.

From this example we obtain the picture that when the integral for ψ_0 converges corresponding to $A \to 0$, while at the same time the integral for ψ_0'' diverges, corresponding to $f_2 \to \infty$ in such a way that $Af_2 \to \infty$, the noise current behaves something like a continuous function which has no derivative. It seems that for physical systems the integrals will always converge since parasitic effects will have the effect of making $w(f)$ tend to zero rapidly enough. The frequency which represents the region where this occurs is of the order of the frequency of the microscopic wiggles.

So far we have been considering the formulas of this section in the most favorable light possible. There are experiments which indicate the possibility of the formulas breaking down in some cases. Prof. Uhlenbeck has pointed out that if a very broad band fluctuation current be forced[29] to flow through a circuit consisting of a condenser, C, in parallel with a series combination of inductance, L, and resistance, R, equation (3.3–11) says that the expected number of zeros per second of the current, I, flowing through R (and L) is independent of R. It is simply $\frac{1}{\pi}(LC)^{-1/2}$. The differential equation for I is the same as that which governs the Brownian motion of a mirror suspended in a gas[30], the gas pressure playing the role of R. Curves are available for this motion and it is seen that their character depends greatly upon the pressure[31]. Unfortunately, it is difficult to tell from the curves whether the expected number of zeros is independent of the pressure. The differences between the curves for various pressures indicates that there may be some dependence*.

3.4 The Distribution of Zeros

The problem of determining the distribution function for the distance between two successive zeros seems to be quite difficult and apparently

[29] For example, by putting the circuit in series with a diode.

[30] This problem in Brownian motion is discussed by G. E. Uhlenbeck and S. Goudsmit, *Phys., Rev.*, 34 (1929), 145–151.

[31] E. Kappler, *Annalen d. Phys.*, 11 (1931) 233–256.

* Since this was written M. Kac and H. Hurwitz have studied the problem of the expected number of zeros using quite a different method of approach which employs the "shot-effect" representation (Sec. 3.11). Their results confirm the correctness of (3.3–11) when the integrals converge. When the integrals diverge the average number of electrons, per sec. producing the shot effect must be considered.

nobody has as yet given a satisfactory solution. Here we shall give some results which are related to the general problem and which give an idea of the form of the distribution for the region of small spacings between the zeros.

We shall show (in the work starting with equation (3.4–12)) that the probability of the noise current, I, passing through zero in the interval $\tau, \tau + d\tau$ with a negative slope, when it is known that I passes through zero at $\tau = 0$ with a positive slope, is

$$\frac{d\tau}{2\pi}\left[\frac{\psi_0}{-\psi_0''}\right]^{1/2}\left[\frac{M_{23}}{H}\right](\psi_0^2 - \psi_\tau^2)^{-3/2}[1 + H \cot^{-1}(-H)] \qquad (3.4\text{–}1)$$

where M_{22} and M_{23} are the cofactors of $\mu_{22} = -\psi_0''$ and $\mu_{23} = -\psi_\tau''$ in the matrix

$$M = \begin{bmatrix} \psi_0 & 0 & \psi_\tau' & \psi_\tau \\ 0 & -\psi_0'' & -\psi_\tau'' & -\psi_\tau' \\ \psi_\tau' & -\psi_\tau'' & -\psi_0'' & 0 \\ \psi_\tau & -\psi_\tau' & 0 & \psi_0 \end{bmatrix}, \qquad (3.4\text{–}2)$$

$$H = M_{23}[M_{22}^2 - M_{23}^2]^{-1/2}.$$

We choose $0 \leq \cot^{-1}(-H) \leq \pi$, the value π being taken at $\tau = 0$, and the value $\pi/2$ being approached as $\tau \to \infty$. It should be remembered that we are writing the arguments of the correlation functions as subscripts, e.g., $-\psi_\tau''$ is really

$$-\psi''(\tau) = 4\pi^2 \int_0^\infty f^2 w(f) \cos 2\pi f\tau \, df \qquad (3.3\text{–}8)$$

As τ becomes larger and larger the behavior of I at τ is influenced less and less by the fact that it goes through zero with a positive slope at $\tau = 0$. Hence (3.4–1) should approach the probability that, for any interval of length $d\tau$ chosen at random, I will go through zero with a negative slope. Because of symmetry, this is half the probability that it will go through zero. Thus (3.4–1) should approach, from (3.3–10),

$$\frac{d\tau}{2\pi}\left[\frac{-\psi_0''}{\psi_0}\right]^{1/2} \qquad (3.4\text{–}3)$$

as $\tau \to \infty$. It actually does this since M approaches a diagonal matrix and both M_{23} and H approach zero with $M_{23}/H \to M_{22} \to -\psi_0^2\psi_0''$. For a low pass filter cutting off at f_b (3.4–3) is

$$d\tau f_b 3^{-1/2} \qquad (3.4\text{–}4)$$

The behavior of (3.4–1) as $\tau \to 0$ is quite a bit more difficult to work out.

M_{22} and M_{23} go to zero as τ^4, $M_{22}^2 - M_{23}^2$ as τ^{10}, and consequently H goes to infinity as τ^{-1}. The final result is that (3.4–1) approaches

$$d\tau \frac{\tau}{8}\left[\frac{\psi_0\psi_0^{(4)} - \psi_0''^2}{-\psi_0\psi_0''}\right] \tag{3.4–5}$$

as $\tau \to 0$, assuming $\psi^{(4)}$ exists. Here the superscript (4) indicates the fourth derivative at $\tau = 0$,

$$\psi_0^{(4)} = 16\pi^4 \int_0^\infty f^4 w(f)\, df \tag{3.4–6}$$

For a low pass filter cutting off at f_b (3.4–5) is

$$d\tau \frac{\tau}{30} (2\pi f_b)^2 \tag{3.4–7}$$

When (3.4–1) is applied to a low pass filter, it turns out that instead of τ the variable

$$\varphi = 2\pi f_b \tau, \qquad d\varphi = 2\pi f_b\, d\tau \tag{3.4–8}$$

is more convenient to handle. Thus, if we write (3.4–1) as $p(\varphi)\, d\varphi$, it follows from (3.4–4) and (3.4–7) that

$$\begin{aligned} p(\varphi) &\to \frac{1}{2\pi\sqrt{3}} = .0919 \quad \text{as} \quad \varphi \to \infty \\ p(\varphi) &\to \frac{\varphi}{30} \quad \text{as} \quad \varphi \to \infty \end{aligned} \tag{3.4–9}$$

$p(\varphi)$ has been computed and plotted on Fig. 1 as a function of φ for the range 0 to 9. From the curve and the theory it is evident that beyond 9 $p(\varphi)$ oscillates about 0.0919 with ever decreasing amplitude.

We may take $p(\varphi)\, d\varphi$ to be the probability that I goes through zero in $\varphi, \varphi + d\varphi$, when it is known that I goes through zero at $\varphi = 0$ with a slope opposite to that at φ. $p(\varphi)\, d\varphi$ exceeds the probability that I goes through zero at $\varphi = 0$ and in $\varphi, \varphi + d\varphi$ with no zeros in between. This is because $p(\varphi)\, d\varphi$ includes all curves of the latter class and in addition those which may have an even number of zeros between 0 and φ. From this it follows that the curve giving the probability density of the intervals between zeros must be underneath the curve of $p(\varphi)$.

A partial check on the curve for $p(\varphi)$ may be obtained by comparing it with a probability density function obtained experimentally by M. E. Campbell for the intervals between 754 successive zeros. He passed thermal noise through a band pass filter, the lower cutoff being around 200 cps and the upper cutoff being around 3000 cps. The upper cutoff was rather gradual and it is difficult to assign a representative value. The crosses on figure 1 are obtained from his data when we assume that his filter behaves like a low pass filter with a cutoff at $f_b = 2850$, this choice being made in order to make the maximum of his curve coincide with that of $p(\varphi)$.

It is seen that some of the crosses lie above $p(\varphi)$. This is probably due to the fact that the actual filter differs somewhat from the assumed low pass filter.

On Fig. 1 there is also plotted a function closely related to (3.4–1). It is the low pass filter form of the following: The probability of I passing

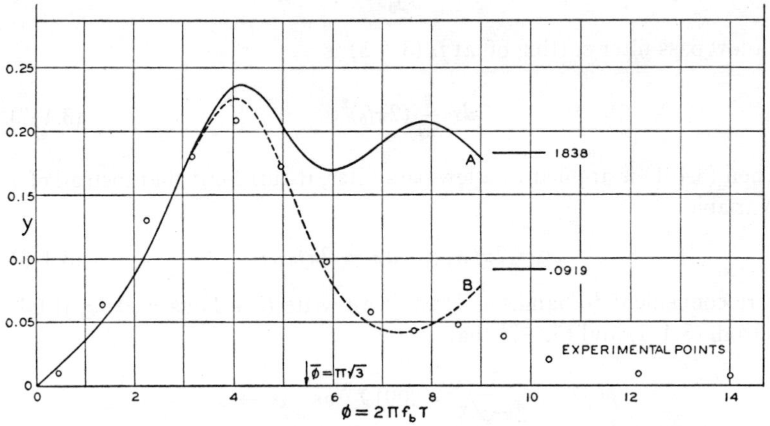

Fig. 1—Distribution of intervals between zeros—low-pass filter
$y_A \Delta\varphi$ is probability of a zero in $\Delta\varphi$ when a zero is at origin.
$y_B \Delta\varphi$ is probability of a zero in $\Delta\varphi$ when a zero is at origin and slopes at zeros are of opposite signs.
$y_B = p(\varphi), f_b =$ filter cutoff, $\tau =$ time between zeros.

through zero in τ, $\tau + d\tau$ when it is known that I passes through zero at $\tau = 0$ is

$$\frac{d\tau}{\pi}\left[\frac{\psi_0}{-\psi_0''}\right]^{1/2}\left[\frac{M_{23}}{H}\right](\psi_0^2 - \psi_\tau^2)^{-3/2}[1 + H \tan^{-1} H] \quad (3.4\text{–}10)$$

where the notation is the same as in (3.4–1) and $-\frac{\pi}{2} \leq \tan^{-1} H \leq \frac{\pi}{2}$. This curve should always lie above $p(\varphi)$ and the small difference between the curves out to $\varphi = 4$ indicates that the true distribution of zeros is given closely by $p(\varphi)$ out to this point.

When (3.4–1) is applied to a relatively narrow band pass filter or some similar device we may make some approximations and obtain an expression somewhat simpler than (3.4–1). As a guide we consider our usual ideal band pass filter whose range extends from f_a to f_b. The correlation function is given by (3.2–5).

$$\psi_\tau = \frac{w_0}{\pi\tau} \sin \pi\tau(f_b - f_a) \cos \pi\tau(f_b + f_a)$$
$$\psi_0 = w_0(f_b - f_a) \tag{3.2–5}$$

From physical considerations we know that in a narrow filter most of the distances between zeros will be nearly equal to

$$\tau_1 = \frac{1}{f_b + f_a}$$

i.e., nearly equal to the distance between the zeros of a sine wave having the mid-band frequency. We therefore expect (3.4–1) to have a peak very close to τ_1. We also expect peaks at $3\tau_1$, $5\tau_1$ etc. but we shall not consider these. We wish to examine the behavior of (3.4–1) near τ_1.

It turns out that M_{23} is nearly equal to M_{22} so that H is large and (3.4–1) becomes approximately

$$\frac{d\tau}{2}\left[\frac{\psi_0}{-\psi_0''}\right]^{1/2} \frac{M_{23}}{[\psi_0^2 - \psi_\tau^2]^{3/2}}$$

where τ is near τ_1.

In order to see that M_{23} is nearly equal to M_{22} we use the expressions

$$M_{22} = -\psi_0''(\psi_0^2 - \psi_\tau^2) - \psi_0\psi_\tau'^2$$

$$M_{23} = \psi_\tau''(\psi_0^2 - \psi_\tau^2) + \psi_\tau\psi_\tau'^2$$

$$M_{22} + M_{23} = (\psi_0 - \psi_\tau)[(\psi_0 + \psi_\tau)(\psi_\tau'' - \psi_0'') - \psi_\tau'^2]$$

$$= (\psi_0 - \psi_\tau)[B + C]$$

$$M_{22} - M_{23} = (\psi_0 + \psi_\tau)[(\psi_0 - \psi_\tau)(-\psi_\tau'' - \psi_0'') - \psi_\tau'^2]$$

$$= (\psi_0 + \psi_\tau)[-B + C]$$

$$B = \psi_0\psi_\tau'' - \psi_\tau\psi_0''$$

$$C = -\psi_0\psi_0'' + \psi_\tau\psi_\tau'' - \psi_\tau'^2$$

From (3.2–5) it is seen that ψ_τ may be written as

$$\psi_\tau = A \cos \beta\tau, \qquad \beta = \pi(f_b + f_a)$$

where $\beta\tau_1 = \pi$ and A is a function of τ which varies slowly in comparison with $\cos \beta\tau$. We see that near τ_1, ψ_τ is nearly equal to $-\psi_0$. Likewise

ψ'_τ hovers around zero and ψ''_τ is nearly equal to $-\psi''_0$. Differentiating with respect to τ gives

$$\psi'_\tau = A' \cos \beta\tau - A\beta \sin \beta\tau$$

$$\psi''_\tau = (A'' - A\beta^2) \cos \beta\tau - 2A'\beta \sin \beta\tau$$

$$\psi''_0 = A''_0 - A_0\beta^2, \qquad \psi_0 = A_0$$

where A_0 and A''_0 are the values of A and its second derivative at τ equal to zero. These lead to

$$B = (A_0A'' - AA''_0) \cos \beta\tau - 2A_0A'\beta \sin \beta\tau$$

$$C = (AA'' - A'^2) \cos^2 \beta\tau - A_0A''_0 + (A_0^2 - A)^2\beta^2$$

We wish to show that $C + B$ and $C - B$ are of the same order of magnitude. If we can do this, it follows that $M_{22} - M_{23}$ is much smaller than $M_{22} + M_{23}$ since $\psi_0 - \psi_{\tau_1}$ is approximately $2\psi_0$ while $\psi_0 + \psi_{\tau 1}$ is quite small. Consequently we will have shown that M_{23} is nearly equal to M_{22}.

So far we have made no approximations. We now express the slowly varying function A as a power series in τ. Since ψ'_0 and ψ'''_0 must be zero for the type of functions we consider, it follows that

$$A = A_0 + \frac{\tau^2}{2} A''_0 + \cdots$$

$$A' = \tau A''_0 + \cdots$$

$$A'' = A''_0 + \frac{\tau^2}{2} A_0^{(4)} + \cdots$$

where we neglect all powers higher than the second. Multiplication and squaring gives

$$A^2 - A_0^2 = \tau^2 A_0 A''_0$$

$$AA'' - A'^2 = A_0 A''_0 + \frac{\tau^2}{2} (A_0 A_0^{(4)} - A_0''^2)$$

$$= A_0 A''_0 + F$$

$$A_0 A'' - AA''_0 = \frac{\tau^2}{2} (A_0 A_0^{(4)} - A_0''^2) = F$$

Since, for small τ, A and A'' are nearly equal to A_0 and A''_0, respectively we see that the difference on the left is small relative to $A_0 A''_0$, i.e.,

$$|F| << |A_0 A''_0|$$

Our expression for B and C become approximately

$$B = F \cos \beta\tau - 2A_0 A_0'' \beta\tau \sin \beta\tau$$

$$C = F \cos^2 \beta\tau - A_0 A_0'' \sin^2 \beta\tau - A_0 A_0'' \beta^2 \tau^2$$

When τ is near τ_1, $\beta\tau$ is approximately π. Hence both $C + B$ and $C - B$ are approximately $-A_0 A_0'' \pi^2$ and are of the same order of magnitude. Consequently M_{22} and M_{23} are both nearly equal and

$$M_{23} = \psi_0[C + B]$$

$$= -A_0^2 A_0'' \pi^2$$

When this expression for M_{23} is used our approximation to (3.4–1) gives us the result: If the correlation function is of the form

$$\psi_\tau = A \cos \beta\tau$$

where A is a slowly varying function of τ, the probability that the distance between two successive zeros lies between τ and $\tau + d\tau$ is approximately

$$\frac{d\tau}{2} \frac{a}{[1 + a^2(\tau - \tau_1)^2]^{3/2}}$$

where a is positive and

$$a^2 = \frac{A_0 \beta^2}{-A_0'' \tau_1^2}, \qquad \tau_1 = \frac{\pi}{\beta}$$

For our ideal band pass filter with the pass band $f_b - f_a$,

$$a = \sqrt{3} \frac{(f_b + f_a)^2}{f_b - f_a}, \qquad \tau_1 = \frac{1}{f_b + f_a}$$

and the average value of $|\tau - \tau_1|$ is a^{-1}. Thus

$$\frac{\text{ave.} |\tau - \tau_1|}{\tau_1} = \frac{1}{a\tau_1} = \frac{f_b - f_a}{\sqrt{3}\,(f_b + f_a)} = \frac{1}{2\sqrt{3}} \frac{\text{band width}}{\text{mid-frequency}}$$

When the correlation function cannot be put in the form assumed above but still behaves like a sinusoidal wave with slowly varying amplitude we may use our first approximation to (3.4–1). Thus, the probability that the distance between two successive zeros lies between τ and $\tau + d\tau$ is approximately

$$\frac{b\, d\tau}{[\psi_0^2 - \psi_\tau^2]^{3/2}}$$

when τ lies near τ_1 where τ_1 is the smallest value of τ which makes ψ_τ approximately equal to $-\psi_0$. This probability is supposed to approach zero rapidly as τ departs from τ_1, and b is chosen so that the integral over the effective region around τ_1 is unity.

It seems to be especially difficult to get an expression for the distribution of zeros for large spacing. One method, suggested by Prof. Goudsmit, is to amend the conditions leading to (3.4–1) by adding conditions that I be positive at equally spaced points along the time axis between 0 and τ. This leads to integrals which are hard to evaluate. For one point between 0 and τ the integral is of the form (3.5-7).

Another method of approach is to use the method of "in and exclusion" of zeros between 0 and τ. Consider the class of curves of I having a zero at $\tau = 0$. Then, in theory, our methods will allow us to compute the functions $p_0(\tau)$, $p_1(r, \tau)$, $p_2(r, s, \tau)$, associated with this class where

$p_0(\tau)\, d\tau$ is probability of curve having zero in $d\tau$

$p_1(r, \tau)\, d\tau\, dr$ is probability of curve having zeros in $d\tau$ and dr

$p_2(r, s, \tau)\, d\tau\, dr\, ds$ is probability of curve having zeros in $d\tau$, dr, and ds

In fact $p_0(\tau)\, d\tau$ is expression (3.4–10). The method of in and exclusion then leads to an expression for $P_0(\tau)\, d\tau$, the probability of having a zero at 0 and a zero in τ, $\tau + d\tau$ but none between 0 and τ. It is

$$P_0(\tau) = p_0(\tau) - \frac{1}{1!}\int_0^\tau p_1(r, \tau)\, dr + \frac{1}{2!}\int_0^\tau\int_0^\tau p_2(r, s, \tau)\, dr\, ds$$
$$- \frac{1}{3!}\int_0^\tau\int_0^\tau\int_0^\tau p_3(r, s, t, \tau)\, dr\, ds\, dt + \dots \tag{3.4–11}$$

Here again we run into difficult integrals. Incidentally, (3.4–11) may be checked for events occurring independently at random. Thus if $\nu\, d\tau$ is the probability of an event happening in $d\tau$, then, if ν is a constant and the events are independent, we have p_0, p_1, p_2, $\cdots$ given by ν, ν^2, ν^3, $\cdots$. From (3.4–11) we obtain the known result $P_0(\tau) = \nu e^{-\nu\tau}$.

We shall now derive (3.4–1). The work is based upon a generalization of (3.3–5): If y is a random curve described by (3.3–1), the probability that y will pass through zero in x_1, $x_1 + dx_1$ with a positive slope and through zero in x_2, $x_2 + dx_2$ with a negative slope is

$$-dx_1\, dx_2 \int_0^{+\infty} d\eta_1 \int_{-\infty}^0 d\eta_2\, \eta_1 \eta_2\, p(0, \eta_1, x_1; 0, \eta_2, x_2) \tag{3.4–12}$$

where $p(\xi_1, \eta_1, x_1; \xi_2, \eta_2, x_2)$ is the probability density function for the four random variables

$$\xi_i = F(a_1, a_2, \cdots, a_N; x_i)$$

$$\eta_i = \left[\frac{\partial F}{\partial x}\right]_{x=x_i}, \qquad i = 1, 2.$$

The x_1 and x_2 play the role of parameters in (3.4–12). This result may be established in much the same way as (3.3–5).

When we identify F with one of our representations, (2.8–1) or (2.8–6), of the noise current $I(t)$ it is seen that p is normal in four dimensions. We may obtain the second moments directly from this representation, as has been done in the equations just below (3.3–7). The same results may be obtained from the definition of $\psi(\tau)$, and for the sake of variety we choose this second method. We set $x_1 = t_1$, $x_2 = t_1 + \tau$. Then

$$\overline{\xi_1^2} = \overline{\xi_2^2} = \overline{I^2(t)} = \psi_0$$

$$\overline{\xi_1 \xi_2} = \overline{I(t)I(t+\tau)} = \psi_\tau \tag{3.4–13}$$

$$\overline{\eta_1 \eta_2} = \overline{\left(\frac{\partial I}{\partial t}\right)_t \left(\frac{\partial I}{\partial t}\right)_{t+\tau}} = \operatorname*{Limit}_{T\to\infty} \frac{1}{T} \int_0^T I'(t+\tau)I'(t)\, dt$$

where primes denote differentiation with respect to the arguments. Integrating by parts:

$$\int_0^T I'(t+\tau)\, dI(t) = [I'(t+\tau)I(t)]_0^T - \int_0^T I''(t+\tau)I(t)\, dt$$

We assume that I and its derivative remains finite so that the integrated portion vanishes, when divided by T, in the limit. Since

$$I''(t+\tau) = \frac{\partial^2}{\partial \tau^2} I(t+\tau)$$

we have

$$\overline{\eta_1 \eta_2} = -\frac{\partial^2}{\partial \tau^2} \psi(\tau) = -\psi_\tau''$$

Setting $\tau = 0$ gives

$$\overline{\eta_1^2} = \overline{\eta_2^2} = -\psi_0''$$

in agreement with the value of μ_{22} obtained from (3.3–7). In the same way

$$\overline{\xi_1 \eta_2} = \operatorname*{Limit}_{T\to\infty} \frac{1}{T} \int_0^T I'(t+\tau)I(t)\, dt = \frac{\partial}{\partial \tau} \psi(\tau)$$

$$= \psi_\tau'$$

$$\overline{\xi_2 \eta_1} = \operatorname*{Limit}_{T\to\infty} \frac{1}{T} \int_0^T I'(t)I(t+\tau)\, dt$$

$$= \text{“} \quad (-)\frac{1}{T} \int_0^T I'(t+\tau)I(t)\, dt$$

$$= -\psi_\tau'$$

where we have integrated by parts in getting $\overline{\xi_2\eta_1}$. Setting $\tau = 0$ and using $\psi_0' = 0$ gives

$$\overline{\xi_1\eta_1} = \overline{\xi_2\eta_2} = 0$$

In order to obtain the matrix M of the second moments μ_{rs} in a form fairly symmetrical about its center we choose the 1, 2, 3, 4 order of our variables to be $\xi_1, \eta_1, \eta_2, \xi_2$. From equations (3.4–13) etc. it is seen that this choice leads to the expression (3.4–2) for M.

When we put ξ_1 and ξ_2 equal to zero, we obtain for the probability density function in (3.4–12) the expression

$$\frac{|M|^{-1/2}}{4\pi^2}\exp\left[-\frac{1}{2|M|}(M_{22}\eta_1^2 + 2M_{23}\eta_1\eta_2 + M_{33}\eta_2^2)\right]$$

Because of the symmetry of M, M_{22} is equal to M_{33}. When, in the integral (3.4–12) we make the change of variable

$$x = \left[\frac{M_{22}}{2|M|}\right]^{1/2}\eta_1, \qquad y = -\left[\frac{M_{22}}{2|M|}\right]^{1/2}\eta_2$$

we obtain

$$\frac{dx_1\,dx_2}{\pi^2}\frac{|M|^{3/2}}{M_{22}^2}\int_0^\infty x\,dx\int_0^\infty dy\;ye^{-x^2-y^2+2(M_{23}/M_{22})xy}$$

The double integral may be evaluated by (3.5–4). Let

$$\varphi = \cos^{-1}\left(-\frac{M_{23}}{M_{22}}\right) = \cot^{-1}(-H), \qquad H = M_{23}[M_{22}^2 - M_{23}^2]^{-1/2}$$

where H is the same as that given in (3.4–2). Our expression now becomes

$$\frac{dx_1\,dx_2}{4\pi^2}\frac{|M|^{3/2}}{M_{22}^2 - M_{23}^2}[1 + H\cot^{-1}(-H)]$$

From a property of determinants

$$M_{22}M_{33} - M_{23}^2 = |M|(\psi_0^2 - \psi_\tau^2)$$

Using this to eliminate $|M|$ and dividing by

$$\frac{dx_1}{2\pi}\left[\frac{-\psi_0''}{\psi_0}\right]^{1/2}$$

which, from (3.3–10), is the probability of going through zero in $x_1, x_1 + dx_1$ with positive slope, gives the probability of going through zero in dx_2 with negative slope when it is known that I goes through zero at x_1 with positive slope:

$$\frac{dx_2}{2\pi}\left[\frac{\psi_0}{-\psi_0''}\right]^{1/2}[M_{22}^2 - M_{23}^2]^{1/2}(\psi_0^2 - \psi_\tau^2)^{-3/2}[1 + H \cot^{-1}(-H)]$$

This is the same as (3.4–1).

The expression (3.4–10) is the same as the probability of I going through zero in $d\tau$ when it is known that I goes through zero at the origin with positive slope. This second probability may be obtained from (3.4–1) by adding the probability that I goes through $d\tau$ with positive slope when it is known to go through zero with positive slope. Thus we must add the expression containing the integral in which the integration in both η_1 and η_2 run from 0 to ∞. In terms of x and y this integral is

$$\int_0^\infty x\, dx \int_0^\infty dy\, y e^{-x^2 - y^2 - 2(M_{23}/M_{22})xy}$$

This is equivalent to a change in the sign of M_{23} and hence of H. After this addition we must consider

$$\begin{aligned} &1 + H \cot^{-1}(-H) + 1 - H \cot^{-1} H \\ &= 2 + H[\cot^{-1}(-H) - \cot^{-1} H] \\ &= 2 + H[\pi - 2 \cot^{-1} H] \\ &= 2[1 + H \tan^{-1} H] \end{aligned}$$

and this leads to (3.4–10).

3.5 Multiple Integrals

We wish to evaluate integrals of the form

$$J = \int_0^\infty dx_1 \int_0^\infty dx_2\, e^{-x_1^2 - 2ax_1x_2 - x_2^2} \tag{3.5–1}$$

Our method of procedure is to first reduce the exponent to the sum of squares by a suitable linear change of variable and then change to polar coordinates. This method appears to work also for triple integrals of the same sort, but when it is applied to a four-fold integral, the last integration apparently cannot be put in closed form.

The reduction of the exponent to the sum of squares is based upon the transformation: If*

$$\begin{aligned} x_1 &= h_1 y_1 + h_2 D_{21} y_2 + h_3 D_{31} y_3 + \cdots + h_n D_{n,1} y_n \\ x_2 &= 0 \quad + h_2 D_{22} y_2 + \cdots \qquad\quad + h_n D_{n,2} y_n \\ &\cdots\cdots\cdots\cdots\cdots\cdots\cdots\cdots\cdots \\ x_n &= 0 \quad + 0 \quad + \cdots \quad + 0 \quad + h_n D_{n,n} y_n \end{aligned} \tag{3.5–2}$$

* T. Fort, Am. Math. Monthly, 43 (1936), pp. 477–481. See also Scott and Mathews, Theory of Determinants, Cambridge (1904), Prob. 63, p. 276.

where $D_0 = 1$, $D_1 = a_{11}$, $D_{r,r} = D_{r-1}$, and D_{rs} is the cofactor of a_{sr} (or of a_{rs} because they are equal) in D_r :

$$D_r = \begin{vmatrix} a_{11} & a_{12} & \cdots & a_{1r} \\ a_{12} & a_{22} & & \\ a_{1r} & \cdots & & a_{rr} \end{vmatrix}, \qquad h_r = [D_{r-1}D_r]^{-1/2},$$

then, if none of the D_r's is zero,

$$\sum_1^n a_{rs} x_r x_s = y_1^2 + y_2^2 + \cdots + y_n^2$$

From (3.5–2); the Jacobian $\partial(x_1, \cdots x_n)/\partial(y_1, \cdots y_n)$ is equal to $D_n^{-1/2}$.

Applying our transformation to the exponent:

$$x_1 = y_1 - aD_2^{-1/2}y_2$$

$$x_2 = 0 + D_2^{-1/2}y_2$$

$$D_2 = 1 - a^2$$

Since x_2 runs from 0 to ∞ so must y_2. The expression for x_1 shows that v_1 runs from $a\, D_2^{-1/2}y_2$ to ∞. The integral is therefore

$$J = D_2^{-1/2} \int_0^\infty dy_2 \int_{aD_2^{-1/2}y_2}^\infty e^{-y_1^2-y_2^2}\, dy_1$$

We now change to polar coordinates:

$$y_1 = \rho \cos \theta \qquad dy_1\, dy_2 = \rho\, d\rho\, d\theta$$

$$y_2 = \rho \sin \theta$$

$y_2 \geq 0$ gives $0 \leq \theta \leq \pi$

$y_1 \geq aD_2^{-1/2}y_2$ gives $\cot\theta \geq aD_2^{-1/2}$

and obtain

$$J = D_2^{-1/2} \int_0^{\cot^{-1} aD_2^{-1/2}} d\theta \int_0^\infty \rho e^{-\rho^2}\, d\rho$$

$$= \tfrac{1}{2} D_2^{-1/2} \cot^{-1}(aD_2^{-1/2})$$

where the arc-cotangent lies between 0 and π. This may be written in the simpler form

$$J = \tfrac{1}{2}(1 - a^2)^{-1/2} \cos^{-1} a = \tfrac{1}{2}\varphi \csc \varphi$$

where

$$a = \cos\varphi,$$

it being understood that $0 \leq \varphi \leq \pi$.

Other integrals may be obtained by differentiation. Thus from

$$\int_0^\infty dx \int_0^\infty dy\, e^{-x^2-y^2-2xy\cos\varphi} = \tfrac{1}{2}\varphi \csc \varphi \tag{3.5–3}$$

we obtain

$$\int_0^\infty dx \int_0^\infty dy\, xy\, e^{-x^2-y^2-2xy\cos\varphi} = \tfrac{1}{4}\csc^2\varphi(1 - \varphi\cot\varphi) \tag{3.5–4}$$

By using the same transformation we may obtain

$$\int_0^\infty dx \int_0^\infty dy\, ye^{-x^2-y^2-2axy} = \frac{\sqrt{\pi}}{4}\,\frac{1}{1+a} \tag{3.5–5}$$

Of course, we may expand part of the exponential in a power series and integrate termwise but this leads to a series which has to be summed in each particular case:

$$\int_0^\infty dx \int_0^\infty dy\, x^n y^m e^{-x^2-y^2-2axy}$$

$$= \frac{1}{4}\sum_{r=0}^{\infty}\frac{(-2a)^r}{r!}\,\Gamma\left(\frac{n+r+1}{2}\right)\Gamma\left(\frac{m+r+1}{2}\right)$$

If we take $-1 < R(m) < -\frac{1}{2}$, $-1 < R(m) < -\frac{1}{2}$, the series may be summed when $a = 1$. The result stated just below equation (3.8–9) is obtained by continuing m and n analytically.

The same methods will work when the limits are $\pm\infty$. We obtain, when m and n are integers,

$$\int_{-\infty}^{+\infty} dx \int_{-\infty}^{+\infty} dy\, x^n y^m e^{-x^2-y^2-2xy\cos\varphi}$$

$$= \begin{cases} 0, & n+m \text{ odd} \\ (-)^n\sqrt{\pi}\,\dfrac{\Gamma\left(\dfrac{m+n+1}{2}\right)}{(\sin\varphi)^{n+m+1}} \\ \quad F\left(-n, -m; \dfrac{1-n-m}{2}; \dfrac{1-\cos\varphi}{2}\right), & n+m \text{ even} \end{cases} \tag{3.5–6}$$

The hypergeometric function may also be written as

$$F\left(-\frac{n}{2}, -\frac{m}{2}; \frac{1-n-m}{2}; \sin^2\varphi\right)$$

By transformations of this we are led to the following expression for the integral

$$0, \; n + m \text{ odd},$$

$$\frac{\Gamma\left(\frac{m+1}{2}\right)\Gamma\left(\frac{n+1}{2}\right)}{(\sin\varphi)^{n+m+1}} F\left(-\frac{n}{2}, -\frac{m}{2}, \frac{1}{2}; \cos^2\varphi\right), \quad m, n \text{ both even},$$

$$-2\frac{\Gamma\left(1+\frac{n}{2}\right)\Gamma\left(1+\frac{m}{2}\right)}{(\sin\varphi)^{n+m+1}} \cos\varphi F\left(\frac{1-m}{2}, \frac{1-n}{2}; \frac{3}{2}; \cos^2\varphi\right),$$

$$m, n \text{ odd}$$

As was mentioned earlier, the method used to evaluate the double integrals may also be applied to similar triple integrals. Here we state two results obtained in this way.

$$\int_0^\infty dx \int_0^\infty dy \int_0^\infty dz \exp\left[-x^2 - y^2 - z^2 - 2cxy - 2bzx - 2ayz\right]$$

$$= \frac{1}{4}\left[\frac{\pi}{D_3}\right]^{1/2} [\alpha + \beta + \gamma - \pi]$$

$$\int_0^\infty dx \int_0^\infty dy \int_0^\infty dz\, yz \exp\left[-x^2 - y^2 - z^2 - 2cxy - 2bzx - 2ayz\right]$$

$$= \frac{\sqrt{\pi}}{8D_3}\left[\frac{1+a-b-c}{1+a} - \frac{a-bc}{D_3^{1/2}}(\alpha+\beta+\gamma-\pi)\right] \tag{3.5–7}$$

where β and γ are obtained by cyclic permutation of a, b, c from

$$\alpha = \cos^{-1}\frac{a - cb}{(1-c^2)^{1/2}(1-b^2)^{1/2}} = \sin^{-1}\left[\frac{D_3}{(1-c^2)(1-b^2)}\right]^{1/2}$$

$$= \cot^{-1}\frac{a-bc}{D_3^{1/2}}$$

where α, β, γ all lie in the range 0, π and where

$$D_3 = \begin{vmatrix} 1 & c & b \\ c & 1 & a \\ b & a & 1 \end{vmatrix} = 1 + 2\,abc - a^2 - b^2 - c^2$$

For reference we state the integrals which arise from the definition of the normal distribution given in section (2.9)

$$\int_{-\infty}^{+\infty} dx_1 \cdots \int_{-\infty}^{+\infty} dx_n \exp\left[-\sum_1^n a_{rs}x_r x_s\right] = \left[\frac{\pi^n}{|a|}\right]^{1/2}$$

$$\int_{-\infty}^{+\infty} dx_1 \cdots \int_{-\infty}^{+\infty} dx_n\, x_t x_u \exp\left[-\sum_1^n a_{rs}x_r x_s\right] = \left[\frac{\pi^n}{|a|^3}\right]^{1/2} \frac{A_{tu}}{2} \tag{3.5–8}$$

where the quadratic form is positive definite and $|a|$ is its determinant. A_{tu} is the cofactor of a_{tu}. Incidentally, these may be regarded as special cases of

$$\int_{-\infty}^{+\infty} dx_1 \cdots \int_{-\infty}^{+\infty} dx_n f\left(\sum_1^n a_{rs} x_r x_s\right) F\left(\sum_1^n b_r x_r\right)$$

$$= \frac{2}{\Gamma\left(\frac{n-1}{2}\right)} \left[\frac{\pi^{n-1}}{|a|}\right]^{1/2} \int_{-\infty}^{+\infty} dx \int_0^{\infty} dy\, y^{n-2} f(x^2 + y^2) \quad (3.5\text{–}9)$$

$$F\left\{x\left[\frac{\sum_1^n A_{rs} b_r b_s}{|a|}\right]^{1/2}\right\},$$

which is a generalization of a result given by Schlömilch.*

3.6 Distribution of Maxima of Noise Current

Here we shall use a result similar to those used in sections 3.3 and 3.4. Let y be a random curve given by (3.3–1),

$$y = F(a_1 \cdots a_N ; x). \quad (3.3\text{–}1)$$

If suitable conditions are satisfied, the probability that y has a maximum in the rectangle $(x_1, x_1 + dx_1, y_1, y_1 + dy_1)$, dx_1 and dy_1 being of the same order of magnitude, is[32]

$$-dx_1\, dy_1 \int_{-\infty}^{0} p(y_1, 0, \zeta)\zeta\, d\zeta \quad (3.6\text{–}1)$$

and the expected number of maxima of y in $a \leq x \leq b$ is obtained by integrating this expression over the range $-\infty \leq y_1 \leq \infty$, $a \leq x_1 \leq b$. $p(\xi, \eta, \zeta)$ is the probability density function for the random variables

$$\xi = F(a_1, \cdots, a_N ; x_1)$$

$$\eta = \left(\frac{\partial F}{\partial x}\right)_{x=x_1} \quad (3.6\text{–}2)$$

$$\zeta = \left(\frac{\partial^2 F}{\partial x^2}\right)_{x=x_1}$$

* Höheren Analysis, Braunschweig (1879), Vol. 2, p. 494, equ. (29).

32 *Am. Jour. Math.*, Vol. 61 (1939) 409–416. A similar problem has been studied by E. L. Dodd, The Length of the Cycles Which Result From the Graduation of Chance Elements, Ann. Math. Stat., Vol. 10 (1939) 254–264. He gives a number of references to the literature dealing with the fluctuations of time series.

In our application of this result we replace x and y by t and I as before. Then

$$\xi = I = \sum_1^N c_n \cos(\omega_n t - \varphi_n)$$

$$\eta = I'$$

$$\zeta = I''$$

where the primes denote differentiation with respect to t. According to the central limit theorem the distribution of ξ, η, ζ approaches a normal law. The second moments defining this law may be obtained either from the above definitions of ξ, η, ζ, or may be obtained from the correlation function as was done in the work following equation (3.4–13).

$$\overline{\xi^2} = \psi_0, \qquad \overline{\eta^2} = -\psi_0'', \qquad \overline{\xi\eta} = 0$$

$$\overline{\eta\zeta} = \overline{I'(t)I''(t)} = \underset{T\to\infty}{\text{Limit}} \frac{1}{T}\int_0^T I'(t)I''(t)\,dt$$

$$= \underset{T\to\infty}{\text{Limit}} \frac{1}{2T}[I'^2(T) - I'^2(0)] = 0$$

$$\overline{\xi\zeta} = \underset{T\to\infty}{\text{Limit}} \frac{1}{T}\int_0^T I(t)I''(t)\,dt$$

$$= \underset{\tau\to 0}{\text{Limit}} \frac{\partial^2\psi(\tau)}{\partial\tau^2} = \psi_0''$$

$$\overline{\zeta^2} = \underset{T\to\infty}{\text{Limit}} \frac{1}{T}\int_0^T I''(t)I''(t)\,dt$$

$$= \underset{T\to\infty}{\text{Limit}} \frac{1}{T}\int_0^T I^{(4)}(t)I(t)\,dt$$

$$= \psi_0^{(4)}$$

where the superscript (4) represents the fourth derivative. The matrix M of the moments is thus

$$M = \begin{bmatrix} \psi_0 & 0 & \psi_0'' \\ 0 & -\psi_0'' & 0 \\ \psi_0'' & 0 & \psi_0^{(4)} \end{bmatrix}$$

The determinant $|M|$ and the cofactors of interest are

$$|M| = -\psi_0''(\psi_0\psi_0^{(4)} - \psi_0''^2) \tag{3.6-3}$$

$$M_{11} = -\psi_0''\psi_0^{(4)}, \qquad M_{13} = \psi_0''^2, \qquad M_{33} = -\psi_0''\psi_0$$

The probability density function in (3.6–1) is

$$p(I, 0, \zeta) = (2\pi)^{-3/2} |M|^{-1/2} \exp \left[-\frac{1}{2|M|} (M_{11}I^2 + M_{33}\zeta^2 + 2M_{13}I\zeta) \right] \tag{3.6–4}$$

and when this is put in (3.6–1) and the integration with respect to ζ performed we get

$$dI\, dt \frac{(2\pi)^{-3/2}}{M_{33}} \left[|M|^{1/2} e^{-M_{11}I^2/2|M|} + M_{13}I \left(\frac{\pi}{2M_{33}}\right)^{1/2} e^{-I^2/2\psi_0} \left(1 + \operatorname{erf} \frac{M_{13}I}{[2|M| M_{33}]^{1/2}}\right) \right] \tag{3.6–5}$$

for the probability of a maximum occurring in the rectangle $dI\, dt$. As is mentioned just below expression (3.6–1), the expected number of maxima in the interval t_1, t_2 may be obtained by integrating (3.6–1) from t_1 to t_2 after replacing x by t, and I from $-\infty$ to $+\infty$ after replacing y by I. When we use (3.6–4) it is easier to integrate with respect to I first. The expected number is then

$$-\int_{t_1}^{t_2} dt \frac{M_{11}^{-1/2}}{2\pi} \int_{-\infty}^{0} \zeta \exp\left[-\frac{\zeta^2}{2|M|} \left(M_{33} - \frac{M_{13}^2}{M_{11}} \right) \right] d\zeta$$

$$= (t_2 - t_1) \frac{\psi_0^{(4)}}{2\pi} M_{11}^{-1/2} = \frac{t_2 - t_1}{2\pi} \left[\frac{\psi_0^{(4)}}{-\psi_0''} \right]^{1/2}$$

Hence the expected number of maxima per second is

$$\frac{1}{2\pi} \left[\frac{\psi_0^{(4)}}{-\psi_0''} \right]^{1/2} = \left[\frac{\int_0^\infty f^4 w(f)\, df}{\int_0^\infty f^2 w(f)\, df} \right]^{1/2} \tag{3.6–6}$$

For a band pass filter, the expected number of maxima per second is

$$\left[\frac{3}{5} \frac{f_b^5 - f_a^5}{f_b^3 - f_a^3} \right]^{1/2} \tag{3.6–7}$$

where f_b and f_a are the cut-off frequencies. Putting $f_a = 0$ so as to get a low pass filter,

$$f_b \left[\frac{3}{5} \right]^{1/2} = .775 f_b \tag{3.6–8}$$

From (3.6–8) and (3.6–5) we may obtain the probability density function for the maxima in the case of a low pass filter. Thus the probability that a maximum selected at random from the universe of maxima will lie in $I, I + dI$ is

$$\frac{dI}{3\sqrt{2\pi\psi_0}}\left[2e^{-9y^2/8} + \left(\frac{5\pi}{2}\right)^{1/2} ye^{-y^2/2}\left(1 + \text{erf } y\left(\frac{5}{8}\right)^{1/2}\right)\right] \tag{3.6–9}$$

where

$$y = \frac{I}{\sqrt{\psi_0}}$$

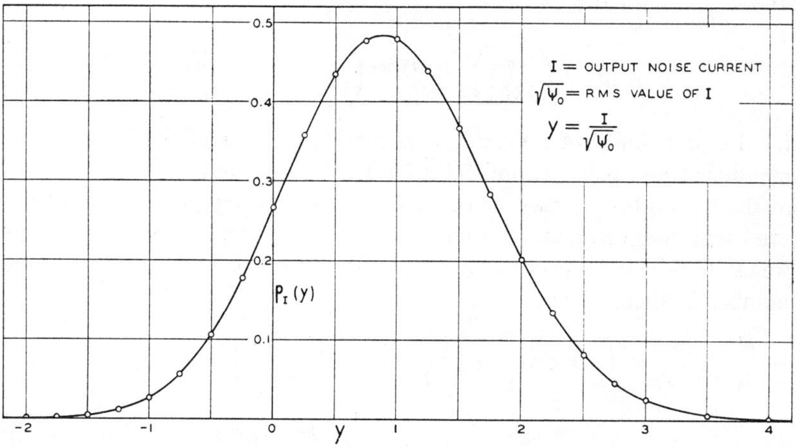

Fig. 2—Distribution of maxima of noise current. Noise through ideal low-pass filter. $\frac{p_I(y)}{\sqrt{\psi_0}} dI$ = probability that a maximum of I selected at random lies between I and $I + dI$.

When y is large and positive (3.6–9) is given asymptotically by

$$\frac{dI}{\sqrt{\psi_0}} \frac{\sqrt{5}}{3} ye^{-y^2/2}$$

If we write (3.6–9) as $p_I(y)\, dy$, the probability density $p_I(y)$ of y may be plotted as a function of y. This plot is shown in Fig. 2. The distribution function $P(I_{\max} < y\sqrt{\psi_0})$ defined by

$$P(I_{\max} < y\sqrt{\psi_0}) = \int_{-\infty}^{y} p_I(y)\cdot dy$$

and which gives the probability that a maximum selected at random is less than a specified $y\sqrt{\psi_0} = I$, is one of the four curves plotted in Fig. 4.

If I is large and positive we may obtain an approximation from (3.6–5). We observe that

$$\frac{M_{11}}{|M|} = \frac{\psi_0^{(4)}}{\psi_0\psi_0^{(4)} - \psi_0''^2} > \frac{1}{\psi_0}$$

so that when I is large and positive

$$e^{-M_{11}I^2/2|M|} \ll e^{-I^2/2\psi_0}$$

Also, in these circumstances the 1 + erf is nearly equal to two. Thus retaining only the important terms and using the definitions of the M's gives the approximation to (3.6–5):

$$\frac{dI\,dt}{2\pi\psi_0}\left[\frac{-\psi_0''}{\psi_0}\right]^{1/2} Ie^{-I^2/2\psi_0} \tag{3.6–10}$$

From this it follows that the expected number of maxima per second lying above the line $I = I_1$ is approximately[33] when I_1 is large,

$$\begin{aligned} &\frac{1}{2\pi}\left[\frac{-\psi_0''}{\psi_0}\right]^{1/2} e^{-I_1^2/2\psi_0} \\ &\quad = e^{-I_1^2/2\psi_0} \times \tfrac{1}{2}[\text{the expected number of zeros of } I \text{ per second}] \end{aligned} \tag{3.6–11}$$

It is interesting to note that the approximation (3.6–11) for the expected number of maxima above I_1 is the same as the exact expression (3.3–14) for the expected number of times I will pass through I_1 with positive slope.

3.7 Results on the Envelope of the Noise Current

The noise current flowing in the output of a relatively narrow band pass filter has the character of a sine wave of, roughly, the midband frequency whose amplitude fluctuates irregularly, the rapidity of fluctuation being of the order of the band width. Here we study the fluctuations of the envelope of such a wave.

First we define the envelope. Let f_m be a representative midband frequency. Then if

$$\omega_m = 2\pi f_m \tag{3.7–1}$$

the noise current may be represented, see (2.8–6), by

$$\begin{aligned} I &= \sum_{n=1}^{N} c_n \cos(\omega_n t - \omega_m t - \varphi_n + \omega_m t) \\ &= I_c \cos \omega_m t - I_s \sin \omega_m t \end{aligned} \tag{3.7–2}$$

where the components I_c and I_s are

$$\begin{aligned} I_c &= \sum_{n=1}^{N} c_n \cos(\omega_n t - \omega_m t - \varphi_n) \\ I_s &= \sum_{n=1}^{N} c_n \sin(\omega_n t - \omega_m t - \varphi_n) \end{aligned} \tag{3.7–3}$$

[33] This expression agrees with an estimate made by V. D. Landon, *Proc. I. R. E.*, 29 (1941), 50–55. He discusses the number of crests exceeding four times the r.m.s. value of I. This corresponds to $I_1^2 = 16\psi_0$.

The envelope, R, is a function of t defined by

$$R = [I_c^2 + I_s^2]^{1/2} \tag{3.7–4}$$

It follows from the central limit theorem and the definitions (3.7–3) of I_c and I_s that these are two normally distributed random variables. They are independent since $\overline{I_c I_s} = 0$. They both have the same standard deviation, namely the square root of

$$\overline{I_c^2} = \overline{I_s^2} = \overline{I^2} = \int_0^\infty w(f)\, df = \psi_0 \tag{3.7–5}$$

Consequently, the probability that the point (I_c, I_s) lies within the elementary rectangle $dI_c dI_s$ is

$$\frac{dI_c\, dI_s}{2\pi\psi_0} \exp\left[-\frac{I_c^2 + I_s^2}{2\psi_0}\right] \tag{3.7–6}$$

In much of the following work it is convenient to introduce another random variable θ where

$$\begin{aligned} I_c &= R\cos\theta \\ I_s &= R\sin\theta \end{aligned} \tag{3.7–7}$$

Since I_c and I_s are random variables so are R and θ. The differentials are related by

$$dI_c dI_s = R d\theta dR \tag{3.7–8}$$

and the distribution function for R and θ is obtainable from (3.7–6) when the change of variables is made:

$$\frac{d\theta}{2\pi} \frac{R\, dR}{\psi_0} e^{-R^2/2\psi_0} \tag{3.7–9}$$

Since this may be expressed as a product of terms involving R only and θ only, R and θ are independent random variables, θ being uniformly distributed over the range 0 to 2π and R having the probability density[34]

$$\frac{R}{\psi_0} e^{-R^2/2\psi_0} \tag{3.7–10}$$

Expression (3.7–10) gives the probability density for the value of the envelope. Like the normal law for the instantaneous value of I, it depends only upon the average total power

$$\psi_0 = \int_0^\infty w(f)\, df$$

[34] See V. D. Landon and K. A. Norton, *I.R.E. Proc.*, 30 (1942), 425–429.

We now study the correlation between R at time t and its value at some later time $t + \tau$. Let the subscripts 1 and 2 refer to the times t and $t + \tau$, respectively. Then from (3.7–3) and the central limit theorem it follows that the four random variables I_{c1}, I_{s1}, I_{c2}, I_{s2} have a four dimensional normal distribution. This distribution is determined by the second moments

$$\overline{I_{c1}^2} = \overline{I_{s1}^2} = \overline{I_{c2}^2} = \overline{I_{s2}^2} = \psi_0 = \mu_{11}$$

$$\overline{I_{c1} I_{s1}} = \overline{I_{c2} I_{s2}} = 0$$

$$\overline{I_{c1} I_{c2}} = \overline{I_{s1} I_{s2}} = \frac{1}{2} \sum_{n=1}^{N} c_n^2 \cos(\omega_n \tau - \omega_m \tau)$$

$$\rightarrow \int_0^\infty w(f) \cos 2\pi(f - f_m)\tau \, df = \mu_{13} \tag{3.7–11}$$

$$\overline{I_{c1} I_{s2}} = -\overline{I_{c2} I_{s1}} = \frac{1}{2} \sum_{n=1}^{N} c_n^2 \sin(\omega_n \tau - \omega_m \tau)$$

$$\rightarrow \int_0^\infty w(f) \sin 2\pi(f - f_m)\tau \, df = \mu_{14}$$

The moment matrix for the variables in the order I_{c1}, I_{s1}, I_{c2}, I_{s2} is

$$M = \begin{bmatrix} \psi_0 & 0 & \mu_{13} & \mu_{14} \\ 0 & \psi_0 & -\mu_{14} & \mu_{13} \\ \mu_{13} & -\mu_{14} & \psi_0 & 0 \\ \mu_{14} & \mu_{13} & 0 & \psi_0 \end{bmatrix}$$

and from this it follows that the cofactors of the determinant $|M|$ are

$$\begin{aligned} M_{11} = M_{22} = M_{33} = M_{44} &= \psi_0(\psi_0^2 - \mu_{13}^2 - \mu_{14}^2) \\ = \psi_0 A, \qquad & A = \psi_0^2 - \mu_{13}^2 - \mu_{14}^2 \end{aligned} \tag{3.7–12}$$

$$M_{12} = M_{34} = 0$$

$$M_{13} = M_{24} = -\mu_{13} A$$

$$M_{14} = -M_{23} = -\mu_{14} A$$

$$|M| = A^2$$

The probability density of the four random variables is therefore

$$\frac{1}{4\pi^2 A} \exp - \frac{1}{2A} [\psi_0(I_1^2 + I_2^2 + I_3^2 + I_4^2)$$

$$- 2\mu_{13}(I_1 I_3 + I_2 I_4) - 2\mu_{14}(I_1 I_4 - I_2 I_3)]$$

where we have written I_1, I_2, I_3, I_4 for $I_{c1}, I_{s1}, I_{c2}, I_{s2}$. We now make the transformation

$$I_1 = R_1 \cos \theta_1 \qquad I_3 = R_2 \cos \theta_2$$
$$I_2 = R_1 \sin \theta_1 \qquad I_4 = R_2 \sin \theta_2$$

and average the resulting probability density over θ_1 and θ_2 in order to get the probability that R_1 and R_2 lie in dR_1 and dR_2. It is

$$\frac{R_1 R_2\, dR_1\, dR_2}{4\pi^2 A} \int_0^{2\pi} d\theta_1 \int_0^{2\pi} d\theta_2 \exp$$
$$-\frac{1}{2A} [\psi_0 R_1^2 + \psi_0 R_2^2 - 2\mu_{13} R_1 R_2 \cos(\theta_2 - \theta_1) - 2\mu_{14} R_1 R_2 \sin(\theta_2 - \theta_1)]$$

Since the integrand is a periodic function of θ_2 we may integrate from $\theta_2 = \theta_1$ to $\theta_2 = \theta_1 + 2\pi$ instead of from 0 to 2π. This integration gives the Bessel function, I_0, of the first kind with imaginary argument. The resulting probability density for R_1 and R_2 is

$$\frac{R_1 R_2}{A} I_0\left(\frac{R_1 R_2}{A} [\mu_{13}^2 + \mu_{14}^2]^{1/2}\right) \exp -\frac{\psi_0}{2A}(R_1^2 + R_2^2) \qquad (3.7\text{–}13)$$

where, from (3.7–12),

$$A = \psi_0^2 - \mu_{13}^2 - \mu_{14}^2$$

μ_{13} and μ_{14} are given by (3.7–11). Of course, R_1 and R_2 are always positive.

For an ideal band pass filter with cut-offs at f_a and f_b we set

$$f_m = \frac{f_b + f_a}{2}, \qquad w(f) = w_0 \quad \text{for} \quad f_a < f < f_b$$

and obtain

$$\psi_0 = w_0(f_b - f_a)$$

$$\mu_{13} = \int_{f_a}^{f_b} w_0 \cos 2\pi(f - f_m)\tau\, df = \frac{w_0 \sin \pi(f_b - f_a)\tau}{\pi\tau}$$

$$\mu_{14} = \int_{f_a}^{f_b} w_0 \sin 2\pi(f - f_m)\tau\, df = 0$$

The I_0 term in (3.7–13), which furnishes the correlation between R_1 and R_2, becomes

$$I_0\left(\frac{R_1 R_2}{\psi_0} \frac{\dfrac{\sin x}{x}}{1 - \dfrac{\sin^2 x}{x^2}}\right)$$

where x is $\pi(f_b - f_a)\tau$. When x is a multiple of π, R_1 and R_2 are independent random variables. When x is zero R_1 and R_2 are equal. Hence we may say, roughly, that the period of fluctuation of R is the time it takes x to increase from 0 to π or $(f_b - f_a)^{-1}$. This is related to the result given in the next section, namely that the expected number of maxima of the envelope is .641 $(f_b - f_a)$ per second.

3.8 Maxima of R

Here we wish to study the distribution of the maxima of R.* Our work is based upon the expression, cf. (3.6–1),

$$-dR\,dt \int_{-\infty}^{0} p(R, 0, R'')R''\,dR'' \tag{3.8–1}$$

for the probability that a maximum of R falls within the elementary rectangle $dR\,dt$. $p(R, R', R'')$ is the probability density for the three dimensional distribution of R, R', R'' where the primes denote differentiation with respect to t.

We shall determine $p(R, R', R'')$ from the probability density of I_c, I_s', I_c'', I_s, I_c', I_s'', which we shall denote by x_1, x_2, $\cdots$ x_6. The interchange of I_s' and I_c' is suggested by the later work. It is convenient to introduce the notation

$$\begin{aligned} b_n &= (2\pi)^n \int_0^\infty w(f)(f - f_m)^n\,df \\ b_0 &= \psi_0 \end{aligned} \tag{3.8–2}$$

where f_m is the mid-band frequency, i.e., the frequency chosen in the definition of the envelope R. b_n is seen to be analogous to the derivatives of $\psi(\tau)$ at $\tau = 0$.

From the definitions (3.7–3) of I_c and I_s we obtain the second moments

$$\begin{aligned} \overline{x_1^2} &= \overline{I_c^2} = \psi_0 = b_0 \\ \overline{x_4^2} &= \overline{I_s^2} = b_0 \\ \overline{x_2^2} &= \overline{I_s'^2} = \sum_1^N w(f_n)\Delta f 4\pi^2(f_n - f_m)^2 = b_2 \\ \overline{x_5^2} &= \overline{I_c'^2} = b_2 \\ \overline{x_3^2} &= \overline{I_c''^2} = b_4 \\ \overline{x_6^2} &= \overline{I_s''^2} = b_4 \end{aligned}$$

* Incidentally, most of the analysis of this section was originally developed in a study of the stability of repeaters in a loaded telephone transmission line. The envelope, R, was associated with the "returned current" produced by reflections from line irregularities. However, the study fell short of its object and the only results which seemed worth salvaging at the time were given in reference[25] cited in Section 3.3.

$$\begin{aligned}
\overline{x_1x_2} &= \overline{I_cI_s'} = \sum_1^N w(f_n)\Delta f 2\pi(f_n - f_m) = b_1 \\
\overline{x_4x_5} &= \overline{I_sI_c'} = -b_1 \\
\overline{x_1x_3} &= \overline{I_cI_c''} = -\sum_1^N w(f)\Delta f 4\pi^2(f_n - f_m)^2 = -b_2 \\
\overline{x_4x_6} &= \overline{I_sI_s''} = -b_2 \\
\overline{x_2x_3} &= \overline{I_s'I_c''} = -b_3 \\
\overline{x_5x_6} &= \overline{I_c'I_s''} = b_3
\end{aligned}$$

All of the other second moments are zero. The moment matrix M is thus

$$M = \begin{bmatrix} b_0 & b_1 & -b_2 & 0 & 0 & 0 \\ b_1 & b_2 & -b_3 & 0 & 0 & 0 \\ -b_2 & -b_3 & b_4 & 0 & 0 & 0 \\ 0 & 0 & 0 & b_0 & -b_1 & -b_2 \\ 0 & 0 & 0 & -b_1 & b_2 & b_3 \\ 0 & 0 & 0 & -b_2 & b_3 & b_4 \end{bmatrix}$$

The adjoint matrix is

$$\begin{bmatrix} B_0 & B_1 & -B_2 & 0 & 0 & 0 \\ B_1 & B_{22} & -B_3 & 0 & 0 & 0 \\ -B_2 & -B_3 & B_4 & 0 & 0 & 0 \\ 0 & 0 & 0 & B_0 & -B_1 & -B_2 \\ 0 & 0 & 0 & -B_1 & B_{22} & B_3 \\ 0 & 0 & 0 & -B_2 & B_3 & B_4 \end{bmatrix}$$

$$\begin{aligned}
B_0 &= (b_2b_4 - b_3^2)B & B_{22} &= (b_0b_4 - b_2^2)B \\
B_1 &= -(b_1b_4 - b_2b_3)B & B_3 &= -(b_0b_3 - b_1b_2)B \\
B_2 &= (b_1b_3 - b_2^2)B & B_4 &= (b_0b_2 - b_1^2)B
\end{aligned} \tag{3.8–3}$$

$$\begin{aligned}
B &= b_0b_2b_4 + 2\,b_1b_2b_3 \\
&\quad - b_2^3 - b_0b_3^2 - b_4b_1^2 \\
|M| &= B^2
\end{aligned}$$

where B is the determinant of the third order matrices in the upper left and lower right corners of M.

As in the earlier work, the distribution of $x_1, \cdots, x_6$ is normal in six dimensions. The exponent is $-[2\,|M|]^{-1}$ times

$$\begin{aligned}
&B_0(x_1^2 + x_4^2) + 2B_1(x_1x_2 - x_4x_5) - 2B_2(x_1x_3 + x_4x_6) \\
&\quad + B_{22}(x_2^2 + x_5^2) \quad - 2B_3(x_2x_3 - x_5x_6) \\
&\quad + B_4(x_3^2 + x_6^2)
\end{aligned} \tag{3.8–4}$$

In line with the earlier work we set

$$x_1 = I_c = R\cos\theta \qquad x_4 = I_s = R\sin\theta$$

$$x_2 = I_s' = R'\sin\theta + R\cos\theta\theta'$$

$$x_5 = I_c' = R'\cos\theta - R\sin\theta\theta'$$

$$x_3 = I_c'' = R''\cos\theta - 2R'\sin\theta\theta' - R\cos\theta\theta'^2 - R\sin\theta\theta''$$

$$x_6 = I_s'' = R''\sin\theta + 2R'\cos\theta\theta' - R\sin\theta\theta'^2 + R\cos\theta\theta''$$

The angle θ varies from 0 to 2π and θ' and θ'' vary from $-\infty$ to $+\infty$. By forming the Jacobian it may be shown that

$$dx_1\,dx_2 \cdots dx_6 = R^3\,dR\,dR'\,dR''\,d\theta\,d\theta'\,d\theta''$$

Also, the quantities in (3.8–4) are

$$x_1^2 + x_4^2 = R^2 \qquad x_1x_3 + x_4x_6 = RR'' - R^2\theta'^2$$

$$x_1x_2 - x_4x_5 = R^2\theta' \qquad x_2^2 + x_5^2 = R'^2 + R^2\theta'^2$$

$$x_2x_3 - x_5x_6 = RR''\theta' - 2R'^2\theta' - R'R\theta'' - R^2\theta'^3$$

$$x_3^2 + x_6^2 = R''^2 - 2RR''\theta'^2 + 4R'^2\theta'^2 + 4RR'\theta'\theta'' + R^2\theta'^4 + R^2\theta''^2$$

The expression for $p(R, 0, R'')$ is obtained when we set these values of the x's in (3.8–4) and integrate the resulting probability density over the ranges of θ, θ', θ'':

$$p(R, 0, R'') = \frac{R^3}{8\pi^3 B}\int_0^{2\pi} d\theta \int_{-\infty}^{+\infty} d\theta' \int_{-\infty}^{+\infty} d\theta'' \tag{3.8–5}$$

$$\exp -\frac{1}{2B^2}[B_0R^2 + 2B_1R^2\theta' - 2B_2(RR'' - R^2\theta'^2) + B_{22}R^2\theta'^2 - 2B_3R\theta'(R'' - R\theta'^2) + B_4(R''^2 - 2RR''\theta'^2 + R^2\theta'^4 + R^2\theta''^2)]$$

The integrations with respect to θ and θ'' may be performed at once leaving $p(R, 0, R'')$ expressed as a single integral which, unfortunately, appears to be difficult to handle. For this reason we assume that $w(f)$ is symmetrical about the mid-band frequency f_m. From (3.8–2), b_1 and b_3 are zero and from (3.8–3), B_1 and B_3 are zero.

With this assumption (3.8–5) yields

$$p(R, 0, R'') = R^2(2\pi)^{-3/2} B_4^{-1/2} \int_{-\infty}^{+\infty} d\theta' \tag{3.8–6}$$

$$\exp -\frac{1}{2B^2}[B_0R^2 + R([B_{22} + 2B_2]R\theta'^2 - 2B_2R'') + B_4(R'' - R\theta'^2)^2]$$

The probability that a maximum occurs in the elementary rectangle dR dt is, from (3.8–1), $p(t, R)\, dR\, dt$ where

$$p(t, R) = -\int_{-\infty}^{0} p(R, 0, R'')R''\, dR'' \tag{3.8–7}$$

We put (3.8–6) in this expression and make the following change of variables.

$$x = \frac{B_4^{1/2}}{\sqrt{2}\,B} R\theta'^2, \qquad y = -\frac{B_4^{1/2}}{\sqrt{2}\,B} R''$$

$$z = -\frac{B_2}{\sqrt{2B_4}\,B} R = \frac{b_2^2}{\sqrt{2B_4}} R \tag{3.8–8}$$

$$b = -\frac{(B_{22} + 2B_2)}{2B\, b_2^2} = \left[\frac{3}{2} - \frac{b_0 b_4}{2b_2^2}\right] = \tfrac{1}{2}(3 - a^2)$$

$$a^2 = \frac{B_0}{2B^2}\frac{2B_4}{b_2^4} = \frac{b_0 b_4}{b_2^2}$$

where we have used the expressions for the B's obtained by setting b_1 and b_3 to zero in (3.8–3). Thus

$$p(t, R) = \frac{4}{b_0 b_2^4}\left(\frac{Bz}{2\pi}\right)^{3/2} \int_0^{\infty} y\, dy \int_0^{\infty} x^{-1/2}\, dx \tag{3.8–9}$$

$$\exp\left[-a^2z^2 + 2bzx + 2zy - (x + y)^2\right]$$

As was to be expected, this expression shows that $p(t, R)$ is independent of t.

A series for $p(t, R)$ may be obtained by expanding $\exp 2z(y + bx)$ and then integrating termwise. We use

$$\int_0^{\infty} dy \int_0^{\infty} dx\, x^{\mu} y^{\gamma} e^{-(x+y)^2} = \frac{\sqrt{\pi}}{2^{\mu+\gamma+2}} \frac{\Gamma(\gamma + 1)\Gamma(\mu + 1)}{\Gamma\left(\dfrac{\mu + \gamma + 3}{2}\right)}$$

which may be evaluated by setting

$$x = \rho^2 \cos^2 \varphi, \qquad y = \rho^2 \sin^2 \varphi$$

The double integral in (3.8–9) becomes

$$e^{-a^2z^2}\sqrt{\frac{\pi}{2}}\sum_{n=0}^{\infty}\frac{(2z)^n}{n!}\sum_{m=0}^{n}\frac{n!\,b^m}{m!\,(n-m)!}\frac{\Gamma(m+\frac{1}{2})\Gamma(n-m+2)}{2^{n+2}\Gamma\left(\frac{n}{2}+\frac{7}{4}\right)}$$

$$= \pi 2^{-5/2}\sum_{n=0}^{\infty}\frac{z^n e^{-a^2z^2}}{\Gamma\left(\frac{n}{2}+\frac{7}{4}\right)}A_n$$

where $A_0 = 1$ and

$$A_n = \sum_{m=0}^{n}\frac{(\frac{1}{2})(\frac{3}{2})\cdots(m-\frac{1}{2})}{m!}(n-m+1)b^m, \qquad 0 < n \qquad (3.8\text{–}10)$$

$$A_n \sim (n+1)(1-b)^{-1/2} - \frac{b}{2}(1-b)^{-3/2}, \qquad n \text{ large}$$

The term corresponding to $m = 0$ in (3.8–10) is $n + 1$.

We thus obtain

$$p(t, R) = \frac{e^{-a^2z^2}}{4b_0b_2^4}\frac{(Bz)^{3/2}}{\sqrt{\pi}}\sum_{n=0}^{\infty}\frac{z^n}{\Gamma\left(\frac{n}{2}+\frac{7}{4}\right)}A_n$$

$$= \frac{e^{-a^2z^2}}{4\sqrt{\pi}}\frac{b_2^{1/2}}{b_0}(a^2-1)^{3/2}z^{3/2}\sum_{n=0}^{\infty}\frac{z^nA_n}{\Gamma\left(\frac{n}{2}+\frac{7}{4}\right)} \qquad (3.8\text{–}11)$$

We are interested in the expected number, N, of maxima per second. From the similar work for I, it follows that N is the coefficient of dt when (3.8–1) is integrated with respect to R from 0 to ∞. Thus from (3.8–7) and

$$dR = \sqrt{2B_4}\,b_2^{-2}\,dz = (2b_0B)^{1/2}b_2^{-3/2}\,dz$$

$$= [2b_0(a^2-1)]^{1/2}\,dz$$

we find

$$N = \int_0^{\infty} p(t, R)\,dR$$

$$= \frac{(a^2-1)^2}{(2a)^{5/2}}\left(\frac{b_2}{\pi b_0}\right)^{1/2}\sum_{n=0}^{\infty}\frac{\Gamma\left(\frac{n}{2}+\frac{5}{4}\right)}{\Gamma\left(\frac{n}{2}+\frac{7}{4}\right)}\frac{A_n}{a^n} \qquad (3.8\text{–}12)$$

Equations (3.8–11) and (3.8–12) have been derived on the assumption that $w(f)$ is symmetrical about f_m, i.e. the band pass filter attenuation is symmetrical about the mid-band frequency. We now go a step further and assume an ideal band pass filter:

$$
\begin{aligned}
w(f) &= w_0 \qquad f_a < f < f_b \\
w(f) &= 0 \qquad \text{otherwise} \\
2f_m &= f_a + f_b
\end{aligned}
\tag{3.8–13}
$$

Putting these in (3.8–2) we obtain zero for b_1 and b_3 and also

$$
\begin{aligned}
b_0 &= w_0(f_b - f_a) = \psi_0 \\
b_2 &= \frac{\pi^2 w_0}{3}(f_b - f_a)^3 \\
b_4 &= \frac{\pi^4 w_0}{5}(f_b - f_a)^5 \\
a^2 &= \tfrac{9}{5} \\
b &= \tfrac{1}{2}(3 - a^2) = \tfrac{3}{5} \\
R &= [2b_0(a^2 - 1)]^{1/2} z = [\tfrac{8}{5}\psi_0]^{1/2} z \\
\left(\frac{b_2}{\pi b_0}\right)^{1/2} &= \left[\frac{\pi}{3}\right]^{1/2}(f_b - f_a), \qquad a^2 z^2 = \frac{9R^2}{8\psi_0}
\end{aligned}
\tag{3.8–14}
$$

n	A_n	n	A_n
0	1	4	6.775
1	2.3	5	8.333
2	3.735	6	9.9002
3	5.238	7	11.4736

$$A_n \sim 1.5811\, n + .3953$$

From (3.8–12) we find that the expected number of maxima per second of the envelope is

$$N = .64110\,(f_b - f_a) \tag{3.8–15}$$

assuming an ideal band pass filter.

The distribution of the maxima of R for an ideal band pass filter may be obtained by placing the results of (3.8–14) in (3.8–11). This gives

$$
p(t, R)\, dR = \frac{dR}{\psi_0^{1/2}} \frac{(f_b - f_a)}{4} \sqrt{\frac{\pi}{3}} \left(\frac{4z}{5}\right)^{3/2} e^{-a^2 z^2} \sum_{n=0}^{\infty} \frac{z^n A_n}{\Gamma\left(\frac{n}{2} + \frac{7}{4}\right)}
$$

It is convenient to define y as the ratio

$$y = \frac{R}{\text{r.m.s. } I(t)} = \frac{R}{\psi_0^{1/2}} = \left(\tfrac{8}{5}\right)^{1/2} z$$

where R is understood to correspond to a maximum of the envelope. Since the value of R corresponding to a maximum of the envelope selected at random is a random variable, y is also a random variable. Its probability density is $p_R(y)$, where

$$p_R(y)\, dy = \frac{p(t, R)\, dR}{0.64110(f_b - f_a)}$$

$p_R(y)$ has been computed and is plotted as a function of y in Fig. 3.

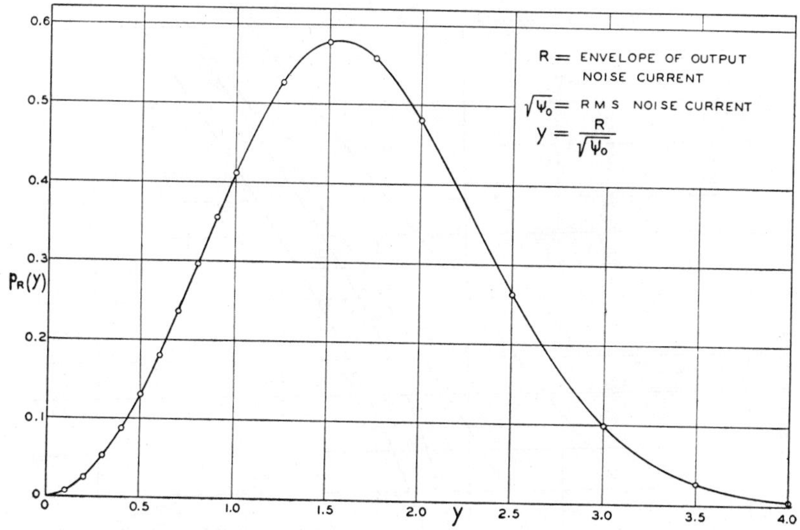

Fig. 3—Distribution of maxima of envelope of noise current. Noise through ideal band-pass filter.
$\frac{p_R(y)}{\sqrt{\psi_0}}\, dR$ = probability that a maximum of R selected at random lies between R and $R + dR$.

The distribution function $P(R_{\max} < y\sqrt{\psi_0})$ defined by

$$P(R_{\max} < y\sqrt{\psi_0}) = \int_0^y p_R(y)\, dy$$

and which gives the probability that a maximum of the envelope selected at random is less than a specified value $y\sqrt{\psi_0} = R$, is plotted in Fig. 4 together with other curves of the same nature.

When y is large, say greater than 2.5,

$$p_R(y) \sim \frac{\sqrt{\frac{\pi}{6}}}{.64110}(y^2 - 1)e^{-y^2/2}$$

$$P(R_{\max} < y\sqrt{\psi_0}) \sim 1 - \frac{\sqrt{\frac{\pi}{6}}}{.64110}ye^{-y^2/2}$$

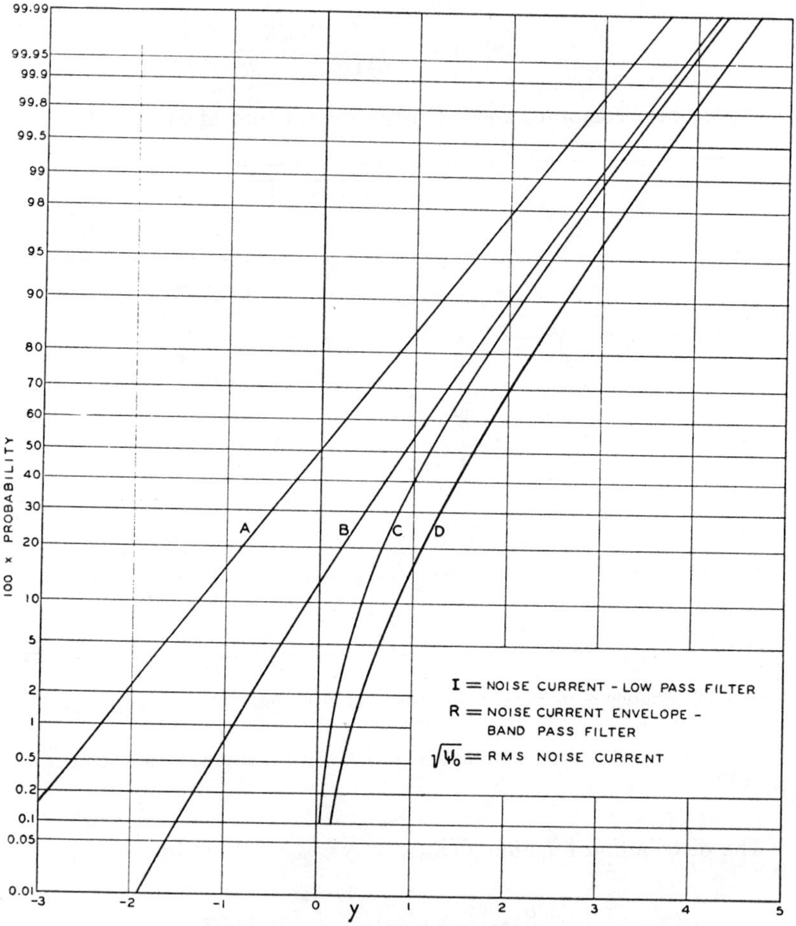

Fig. 4—Distribution of maxima

$A = P(I < y\sqrt{\psi_0})$ = probability of I being less than $y\sqrt{\psi_0}$. Similarly $C = P(R < y\sqrt{\psi_0})$.

$B = P(I \max < y\sqrt{\psi_0})$ = probability of random maximum of I being less than $y\sqrt{\psi_0}$. Similarly $D = P(R \max < y\sqrt{\psi_0})$.

The asymptotic expression for $p_R(y)$ may be obtained from the integral (3.8–9) for $p(t, R)$. Indeed, replacing the variables of integration x, y in (3.8–9) by

$$x' = x$$

$$y' = x + y,$$

integrating a portion of the y' integral by parts, and assuming $b < 1$ ($a^2 \geq 1$, by Schwarz's inequality, so that $b \leq 1$ always) leads to

$$p(t, R) \sim \left(\frac{b_2}{2\pi}\right)^{\frac{1}{2}} \frac{e^{-R^2/2\psi_0}}{\psi_0} \left(\frac{R^2}{\psi_0} - 1\right)$$

when R is large.

If, instead of an ideal band pass filter, we assume that $w(f)$ is given by

$$w(f) = \frac{1}{\sigma\sqrt{2\pi}} e^{-(f-f_m)^2/2\sigma^2}, \qquad f_m \gg \sigma \qquad (3.8\text{–}16)$$

we find that

$$b_0 = 1$$

$$b_2 = 4\pi^2\sigma^2$$

$$b_4 = 16\pi^4 \cdot 3\sigma^4$$

$$a^2 = 3, \ b = 0$$

$$A_n = (n + 1)$$

Some rough work indicates that the sum of the series in (3.8–12) is near 3.97. This gives the expected number of maxima of the envelope as

$$N = 2.52\sigma \qquad (3.8\text{–}17)$$

per second.

The pass band is determined by σ. It appears difficult to compare this with an ideal band pass filter. If we use the fact that the filter given by

$$w(f) = w_0 \exp\left[-\pi\left(\frac{f - f_m}{f_b - f_a}\right)^2\right]$$

passes the same average amount of power as does an ideal band pass filter whose pass band is $f_b - f_a$, we have

$$f_b - f_a = \sigma\sqrt{2\pi}$$

and the expression for N becomes $1.006\ (f_b - f_a)$.

3.9 Energy Fluctuation

Some information regarding the statistical behavior of the random variable

$$E = \int_{t_1}^{t_1+T} I^2(t)\, dt \qquad (3.9\text{–}1)$$

where $I(t)$ is a noise current and t_1 is chosen at random, has been given in a recent article.[35] Here we study this behavior from a somewhat different point of view.

If we agree to use the representations (2.8–1) or (2.8–6) we may write, as in the paper, the random variable E as

$$E = \int_{-T/2}^{T/2} I^2(t)\, dt \tag{3.9–2}$$

where the randomness on the right is due either to the a_n's and b_n's if (2.8–1) is used or to the φ_n's if (2.8–6) is used.

The average value of E is m_T where, from (3.1–2),

$$\begin{aligned} \bar{E} = m_T &= \int_{-T/2}^{T/2} \overline{I^2(t)}\, dt = \int_{-T/2}^{T/2} \psi(0)\, dt = T\psi_0 \\ &= T \int_0^{\infty} w(f)\, df \end{aligned} \tag{3.9–3}$$

The second moment of E is

$$\overline{E^2} = \int_{-T/2}^{T/2} dt_1 \int_{-T/2}^{T/2} dt_2\, \overline{I^2(t_1) I^2(t_2)} \tag{3.9–4}$$

If, for the time being, we set t_2 equal to $t_1 + \tau$, it is seen from section 3.2 that we have an expression for the probability density of $I(t_1)$ and $I(t_1 + \tau)$ and hence we may obtain the required average:

$$\begin{aligned} \overline{I_1^2 I_2^2} = \frac{1}{2\pi A} \int_{-\infty}^{+\infty} dI_1 \int_{-\infty}^{+\infty} dI_2\, I_1^2 I_2^2 \exp \\ \left(-\frac{1}{2A^2} (\psi_0 I_1^2 + \psi_0 I_2^2 - 2\psi_\tau I_1 I_2) \right) \end{aligned} \tag{3.9–5}$$

$$A^2 = \psi_0^2 - \psi_\tau^2, \quad I_1 = I(t_1), \quad I_2 = I(t_1 + \tau) = I(t_2)$$

The integral may be evaluated by (3.5–6) when we set

$$\begin{aligned} I_1 = Ax\sqrt{\frac{2}{\psi_0}}, \qquad & I_2 = Ay\sqrt{\frac{2}{\psi_0}} \\ \psi_\tau &= -\psi_0 \cos \varphi \\ A &= \psi_0 \sin \varphi \end{aligned} \tag{3.9–6}$$

35 "Filtered Thermal Noise—Fluctuation of Energy as a Function of Interval Length", *Jour. Acous. Soc. Am.*, 14 (1943), 216–227.

Thus

$$\overline{I_1^2 I_2^2} = \psi_0^2(1 + 2\cos^2\varphi) = \psi_0^2 + 2\psi_\tau^2 \tag{3.9-7}$$

Incidentally, this gives an expression for the correlation function of $I^2(t)$. Replacing τ by its value of $t_2 - t_1$ and returning to (3.9–4),

$$\overline{E^2} = T^2\psi_0^2 + 2\int_{-T/2}^{T/2} dt_1 \int_{-T/2}^{T/2} dt_2\, \psi^2(t_2 - t_1) \tag{3.9-8}$$

When we introduce σ_T, the standard deviation of E, and use

$$\sigma_T^2 = \overline{E^2} - m_T^2$$

we obtain

$$\sigma_T^2 = \overline{(E - \overline{E})^2} = 2\int_{-T/2}^{T/2} dt_1 \int_{-T/2}^{T/2} dt_2\, \psi^2(t_2 - t_1) = 4\int_0^T (T - x)\psi^2(x)\, dx$$

where the second line may be obtained from the first either by changing the variables of integration, as in (3.9–27), or by the method used below in dealing with $\overline{E^3}$. I am indebted to Prof. Kac for pointing out the advantage obtained by reducing the double integral to a single integral. It should be noted that the limits of integration $-T/2$, $T/2$ in the double integral may be replaced by 0, T by making the change of variable $t = t' - T/2$ for both t_1 and t_2.

When we use

$$\psi(\tau) = \int_0^\infty w(f) \cos 2\pi f\tau\, df \tag{2.1-6}$$

we obtain the result stated in the paper, namely,

$$\sigma_T^2 = \int_0^\infty w(f_1)\, df_1 \int_0^\infty w(f_2)\, df_2 \left[\frac{\sin^2 \pi(f_1 + f_2)T}{\pi^2(f_1 + f_2)^2} + \frac{\sin^2 \pi(f_1 - f_2)T}{\pi^2(f_1 - f_2)^2}\right] \tag{3.9-9}$$

If this formula is applied to a relatively narrow band-pass filter and if $T(f_b - f_a) >> 1$ the contribution of the $f_1 + f_2$ term may be neglected and we have the approximation

$$\sigma_T^2 \approx \int_{f_a}^{f_b} w_0\, df_1 \int_{-\infty}^{+\infty} w_0\, df_2 \frac{\sin^2 \pi(f_1 - f_2)T}{\pi^2(f_1 - f_2)^2} = w_0^2 T(f_b - f_a) = w_0 m_T \tag{3.9-10}$$

where, from (3.9–3)

$$m_T = w_0 T(f_b - f_a) \tag{3.9–11}$$

The third moment $\overline{E^3}$ may be computed in the same way. However, in this case it pays to introduce the characteristic function for the distribution of $I(t_1)$, $I(t_2)$, $I(t_3)$. Since this distribution is normal its characteristic function is

$$\begin{aligned} &\text{Average exp } [iz_1 I_1 + iz_2 I_2 + iz_3 I_3] \\ &\quad = \exp -\left[\frac{\psi_0}{2}(z_1^2 + z_2^2 + z_3^2) + \psi(t_2 - t_1) z_1 z_2 \right. \\ &\qquad\qquad \left. + \psi(t_3 - t_1) z_1 z_3 + \psi(t_3 - t_2) z_2 z_3 \right] \end{aligned} \tag{3.9–12}$$

From the definition of the characteristic function it follows that

$$\begin{aligned} \overline{I_1^2 I_2^2 I_3^2} &= -\text{coeff. of } \frac{z_1^2 z_2^2 z_3^2}{2!2!2!} \text{ in ch. f.} \\ &= \psi_0^3 + 2\psi_0(\psi_{21}^2 + \psi_{31}^2 + \psi_{32}^2) \\ &\qquad + 8\psi_{21}\psi_{31}\psi_{32} \end{aligned} \tag{3.9–13}$$

where we have written ψ_{21} for $\psi(t_2 - t_1)$, etc. When (3.9–13) is multiplied by $dt_1\, dt_2\, dt_3$, the variables integrated from 0 to T, and the above double integral expression for σ_T^2 used, we find

$$\overline{(E - \overline{E})^3} = 2!2^2 \int_0^T dt_1 \int_0^T dt_2 \int_0^T dt_3 \psi_{21}\psi_{31}\psi_{32} \,.$$

Denoting the triple integral on the right by J and differentiating,

$$\begin{aligned} \frac{dJ}{dT} &= 3 \int_0^T dt_1 \int_0^T dt_2 \psi(t_2 - t_1)\psi(T - t_1)\psi(T - t_2) \\ &= 3 \int_0^T dx \int_0^T dy \psi(x - y)\psi(x)\psi(y) \\ &= 6 \int_0^T dx \int_0^x dy \psi(x - y)\psi(x)\psi(y) \end{aligned}$$

In going from the first line to the second t_1 and t_2 were replaced by $T - x$ and $T - y$, respectively. In going from the second to the third use was made of the relations symbolized by

$$\begin{aligned} \int_0^T dx \int_0^T dy &= \int_0^T dx \int_0^x dy + \int_0^T dx \int_x^T dy \\ &= \int_0^T dx \int_0^x dy + \int_0^T dy \int_0^y dx \end{aligned}$$

and of the fact that the integrand is symmetrical in x and y. Integrating dJ/dT with respect to T from 0 to T_1, using the formula

$$\int_0^{T_1} dT \int_0^T f(x)\,dx = \int_0^{T_1} (T_1 - x)f(x)\,dx,$$

noting that J is zero when T is zero, and dropping the subscript on T_1 finally gives

$$\overline{(E - \overline{E})^3} = 48 \int_0^T dx \int_0^x dy(T - x)\psi(x)\psi(y)\psi(x - y).$$

E^4 may be treated in a similar way. It is found that

$$\overline{(E - \overline{E})^4} - 3\overline{(E - \overline{E})^2}^2 = 3!2^3 \int_0^T dt_1 \int_0^T dt_2 \int_0^T dt_3 \int_0^T dt_4 \psi_{21}\psi_{31}\psi_{42}\psi_{43}$$

which may be reduced to the sum of two triple integrals. It is interesting to note that the expression on the left is the fourth semi-invariant of the random variable E and gives us a measure of the peakedness of the distribution (kurtosis). Likewise, the second and third moments about the mean are the second and third semi-invariants of E. This suggests that possibly the higher semi-invariants may also be expressed as similar multiple integrals.

So far, in this section, we have been speaking of the statistical constants of E. The determination of an exact expression for the probability density of E, in which T occurs as a parameter, seems to be quite difficult.

When T is very small E is approximately $I^2(t)T$. The probability that E lies in dE is the probability that the current lies in $-I, -I - dI$ plus the probability that the current lies in $I, I + dI$:

$$\frac{2dI}{\sqrt{2\pi\psi_0}} \exp - \frac{I^2}{2\psi_0} = (2\pi\psi_0 ET)^{-1/2} \exp - \frac{E}{2\psi_0 T}\,dE \quad (3.9\text{–}14)$$

where E is positive,

$$I = \left(\frac{E}{T}\right)^{1/2}, \qquad dI = \frac{1}{2}(ET)^{-1/2}\,dE$$

and T is assumed to be so small that $I(t)$ does not change appreciably during an interval of length T.

When T is very large we may divide it into a number of intervals, say n, each of length T/n. Let E_r be the contribution of the r th interval. The energy E for the entire interval is then

$$E = E_1 + E_2 + \cdots + E_n$$

If the sub-intervals are large enough the E_r's are substantially independent random variables. If in addition n is large enough E is distributed normally, approximately. Hence when T is very large the probability that E lies in dE is

$$\frac{dE}{\sigma_T\sqrt{2\pi}} \exp - \frac{(E - m_T)^2}{2\sigma_T^2} \tag{3.9–15}$$

where

$$\begin{aligned} m_T &= T \int_0^\infty w(f)\, df \\ \sigma_T^2 &= T \int_0^\infty w^2(f)\, df \end{aligned} \tag{3.9–16}$$

the second relation being obtained by letting $T \to \infty$ in (3.9–9). The analogy with Campbell's theorem, section 1.2, is evident. When we deal with a band pass filter we may use (3.9–10) and (3.9–11).

Consider a relatively narrow band pass filter such that we may find a T for which $Tf_a >> 2\pi$ but $T(f_b - f_a) << .64$. Thus several cycles of frequency f_a are contained in T but, from (3.8–15), the envelope does not change appreciably during this interval. Thus throughout this interval $I(t)$ may be considered to be a sine wave of amplitude R. The corresponding value of E is approximately

$$E = T\frac{R^2}{2}$$

where the distribution of the envelope R is given by (3.7–10). From this it follows that the probability of E lying in dE is

$$\frac{dE}{\psi_0 T} \exp - \frac{E}{\psi_0 T} = \frac{dE}{m_T} e^{-E/m_T} \tag{3.9–17}$$

when E is small but not too small.

When we look at (3.9–14) and (3.9–17) we observe that they are of the form

$$\frac{a^{n+1}E^n}{\Gamma(n+1)} e^{-aE}\, dE \tag{3.9–18}$$

Moreover, the normal law (3.9–15), may be obtained from this by letting n become large. This suggests that an approximate expression for the distribution of E is given by (3.9–18) when a and n are selected so as to give the values of m_T and σ_T obtained from (3.9–3) and (3.9–9). This gives

$$a = \frac{m_T}{\sigma_T^2}, \qquad n + 1 = \frac{m_T^2}{\sigma_T^2} \tag{3.9–19}$$

and if we drop the subscript T and substitute the value of a in (3.9–18) we get

$$\frac{\left(\frac{mE}{\sigma^2}\right)^n}{\Gamma(n+1)} \exp\left(-\frac{mE}{\sigma^2}\right) d\left(\frac{mE}{\sigma^2}\right), \qquad n = \frac{m^2}{\sigma^2} - 1 \qquad (3.9\text{–}20)$$

An idea of how this distribution behaves may be obtained from the following table:

n	$T(f_b - f_a)$	$x_{.25}$	$x_{.50}$	$x_{.75}$	$\frac{x_{.25}}{x_{.50}}$	$\frac{x_{.75}}{x_{.50}}$
0	0	.29	.695	1.39	.415	2.00
1	1.45	.96	1.68	2.69	.572	1.60
2	2.4	1.73	2.67	3.94	.647	1.47
3	3.4	2.54	3.67	5.12	.692	1.39
5	5.4	4.22	5.67	7.42	.744	1.31
10	10.5	8.63	10.67	13.02	.808	1.22
24	25	21.47	24.67	28.17	.870	1.14
48	50	44.1	48.7	53.5	.905	1.10

where n is the exponent in (3.9–20). The column $T(f_b - f_a)$ holds only for a narrow band pass filter and was obtained by reading the curve y_A in Fig. 1 of the above mentioned paper. The figures in this column are not very accurate. The next three columns give the points which divide the distribution into four intervals of equal probability:

$$x_{.25} = \frac{mE_{.25}}{\sigma^2}, \quad E_{.25} = \text{energy exceeded } 75\% \text{ of time}$$

$$x_{.50} = \frac{mE_{.50}}{\sigma^2}, \quad E_{.50} = \text{energy exceeded } 50\% \text{ of time}$$

$$x_{.75} = \frac{mE_{.75}}{\sigma^2}, \quad E_{.75} = \text{energy exceeded } 25\% \text{ of time}$$

The values in these columns were obtained from Pearson's table of the incomplete gamma function. The last two columns show how the distribution clusters around the average value as the normal law is approached.

For the larger values of n we expected the normal law (3.9–15) to be approached. Since, for this law the 25, 50, and 75 per cent points are at $m - .675\sigma$, m, and $m + .675\sigma$ we have to a first approximation

$$\begin{aligned} x_{.50} &= \frac{m^2}{\sigma^2} = (n+1) \approx T(f_b - f_a) \\ x_{.25} &= \frac{m}{\sigma^2}(m - .675\sigma) = x_{.50} - .675\sqrt{x_{.50}} \\ x_{.75} &= x_{.50} + .675\sqrt{x_{.50}} \end{aligned} \qquad (3.9\text{–}21)$$

This agrees with the table.

Thiede[36] has studied the mean square value of the fluctuations of the integral

$$A(t) = \int_{-\infty}^{t} I^2(\tau)e^{-\alpha(t-\tau)}\,d\tau \tag{3.9–22}$$

The reading of a hot wire ammeter through which a current I is passing is proportional to $A(t)$. α is a constant of the meter. Here we study $A(t)$ by

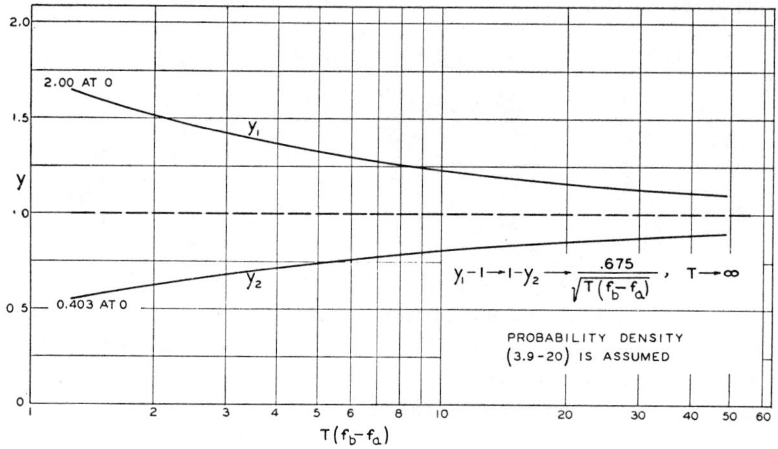

Fig. 5*—Filtered thermal noise—spread of energy fluctuation

$E = \int_{t_1}^{t_1+T} I^2(t)\,dt$, t_1 random, I is noise current.

$y_1 = E_{.75}/E_{.50}$, $y_2 = E_{.25}/E_{.50}$.

$f_b - f_a$ = band width of filter.

first obtaining its correlation function. This method of approach enables us to extend Thiede's results

The distributed portion of the power spectrum of $A(t)$ is given by (3.9–30). When the power spectrum $w(f)$ of $I(t)$ is zero except over the band $f_a < f < f_b$ where it is w_0, the power spectrum of $A(t)$ is

$$\frac{2w_0^2(f_b - f_a - f)}{\alpha^2 + 4\pi^2 f^2} \quad \text{for} \quad 0 < f < f_b - f_a$$

and is zero from $f_b - f_a$ up to $2f_a$. The spectrum from $2f_a$ to $2f_b$ is not zero, and may be obtained from (3.9–34). The mean square fluctuation of $A(t)$ is given, in the general case, by (3.9–28) and (3.9–32). For the band pass case, when $(f_b - f_a)/\alpha$ is large,

$$\text{r.m.s.}\ \frac{A(t) - \bar{A}}{\bar{A}} = \left[\frac{\alpha}{2(f_b - f_a)}\right]^{1/2}$$

[36] *Elec. Nachr. Tek.*, 13 (1936), 84–95. This is an excellent article.

* Note added in proof. The value of y_2 at 0 should be .415 instead of .403.

We start by setting $\tau = t - u$ which transforms the integral for $A(t)$ into

$$A(t) = \int_0^\infty I^2(t - u)e^{-\alpha u}\, du \tag{3.9-23}$$

In order to obtain the correlation function $\Psi(\tau)$ for $A(t)$ we multiply $A(t)$ by $A(t + \tau)$ and average over all the possible currents

$$\begin{aligned}\Psi(\tau) &= \overline{A(t)A(t + \tau)} \\ &= \int_0^\infty e^{-\alpha u}\, du \int_0^\infty e^{-\alpha v}\, dv \text{ ave. } I^2(t - u)I^2(t + \tau - v)\end{aligned}$$

Just as in (3.9–4) the average in the integrand is the correlation function of $I^2(t)$, the argument being $t + \tau - v - t + u = \tau + u - v$. From (3.9–7) it is seen that this is

$$\psi_0^2 + 2\psi^2(\tau + u - v)$$

where $\psi(\tau)$ is the correlation function of $I(t)$. Hence

$$\Psi(\tau) = \frac{\psi_0^2}{\alpha^2} + 2\int_0^\infty du \int_0^\infty dv\, e^{-\alpha u - \alpha v}\psi^2(\tau + u - v) \tag{3.9-24}$$

From the integral (3.9–23) for $A(t)$ it is seen that the average value of $A(t)$ is

$$\bar{A} = \frac{\overline{I^2}}{\alpha} = \frac{\psi_0}{\alpha} \tag{3.9-25}$$

where we have used

$$\psi_0 = \psi(0) = \int_0^\infty w(f)\, df = \overline{I^2}$$

Using this result again, only this time applying it to $A(t)$, gives

$$\begin{aligned}\overline{A^2(t)} &= \Psi(0) \\ &= \bar{A}^2 + 2\int_0^\infty du \int_0^\infty dv\, e^{-\alpha u - \alpha v}\psi^2(u - v)\end{aligned} \tag{3.9-26}$$

The double integrals may be transformed by means of the change of variable $u + v = x$, $u - v = y$. Then (3.9–24) becomes

$$\begin{aligned}\Psi(\tau) &= \bar{A}^2 + \left[\int_0^\infty dy \int_y^\infty dx + \int_{-\infty}^0 dy \int_{-y}^\infty dx\right] e^{-\alpha x}\psi^2(\tau + y) \\ &= \bar{A}^2 + \frac{1}{\alpha}\int_0^\infty e^{-\alpha y}[\psi^2(\tau + y) + \psi^2(\tau - y)]\, dy\end{aligned} \tag{3.9-27}$$

When we make use of the fact that $\psi(y)$ is an even function of y we see, from (3.9–26), that the mean square fluctuation of $A(t)$ is

$$\overline{(A(t) - \bar{A})^2} = \overline{A^2(t)} - \bar{A}^2 = \frac{2}{\alpha}\int_0^\infty e^{-\alpha y}\psi^2(y)\,dy \qquad (3.9\text{–}28)$$

$\Psi(\tau)$ may be expressed in terms of integrals involving the power spectrum $w(f)$ of $I(t)$. The work starts with (3.9–24) and is much the same as in going from (3.9–8) to (3.9–9). The result is

$$\Psi(\tau) = \bar{A}^2 + \int_0^\infty df_1 \int_0^\infty df_2\, w(f_1)w(f_2)$$
$$\left[\frac{\cos 2\pi(f_1 + f_2)\tau}{\alpha^2 + [2\pi(f_1 + f_2)]^2} + \frac{\cos 2\pi(f_1 - f_2)\tau}{\alpha^2 + [2\pi(f_1 - f_2)]^2}\right]$$

It is convenient to define $w(-f)$ for negative frequencies to be equal to $w(f)$. The integration with respect to f_2 may then be taken from $-\infty$ to $+\infty$ and we get

$$\Psi(\tau) = \bar{A}^2 + \int_0^\infty df_1 \int_{-\infty}^{+\infty} df_2\, w(f_1)w(f_2)\,\frac{\cos 2\pi(f_1 - f_2)\tau}{\alpha^2 + [2\pi(f_1 - f_2)]^2} \qquad (3.9\text{–}29)$$

The power spectrum $W(f)$ of $A(t)$ may be obtained by integrating $\Psi(\tau)$:

$$W(f) = 4\int_0^\infty \Psi(\tau)\cos 2\pi f\tau\, d\tau$$

Let us concern ourselves with the fluctuating portion $A(t) - \bar{A}$ of $A(t)$. Its power spectrum $W_c(f)$ is

$$W_c(f) = 4\int_0^\infty (\Psi(\tau) - \bar{A}^2)\cos 2\pi f\tau\, d\tau$$

The integration is simplified by using Fourier's integral formula in the form

$$\int_0^\infty d\tau \int_{-\infty}^{+\infty} df_2\, F(f_2)\cos 2\pi(u - f_2)\tau = \tfrac{1}{2}F(u)$$

We get

$$\begin{aligned} W_c(f) &= \frac{1}{\alpha^2 + 4\pi^2 f^2}\int_0^\infty df_1[w(f_1)w(f + f_1) + w(f_1)w(-f + f_1)] \\ &= \frac{1}{\alpha^2 + 4\pi^2 f^2}\int_{-\infty}^{+\infty} w(f_1)w(f - f_1)\,df_1 \end{aligned} \qquad (3.9\text{–}30)$$

The simplicity of this result suggests that a simpler derivation may be found. If we attempt to use the result

$$\bar{w}(f) = \operatorname*{Limit}_{T\to\infty} \frac{\overline{2|S(f)|^2}}{T} \qquad (2.5\text{–}3)$$

where $S(f)$ is given by (2.1–2) we find that we need the result

$$\underset{T\to\infty}{\text{Limit}} \frac{2}{T}\int_0^T dt_1 \int_0^T dt_2 e^{2\pi i f(t_2-t_1)} I^2(t_1)I^2(t_2)$$
$$= \int_{-\infty}^{+\infty} w(f_1)w(f-f_1)\,df_1 \tag{3.9–31}$$

where $f > 0$ and $I(t)$ is a noise current with $w(f)$ as its power spectrum. This may be proved by using (3.9–7) and

$$8\int_0^\infty \psi^2(\tau)\cos 2\pi f\tau\, d\tau = \int_{-\infty}^{+\infty} w(x)w(f-x)\,dx$$

which is given by equation (4C-6) in Appendix 4C.

An expression for the mean square fluctuation of $A(t)$ in terms of $w(f)$ may be obtained by setting τ equal to zero in (3.9–29)

$$\overline{(A(t)-\bar{A})^2} = \Psi(0) - \bar{A}^2$$
$$= \int_0^\infty df_1 \int_{-\infty}^{+\infty} df_2 \frac{w(f_1)w(f_2)}{\alpha^2 + 4\pi^2(f_1-f_2)^2} \tag{3.9–32}$$

The same result may be obtained by integrating $W_c(f)$, (3.9–30), from 0 to ∞:

$$\int_0^\infty \frac{df}{\alpha^2+4\pi^2f^2}\int_{-\infty}^{+\infty} df_1\, w(f_1)w(f-f_1) \tag{3.9–33}$$

Although this differs in appearance from (3.9–32) it may be transformed into that expression by making use of $w(-f) = w(f)$.

Suppose that $I(t)$ is the current through an ideal band pass filter so that $w(f)$ is zero except in the band $f_a < f < f_b$ where it is w_0. Then, if $3f_a > f_b$,

$$\bar{A} = \frac{w_0}{\alpha}(f_b - f_a) \tag{3.9–34}$$

$$\int_{-\infty}^{+\infty} w(x)w(f-x)\,dx = \begin{cases} 2w_0^2(f_b - f_a - f) & 0 < f \leq f_b - f_a \\ w_0^2(f-2f_a) & 2f_a \leq f \leq f_b + f_a \\ w_0^2(2f_b - f) & f_b + f_a \leq f \leq 2f_b \end{cases}$$

and is zero outside these ranges. The power spectrum $W_c(f)$ may be obtained immediately from (3.9–30) by dividing these values by $\alpha^2 + 4\pi^2f^2$.

From (3.9–33)

$$\overline{(A(t)-\bar{A})^2} = 2w_0^2\int_0^{f_b-f_a}\frac{(f_b-f_a-f)\,df}{\alpha^2+4\pi^2f^2}$$
$$+ w_0^2\int_{2f_a}^{f_b+f_a}\frac{(f-2f_a)}{\alpha^2+4\pi^2f^2}\,df + w_0^2\int_{f_b+f_a}^{2f_b}\frac{(2f_b-f)}{\alpha^2+4\pi^2f^2}\,df$$

If an exact answer is desired the integrations may be performed. When we assume that $f_b - f_a << f_b + f_a$ we may obtain approximations for the last two integrals.

$$\overline{(A(t) - \bar{A})^2} \approx w_0^2 \left[\frac{f_b - f_a}{\pi\alpha} \tan^{-1} \frac{2\pi(f_b - f_a)}{\alpha} - \frac{1}{4\pi^2} \log \frac{\alpha^2 + 4\pi^2(f_b - f_a)^2}{\alpha^2} \cdot + \frac{(f_b - f_a)^2}{\alpha^2 + 4\pi^2(f_b + f_a)^2} \right]$$

Furthermore, if $2\pi(f_b - f_a)/\alpha$ is large we have

$$\overline{(A(t) - \bar{A})^2} \approx w_0^2 \frac{f_b - f_a}{2\alpha}$$

and the relative r.m.s. fluctuation is

$$\text{r.m.s. of } \left[\frac{(A(t) - \bar{A})}{\bar{A}} \right] \approx \left[\frac{\alpha}{2(f_b - f_a)} \right]^{1/2}$$

This result may also be obtained from (3.9–10) and (3.9–11) by assuming α so small that the integral for $A(t)$ may be broken into a great many integrals each extending over an interval T. αT is assumed so small that $e^{-\alpha u}$ is substantially constant over each interval.

3.10 Distribution of Noise Plus Sine Wave

Suppose we have a steady sinusoidal current

$$I_p = I_p(t) = P \cos(\omega_p t - \varphi_p) \tag{3.10–1}$$

We pick times $t_1, t_2, \cdots$ at random and note the corresponding values of the current. How are these values distributed? Picking the times at random in (3.10–1) is the same, statistically, as holding t constant and picking the phase angles φ_p at random from the range 0 to 2π. If I_p be regarded as a random variable defined by the random variable φ_p, its characteristic function is

$$\begin{aligned} \text{ave. } e^{izI_p} &= \frac{1}{2\pi} \int_0^{2\pi} e^{izP \cos(\omega_p t - \varphi)} \, d\varphi \\ &= J_0(Pz) \end{aligned} \tag{3.10–2}$$

and its probability density is

$$\frac{1}{2\pi} \int_{-\infty}^{+\infty} e^{-izI_p} J_0(Pz) \, dz = \begin{cases} \dfrac{1}{\pi} (P^2 - I_p^2)^{-1/2} & |I_p| < P \\ 0 & |I_p| > P \end{cases} \tag{3.10–3}$$

In this case it is simpler to obtain the probability density directly from (3.10–1) instead of from the characteristic function.

Now suppose that we have a noise current I_N plus a sine wave. By combining our representation (2.8–6) for I_N with the idea of φ_p being random mentioned above we are led to the representation

$$\begin{aligned} I(t) &= I = I_p + I_N \\ &= P \cos(\omega_p t - \varphi_p) + \sum_1^M c_n \cos(\omega_n t - \varphi_n), \qquad (3.10\text{–}4) \\ c_n^2 &= 2w(f_n)\Delta f \end{aligned}$$

where φ_p and $\varphi_1, \cdots \varphi_M$ are independent random angles.

If we note I at the random times $t_1, t_2 \cdots$ how are the observed values distributed? Since I_p and I_N may be regarded as independent random variables and since the characteristic function for the sum of two such variables is the product of their characteristic functions we have from (3.1–6) and (3.10–2)

$$\begin{aligned} \text{ave. } e^{izI} &= \text{ave. } e^{iz(I_p+I_N)} \\ &= J_0(Pz) \exp\left(\frac{-\psi_0 z^2}{2}\right) \end{aligned} \qquad (3.10\text{–}5)$$

which gives the characteristic function of I. The probability density of I is[37]

$$\frac{1}{2\pi}\int_{-\infty}^{+\infty} e^{-izI-(\psi_0 z^2/2)} J_0(Pz)\, dz = \frac{1}{\pi\sqrt{2\pi\psi_0}}\int_0^\pi e^{-(I-P\cos\theta)^2/2\psi_0}\, d\theta \qquad (3.10\text{–}6)$$

In the same way the two-dimensional probability density of (I_1, I_2), where $I_1 = I(t)$ is a sine wave plus noise (3.10–4) and $I_2 = I(t + \tau)$ is its value at a constant interval τ later, may be shown to be

$$\frac{(\psi_0^2 - \psi_\tau^2)^{-1/2}}{2\pi}\int_0^{2\pi} d\theta \exp\left[-\frac{B(\theta)}{2(\psi_0^2 - \psi_\tau^2)}\right] \qquad (3.10\text{–}7)$$

where

$$\begin{aligned} B(\theta) &= \psi_0[(I_1 - P\cos\theta)^2 + (I_2 - P\cos(\theta + \omega_p\tau))^2] \\ &\quad - 2\psi_\tau(I_1 - P\cos\theta)(I_2 - P\cos(\theta + \omega_p\tau)) \end{aligned}$$

The characteristic function for I_1 and I_2 is

$$\begin{aligned} \text{ave. } e^{iuI_1+ivI_2} &= J_0(P\sqrt{u^2 + v^2 + 2uv\cos\omega_p\tau}) \\ &\quad \times \exp\left[-\frac{\psi_0}{2}(u^2 + v^2) - \psi_\tau uv\right] \end{aligned} \qquad (3.10\text{–}8)$$

[37] A different derivation of this expression is given by W. R. Bennett, *Jour. Acous. Soc. Amer.*, Vol. 15, p. 165 (Jan. 1944); *B.S.T.J.*, Vol. 23, p. 97 (Jan. 1944).

Sometimes the distribution of the envelope of

$$I = P \cos pt + I_N \tag{3.10–9}$$

is of interest. Here we have replaced ω_p by p and have set φ_p to zero. By the envelope we mean $R(t)$ given by

$$R^2(t) = R^2 = (P + I_c)^2 + I_s^2 \tag{3.10–10}$$

where I_c is the component of I_N "in phase" with cos pt and I_s is the component "in phase" with sin pt:

$$\begin{aligned} I_c &= \sum c_n \cos[(\omega_n - p)t - \varphi_n] \\ I_s &= \sum c_n \sin[(\omega_n - p)t - \varphi_n] \\ I_N &= I_c \cos pt - I_s \sin pt \\ \overline{I_N^2} &= \overline{I_c^2} = \overline{I_s^2} = \psi_0 \end{aligned}$$

Since I_c and I_s are distributed normally about zero with a variance of ψ_0, the probability densities of the variables

$$x = P + I_c$$

$$y = I_s$$

are

$$(2\pi\psi_0)^{-1/2} \exp - \frac{(x - P)^2}{2\psi_0}$$

$$(2\pi\psi_0)^{-1/2} \exp - \frac{y^2}{2\psi_0}$$

respectively. Setting

$$x = R \cos \theta$$

$$y = R \sin \theta$$

and using these distributions shows that the probability of a point (x, y) lying in the ring R, $R + dR$ is

$$\frac{R\,dR}{2\pi\psi_0} \int_0^{2\pi} \exp\left[-\frac{1}{2\psi_0}(R^2 + P^2 - 2RP\cos\theta)\right] d\theta$$

$$= \frac{R\,dR}{\psi_0} \exp\left[-\frac{R^2 + P^2}{2\psi_0}\right] I_0\left(\frac{RP}{\psi_0}\right) \tag{3.10–11}$$

where I_0 is the Bessel function with imaginary argument.

$$I_0(z) = \sum_{n=0}^{\infty} \frac{z^{2n}}{2^{2n}\, n!\, n!}$$

and is a tabulated function. Thus (3.10–11) gives the probability density of the envelope R.

The average value of R^n may be obtained by multiplying (3.10–11) by R^n and integrating from 0 to ∞. Expansion of the Bessel function and termwise integration gives

$$\overline{R^n} = (2\psi_0)^{n/2}\Gamma\left(\frac{n}{2}+1\right)e^{-P^2/2\psi_0}\,{}_1F_1\left(\frac{n}{2}+1;\,1;\,\frac{P^2}{2\psi_0}\right)$$

$$= (2\psi_0)^{n/2}\Gamma\left(\frac{n}{2}+1\right){}_1F_1\left(-\frac{n}{2};\,1;\,-\frac{P^2}{2\psi_0}\right) \qquad (3.10\text{–}12)$$

where ${}_1F_1$ is a hypergeometric function.[38] In going from the first line to the second we have used Kummer's first transformation of this function. A special case is

$$\overline{R^2} = P^2 + 2\psi_0 \qquad (3.10\text{–}13)$$

When only noise is present, $P = 0$ and

$$\bar{R} = (2\psi_0)^{1/2}\Gamma(\tfrac{3}{2}) = \left(\frac{\psi_0\pi}{2}\right)^{1/2}$$
$$\overline{R^2} = 2\psi_0 \qquad (3.10\text{–}14)$$

Before going further with (3.10–11) it is convenient to make the following change of notation

$$v = \frac{R}{\psi_0^{1/2}}, \qquad dv = \frac{dR}{\psi_0^{1/2}}, \qquad a = \frac{P}{\psi_0^{1/2}} \qquad (3.10\text{–}15)$$

"a" is the ratio (sine wave amplitude)/(r.m.s. noise current).
Instead of the random variable R we now have the random variable v whose probability density is

$$p(v) = v\exp\left[-\frac{v^2+a^2}{2}\right]I_0(av) \qquad (3.10\text{–}16)$$

Curves of $p(v)$ versus v are plotted in Fig. 6 for the values 0, 1, 2, 3, 5 of a. Curves showing the probability that v is less than a stated amount, i.e., distribution curves for v, are given in Fig. 7. These curves were obtained by integrating $p(v)$ numerically. The following useful expression for this probability has been given by W. R. Bennett in some unpublished work.

$$\int_0^v p(u)\,du = \exp\left[-\frac{v^2+a^2}{2}\right]\sum_{n=1}^{\infty}\left(\frac{v}{a}\right)^n I_n(av) \qquad (3.10\text{–}17)$$

[38] Curves of this function are given in "Tables of Functions", Jahnke and Emde (1938), p. 275, and some of its properties are stated in Appendix 4C.

This is obtained by integration by parts using

$$\int u^n I_{n-1}(au)\, du = u^n I_n(au)/a$$

When $av >> 1$ but $1 << a - v$, Bennett has shown that (3.10–17) leads to

$$\int_0^v p(u)\, du \approx \left(\frac{v}{2\pi a}\right)^{1/2} \frac{1}{a - v} \exp\left[-\frac{(v-a)^2}{2}\right] \left(1 - \frac{3(a+v)^2 - 4v^2}{8av(a-v)^2} \cdots\right) \tag{3.10–18}$$

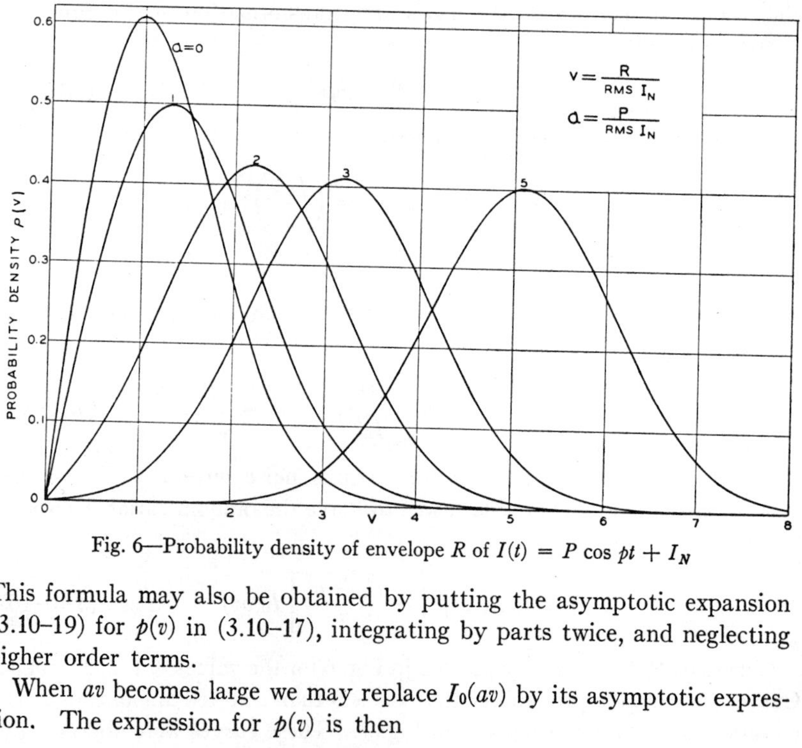

Fig. 6—Probability density of envelope R of $I(t) = P \cos pt + I_N$

This formula may also be obtained by putting the asymptotic expansion (3.10–19) for $p(v)$ in (3.10–17), integrating by parts twice, and neglecting higher order terms.

When av becomes large we may replace $I_0(av)$ by its asymptotic expression. The expression for $p(v)$ is then

$$p(v) \sim \left(1 + \frac{1}{8av}\right)\left(\frac{v}{2\pi a}\right)^{1/2} \exp\left[-\frac{(v-a)^2}{2}\right] \tag{3.10–19}$$

Thus when either a becomes large or v is far out on the tail of the probability density curve, the distribution behaves like a normal law. In terms of the original quantities, the normal law has an average of P and a standard deviation of $\psi_0^{1/2}$. This standard deviation is the same as the standard deviation of the instantaneous values of I_N. When $av \gg 1$ and $a \gg |v - a|$ we may expand the coefficient of the exponential term in (3.10–19) in powers of

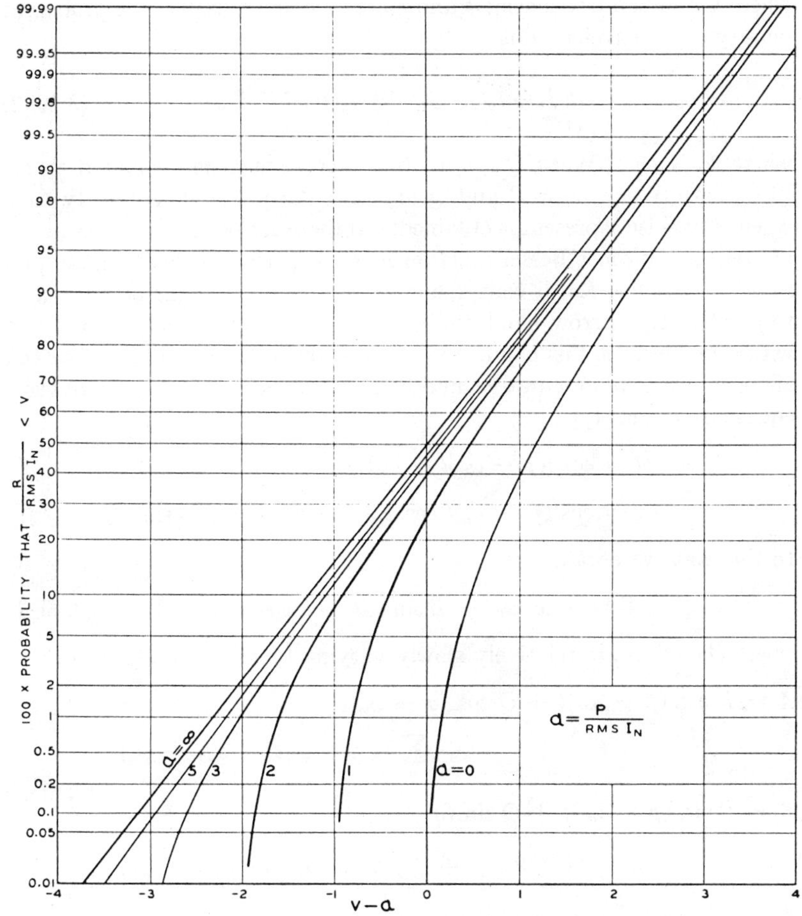

Fig. 7—Distribution function of envelope R of $I(t) = P \cos pt + I_N$

$(v - a)/a$. Integrating this expansion termwise gives, when terms of magnitude less than a^{-3} are neglected,

$$\int_0^v p(u)\,du \approx \frac{1}{2} + \frac{1}{2}\,erf\,\frac{v-a}{\sqrt{2}}$$

$$- \frac{1}{2a\sqrt{2\pi}}\left[1 - \frac{v-a}{4a} + \frac{1+(v-a)^2}{8a^2}\right]\exp\left[-\frac{(v-a)^2}{2}\right]$$

When I consists of two sine waves plus noise

$$I = P \cos pt + Q \sin qt + I_N , \tag{3.10–20}$$

where the radian frequencies p and q are incommensurable, the probability density of the envelope R is

$$R \int_0^\infty rJ_0(Rr)J_0(Pr)J_0(Qr)e^{-\psi_0 r^2/2}\, dr \tag{3.10–21}$$

where ψ_0 is $\overline{I_N^2}$. When Q is zero the integral may be evaluated to give (3.10–11). When both P and Q are zero the probability density for R when only noise is present is obtained. If there are three sine waves instead of two then another Bessel function must be placed in the integrand, and so on. To define R it is convenient to think of the noise as being confined to a relatively narrow band and the frequencies of the sine waves lying within, or close to, this band. As in equations (3.7–2) to (3.7–4), we refer all terms to a representative mid-band frequency $f_m = \omega_m/2\pi$ by using equations of the type

$$\begin{aligned} \cos pt &= \cos\, [(p - \omega_m)t + \omega_m t] \\ &= \cos\, (p - \omega_m)t \cos \omega_m t - \sin\, (p - \omega_m)t \sin \omega_m t. \end{aligned}$$

In this way we obtain

$$V = A \cos \omega_m t - B \sin \omega_m t = R \cos\, (\omega_m t + \theta) \tag{3.10–22}$$

where A and B are relatively slowly varying functions of t given by

$$\begin{aligned} A &= P \cos\, (p - \omega_m)t + Q \cos\, (q - \omega_m)t \\ &\qquad + \sum_n c_n \cos\, (\omega_n t - \omega_m t - \varphi_n) \\ B &= P \sin\, (p - \omega_m)t + Q \sin\, (q - \omega_m)t \\ &\qquad + \sum_n c_n \sin\, (\omega_n t - \omega_m t - \varphi_n) \end{aligned} \tag{3.10–23}$$

and

$$\begin{aligned} R^2 &= A^2 + B^2, \qquad R > 0 \\ \tan \theta &= B/A \end{aligned} \tag{3.10–24}$$

As might be expected, (3.10–21) is closely associated with the problem of random flights and may be obtained from Kluyver's result[39] by assuming

[39] G. N. Watson, "Theory of Bessel Functions" (Cambridge, 1922), p. 420.

the noise to correspond to a very large number of very small random displacements.

Another way of deriving (3.10–21) is to assume $(p - \omega_m)t$, $(q - \omega_m)t$, $\varphi_1, \varphi_2, \cdots$ are independent random angles. The characteristic function of A, B is

$$\text{ave. } e^{iuA+ivB} = J_0(P\sqrt{u^2+v^2})J_0(Q\sqrt{u^2+v^2})e^{-(\psi_0/2)(u^2+v^2)}$$

The probability density of A, B is

$$\left(\frac{1}{2\pi}\right)^2 \int_{-\infty}^{+\infty} du \int_{-\infty}^{+\infty} dv\, e^{-iuA-ivB} \text{ ave. } e^{iuA+ivB}$$

When the change of variables

$$A = R\cos\theta \qquad u = r\cos\varphi$$

$$B = R\sin\theta \qquad v = r\sin\varphi$$

is made the integration with respect to φ may be performed. The double integral becomes

$$\frac{1}{2\pi}\int_0^\infty rJ_0(Pr)J_0(Qr)J_0(Rr)e^{-(\psi_0/2)r^2}\, dr$$

This leads directly to (3.10–21) when we observe that $dAdB = RdRd\theta$.

Incidentally, if

$$I = Q(1 + k\cos pt)\cos qt + I_N$$

in which $p << q$, similar considerations show that the probability density of R is

$$\frac{R}{2\pi}\int_0^{2\pi} d\alpha \int_0^\infty rJ_0(Rr)J_0[Qr(1 + k\cos\alpha)]e^{-(\psi_0/2)r^2}\, dr$$

when ω_m is taken to be q. The integration with respect to r may be performed. This relation is closely connected with (3.10–11).

Returning now to the case in which I is the sum of two sine waves plus noise, we may show from (3.10–21) and

$$\int_0^\infty R^{n+1}J_0(Rr)\, dR = \frac{2^{n+1}\Gamma\left(1+\frac{n}{2}\right)}{r^{n+2}\Gamma\left(-\frac{n}{2}\right)}$$

that the average value of R^n is, when $-2 < re\,(n) < -\frac{3}{2}$,

$$\overline{R^n} = \frac{2^{n+1}\Gamma\left(1+\frac{n}{2}\right)}{\Gamma\left(-\frac{n}{2}\right)} \int_0^\infty r^{-n-1} J_0(Pr) J_0(Qr) e^{-\psi_0 r^2/2}\, dr$$

$$= (2\psi_0)^{n/2}\Gamma\left(\frac{n}{2}+1\right)\sum_{k=0}^{\infty}\sum_{m=0}^{\infty} \frac{\left(-\frac{n}{2}\right)_{k+m}(-x)^k(-y)^m}{k!\,k!\,m!\,m!} \tag{3.10–25}$$

$$= (2\psi_0)^{n/2}\Gamma\left(\frac{n}{2}+1\right)\sum_{k=0}^{\infty} \frac{\left(-\frac{n}{2}\right)_k (y-x)^k}{k!\,k!} P_k\left(\frac{x+y}{x-y}\right)$$

It appears very probable that this result could be extended, by analytic continuation, to positive integer values of n. We have used the notation

$$(\alpha)_0 = 1, \qquad (\alpha)_k = \alpha(\alpha+1)\cdots(\alpha+k-1)$$

$$x = \frac{P^2}{2\psi_0}, \qquad y = \frac{Q^2}{2\psi_0} \tag{3.10–26}$$

and have denoted the Legendre polynomial by $P_k(z)$. The series converge for all values of P, Q, and ψ_0 and terminate when n is an even positive integer.

When x or y, or both, are large in comparison with unity we may use the integral for $\overline{R^n}$ to obtain the asymptotic expansion, assuming $Q < P$ so that $y < x$,

$$\overline{R^n} \sim P^n \sum_{k=0}^{\infty} \frac{\left(-\frac{n}{2}\right)_k\left(-\frac{n}{2}\right)_k}{k!\,x^k}\, {}_2F_1\left(k-\frac{n}{2}, k-\frac{n}{2}; 1; \frac{y}{x}\right) \tag{3.10–27}$$

When n is an even positive integer this series terminates and gives the same expression as (3.10–25). When n is an odd integer the ${}_2F_1$ may be expressed in terms of the complete elliptic functions E and K of modulus $y^{1/2}x^{-1/2}$:

$${}_2F_1\left(-\tfrac{1}{2}, -\tfrac{1}{2}; 1; \frac{y}{x}\right) = \frac{4}{\pi}E - \frac{2}{\pi}\left(1-\frac{y}{x}\right)K$$

$${}_2F_1\left(\tfrac{1}{2}, \tfrac{1}{2}; 1; \frac{y}{x}\right) = \frac{2}{\pi}K \tag{3.10–28}$$

The higher terms may be computed from

$$a(1-z)^2\,{}_2F_1(a+1, a+1; 1; z) = (2a-1)(1+z)\,{}_2F_1(a, a; 1; z) + (1-a)\,{}_2F_1(a-1, a-1; 1; z) \tag{3.10–29}$$

which is a special case of

$$ab(\gamma + 1)(1 - z)^2 {}_2F_1(a + 1, b + 1; c; z) = A\, {}_2F_1(a, b; c; z)$$
$$- (\gamma - 1)(c - a)(c - b)\, {}_2F_1(a - 1, b - 1; c; z) \quad (3.10\text{–}30)$$

where $\gamma = c - a - b$ and

$$A = (\gamma^2 - 1)\gamma + (1 - z)[(\gamma - 1)(c - b)(b - 1) + (\gamma + 1)a(c - a - 1)]$$

Although this expression does not show it, A is really symmetrical in a and b. A symmetrical form may be obtained by using the expression obtained by putting $z = 0$ in (3.10–30).

3.11 Shot Effect Representation

In most of the work in this part the representations (2.8–1) or (2.8–6) have been used as a starting point. Here we point out that the shot effect representation used in Part I may also be used as a starting point.

For example, suppose we wish to find the two dimensional distribution of $I(t)$ and $I(t + \tau)$ discussed in Section 3.2. This is a special case of the distribution of the two variables

$$\begin{aligned} I(t) &= \sum_{k=-\infty}^{+\infty} F(t - t_k) \\ J(t) &= \sum_{k=-\infty}^{+\infty} G(t - t_k) \end{aligned} \quad (3.11\text{–}1)$$

where we now assume

$$\int_{-\infty}^{+\infty} F(t)\, dt = \int_{-\infty}^{+\infty} G(t)\, dt = 0 \quad (3.11\text{–}2)$$

in order that the average values of I and J may be zero. In fact, to get $I(t + \tau)$ from $J(t)$ we set $G(t)$ equal to $F(t + \tau)$.

The distribution of I and J may be obtained in much the same manner as was the distribution of I alone in section 1.4. The characteristic function of the distribution is

$$\begin{aligned} f(u, v) &= \text{ave. } e^{iuI+ivJ} \\ &= \exp \nu \int_{-\infty}^{+\infty} [e^{iuF(t)+ivG(t)} - 1]\, dt \end{aligned} \quad (3.11\text{–}3)$$

where ν is the expected number of events (electron arrivals in the shot effect) per second. The probability density of I and J is

$$\frac{1}{4\pi^2} \int_{-\infty}^{+\infty} du \int_{-\infty}^{+\infty} dv\, e^{-iuI-ivJ} f(u, v) \quad (3.11\text{–}4)$$

The semi-invariants $\lambda_{m,n}$ are given by the generating function

$$\log f(u, v) = \sum_{m,n=1}^{k} \frac{\lambda_{m,n}}{m!n!} (iu)^m (iv)^n + o[(iu)^k, (iv)^k]$$

and are

$$\lambda_{m,n} = \nu \int_{-\infty}^{+\infty} F^m(t) G^n(t)\, dt \tag{3.11–5}$$

As $\nu \to \infty$ the distribution of I and J approaches a two dimensional normal law. The approximation to this normal law may be obtained in much the same manner as in section 1.6. From our assumption (3.11–2) it follows that λ_{10} and λ_{01} are zero. From the relation between the second moments and semi-invariants λ we have

$$\begin{aligned} \mu_{11} &= \lambda_{20} + \lambda_{10}^2 = \nu \int_{-\infty}^{+\infty} F^2(t)\, dt \\ \mu_{12} &= \lambda_{11} + \lambda_{10}\lambda_{01} = \nu \int_{-\infty}^{+\infty} F(t)G(t)\, dt \\ \mu_{22} &= \lambda_{02} + \lambda_{01}^2 = \nu \int_{-\infty}^{+\infty} G^2(t)\, dt \end{aligned} \tag{3.11–6}$$

where the notation in the subscripts of the μ's differs from that of the λ's, the change being made to bring it in line with sections 2.9 and 2.10 so that we may write down the normal distribution at once.

The formulas (3.11–6) are closely related to Rowland's generalization of Campbell's theorem mentioned just below equation (1.5–9).

PART IV

NOISE THROUGH NON-LINEAR DEVICES

4.0 Introduction

We shall consider two problems which concern noise passing through detectors or other non-linear devices. The first deals with the statistical properties of the output of a non-linear device, that is, with its average value, its fluctuation about this average and so on. The second problem may be stated more definitely: Given a non-linear device and an input consisting of noise alone, or of noise plus a signal. What is the power spectrum of the output?

There does not seem to be much published material on the first problem. However, from conversation with other people, I have learned that it has been studied independently by several investigators. The same is probably true of the second problem although here the published material is somewhat more plentiful. This makes it difficult to assign credit where credit is due. Much of the material given here had its origin in discussions with friends, especially with W. R. Bennett, J. H. Van Vleck, and David Middleton. Help was obtained from the recent paper[37] by Bennett, and also from the manuscript of a forthcoming paper by Middleton.[40]

4.1 Low Frequency Output of a Square Law Device

Let the output current I of the device be related to the input voltage V by

$$I = \alpha V^2 \qquad (4.1\text{–}1)$$

where α is a constant. When the power spectrum of V is confined to a relatively narrow band, the power spectrum of I consists of two portions. One portion clusters around twice the mid-band frequency of V and the other around zero frequency. We are interested in the low frequency portion. The current corresponding to this portion will be denoted by $I_{t\ell}$, and is the current which would flow if a low pass filter were inserted in the output to remove the upper portion of the spectrum. It is convenient to divide $I_{t\ell}$ into two components:

$$I_{t\ell} = I_{dc} + I_{\ell f} \qquad (4.1\text{–}2)$$

[37] Loc. cit. (Section 3.10).

[40] Cruft Laboratory and the Research Laboratory of Physics, Harvard University, Cambridge, Mass. In the following sections references to Bennett's paper and Middleton's manuscript are made by simply giving the authors' names.

where the subscripts stand for "total low" frequency, "direct current," and "low frequency," respectively. We have

$$I_{dc} = \text{average } I_{t\ell} = \bar{I}_{t\ell} \tag{4.1-3}$$

$$\text{Mean Square } I_{\ell f} = \text{average } (I_{t\ell} - I_{dc})^2 = \overline{I_{t\ell}^2} - I_{dc}^2$$

Probably the simplest method of obtaining I_{dc} is to square the given expression for V and pick out the terms independent of time. Thus if

$$V = P \cos pt + Q \cos qt + V_N \tag{4.1-4}$$

we have

$$I_{dc} = \alpha\left(\frac{P^2}{2} + \frac{Q^2}{2} + \overline{V_N^2}\right) \tag{4.1-5}$$

$I_{\ell f}$ may also be obtained by picking out the low frequency terms. However, here we wish to use the square law device, and the linear rectifier in the next section, to illustrate a general method of dealing with the statistical properties of the output of a non-linear device when the input voltage is restricted to a relatively narrow band.

If none of the low frequency spectrum is removed by filters,

$$I_{t\ell} = \alpha \frac{R^2}{2} \tag{4.1-6}$$

where R is the envelope of V. The probability density and the statistical properties of $I_{t\ell}$ may be derived from this relation when the distribution function of R is known.[41] Before discussing these properties we shall establish (4.1-6).

Equation (4.1-6) is a special case of a more general result established in Section 4.3. However, its truth may be seen by taking the example

$$V = P \cos pt + Q \cos qt + V_N \tag{4.1-4}$$

where $f_p = p/2\pi$ and $f_q = q/2\pi$ lie within, or close to, the band of the noise voltage V_N.

By using formulas of the type

$$\begin{aligned} \cos pt &= \cos [(p - \omega_m)t + \omega_m t] \\ &= \cos (p - \omega_m)t \cos \omega_m t - \sin (p - \omega_m)t \sin \omega_m t \end{aligned} \tag{4.1-7}$$

[41] When part of the low-frequency spectrum is removed, the problem becomes much more difficult. I_{dc} may be obtained as above, but to get $\overline{I_{\ell f}^2}$ it is necessary to first determine the power spectrum of I (Section 4.5) and then integrate over the appropriate portion of it. Concerning the distribution of $I_{\ell f}$, our present knowledge tells us only that it lies between the one given by (4.1-6) and the normal law which it approaches when only a narrow portion of the low frequency spectrum is passed by the audio frequency filter (Section 4.3).

we may refer all terms to the mid-band frequency $f_m = \omega_m/2\pi$, as is done in equations (3.7–2) to (3.7–4).

In this way we obtain

$$V = A \cos \omega_m t - B \sin \omega_m t = R \cos (\omega_m t + \theta), \tag{4.1–8}$$

where A and B are relatively slowly varying functions of t given by

$$A = P \cos (p - \omega_m)t + Q \cos (q - \omega_m)t + \sum_n c_n \cos (\omega_n t - \omega_m t - \varphi_n),$$

$$B = P \sin (p - \omega_m)t + Q \sin (q - \omega_m)t + \sum_n c_n \sin (\omega_n t - \omega_m t - \varphi_n)$$

and

$$R^2 = A^2 + B^2, \qquad R > 0$$
$$\tan \theta = B/A. \tag{4.1–9}$$

This definition of R has also been given in equations (3.10–22, 23, 24). The envelope of V is R and the output current is

$$I = \alpha R^2 \left[\frac{1}{2} + \frac{1}{2} \cos (2\omega_m t + 2\theta)\right] \tag{4.1–10}$$

Since R is a slowly varying function of time, so is R^2. The power spectrum of R^2 is confined to frequencies much lower than $2f_m$ and consequently the power spectrum of $R^2 \cos (2\omega_m t + 2\theta)$ is clustered around $2f_m$. Thus the only term in I contributing to the low frequency output is $\alpha R^2/2$ which is what we wished to show.

We now return to the statistical properties of $I_{\ell\ell}$. First, consider the case in which V consists of noise only, $V = V_N$, so that the probability density of the envelope R is

$$\frac{R}{\psi_0} e^{-R^2/2\psi_0} \tag{3.7–10}$$

where

$$\psi_0 = [rms\ V_N]^2 = \overline{V_N^2} \tag{4.1–11}$$

Hence

$$I_{dc} = \overline{I}_{\ell\ell} = \frac{\alpha \overline{R^2}}{2}$$
$$= \int_0^\infty \frac{\alpha R^2}{2} \frac{R}{\psi_0} e^{-R^2/2\psi_0} \, dR$$
$$= \alpha \psi_0$$

$$\overline{I_{\ell f}^2} = \overline{I_{\ell\ell}^2} - I_{dc}^2 = \int_0^\infty \frac{\alpha^2 R^5}{4\psi_0} e^{-R^2/2\psi_0} \, dR - I_{dc}^2$$
$$= \alpha^2 \psi_0^2 \tag{4.1–12}$$

Second, consider the case in which

$$V = V_N + P \cos pt \tag{4.1–13}$$

where $p/2\pi$ lies near the noise band of V_N. The probability density of the envelope R is

$$\frac{R}{\psi_0} \exp\left[-\frac{R^2 + p^2}{2\psi_0}\right] I_0\left(\frac{RP}{\psi_0}\right) \tag{3.10–11}$$

From this and equations (3.10–12), (3.10–13), we find

$$I_{dc} = \frac{\alpha \overline{R^2}}{2} = \alpha\psi_0 + \frac{\alpha P^2}{2} \tag{4.1–14}$$

$$\overline{I_{tl}^2} = \frac{\alpha^2}{4}\overline{R^4} = \alpha^2\left[2\psi_0^2 + 2P^2\psi_0 + \frac{P^4}{4}\right]$$

$$\overline{I_{lf}^2} = \overline{I_{tl}^2} - I_{dc}^2 = \alpha^2[\psi_0 + P^2]\psi_0 \tag{4.1–15}$$

In (4.1–14) ψ_0 is the mean square value of V_N and $P^2/2$ is the mean square value of the signal. These two equations show that I_{dc} and the rms value of I_{lf} are independent of the distribution of the noise power spectrum in V_N as long as the input V is confined to a relatively narrow band. In other words, although this distribution does affect the power spectrum of the output, it does not affect the d.c. and rms I_{lf} when ψ_0 and P are given. That the same is also true for a large class of non-linear devices was first pointed out by Middleton (see end of Section 4.9).

When the voltage is[42]

$$V = V_N + P \cos pt + Q \cos qt, \tag{4.1–4}$$

$p \neq q$, we obtain from equation (3.10–25)

$$I_{dc} = \frac{\alpha}{2}\overline{R^2} = \alpha\left(\psi_0 + \frac{P^2}{2} + \frac{Q^2}{2}\right)$$

$$\overline{I_{tl}^2} = \frac{\alpha^2}{4}\overline{R^4} \tag{4.1–16}$$

$$\overline{I_{lf}^2} = \alpha^2\left[\psi_0^2 + P^2\psi_0 + Q^2\psi_0 + \frac{P^2Q^2}{2}\right]$$

[42] These results are special cases, obtained by assuming no audio frequency filter, of formulas given by F. C. Williams, *Jour. Inst.* of *E. E.*, 80 (1937), 218–226. Williams also discusses the response of a linear rectifier to (4.1–4) when $P \gg Q + V_N$. An account of Williams' work is given by E. B. Moullin, "Spontaneous Fluctuations of Voltage," Oxford (1938), Chap. 7.

4.2 Low Frequency Output of a Linear Rectifier

In the case of the linear rectifier

$$I = \begin{cases} 0, & V < 0 \\ \alpha V, & V > 0 \end{cases} \tag{4.2–1}$$

the low frequency output current, assuming no audio frequency filter, is

$$I_{\ell f} = \frac{\alpha R}{\pi} \tag{4.2–2}$$

This formula, like its analogue (4.1–6) for the square law device, assumes that the applied signal and noise lie within a relatively narrow band. It may be used to compute the probability density and statistical properties of $I_{\ell f}$ when the corresponding information regarding the envelope R of the applied voltage is known.

The truth of (4.2–2) may be seen by considering the output I. It consists of the positive halves of the oscillations of αV. The envelope of I is the same as that of αV. However, the area under the loops of I is only about $1/\pi$ of the area under αR, this being the ratio of the area under a loop of $\sin x$ to the area of a rectangle of unit height and length 2π. From the low frequency point of view these loops of I merge into a current which varies as $\alpha R/\pi$.

When V is a sine wave plus noise,

$$V = V_N + P \cos pt \tag{4.1–13}$$

the average value of $I_{\ell f}$ is[43]

$$\begin{aligned} I_{dc} &= \frac{\alpha}{\pi} \overline{R} = \alpha \left(\frac{\psi_0}{2\pi}\right)^{1/2} {}_1F_1\left(-\frac{1}{2}; 1; -\frac{P^2}{2\psi_0}\right) \\ &= \alpha \left(\frac{\psi_0}{2\pi}\right)^{1/2} e^{-x/2} \left[(1 + x) I_0\left(\frac{x}{2}\right) + x I_1\left(\frac{x}{2}\right)\right] \end{aligned} \tag{4.2–3}$$

where I_0, I_1 are Bessel functions of imaginary argument and

$$x = \frac{P^2}{2\psi_0} = \frac{\text{ave. sine wave power}}{\text{ave. noise power}} \tag{4.2–4}$$

[43] This result was discovered independently by several investigators, among whom we may mention W. R. Bennett and D. O. North. The latter has applied it to noise measurement work. He has found that the diode detector, when adapted to noise metering, is a great improvement over the thermocouple, and has used noise meters of this type satisfactorily since 1940. See D. O. North, "The Modification of Noise by Certain Non-Linear Devices", Paper read before I.R.E., Jan. 28, 1944.

ψ_0 being the average value of V_N^2. Equation (4.2–3) follows from the formulas (3.10–12) and (4B–9). When x is large the asymptotic expansion (4B–3) of the ${}_1F_1$ gives

$$I_{dc} \sim \frac{\alpha}{\pi}\left[P + \frac{\psi_0}{2P} + \frac{\psi_0^2}{8P^3} + \cdots\right] \tag{4.2–5}$$

Similarly, the mean square value of $I_{t\ell}$ is

$$\overline{I_{t\ell}^2} = \frac{\alpha^2}{\pi^2}\overline{R^2} = \frac{\alpha^2}{\pi^2}(P^2 + 2\psi_0) \tag{4.2–6}$$

and the mean square value of the low frequency current $I_{\ell f}$, excluding the d.c., is given by

$$\overline{I_{\ell f}^2} = \overline{I_{t\ell}^2} - I_{dc}^2$$

When x is large we have

$$\overline{I_{\ell f}^2} \sim \frac{\alpha^2}{\pi^2}\left[\psi_0 - \frac{\psi_0^2}{2P^2}\cdots\right] = \frac{\alpha^2}{\pi^2}\psi_0\left[1 - \frac{1}{4x}\cdots\right] \tag{4.2–7}$$

and when $x = 0$,

$$\overline{I_{\ell f}^2} = \frac{\alpha^2}{\pi^2}\psi_0\left(2 - \frac{\pi}{2}\right) \tag{4.2–8}$$

Curves for I_{dc} are given in Figures 1, 2 and 3 of Bennett's paper. He also gives curves, in Fig. 4, showing $\overline{I_{\ell f}^2}$ versus x. These show that the effect of the higher order modulation terms is small when $I_{\ell f}$ is computed by adding low frequency modulation products.

When V consists of two sine waves plus noise,

$$V = V_N + P\cos pt + Q\cos qt, \tag{4.1–4}$$

the average value of $I_{t\ell}$ is, from (3.10–25), a sort of double ${}_1F_1$ function:

$$\begin{aligned} I_{dc} &= \frac{\alpha}{\pi}\overline{R} = \alpha\left(\frac{\psi_0}{2\pi}\right)^{1/2}\sum_{k=0}^{\infty}\sum_{m=0}^{\infty}\frac{(-\frac{1}{2})_{k+m}}{k!k!m!m!}(-x)^k(-y)^m \\ &= \alpha\left(\frac{\psi_0}{2\pi}\right)^{1/2}\sum_{k=0}^{\infty}\frac{(-\frac{1}{2})_k}{k!k!}(y-x)^k P_k\left(\frac{x+y}{x-y}\right) \end{aligned} \tag{4.2–9}$$

where

$$x = \frac{P^2}{2\psi_0}, \qquad y = \frac{Q^2}{2\psi_0}, \qquad P_k(z) = \text{Legendre polynomial} \tag{4.2–10}$$

If x is large and $y < x$, we have from (3.10–27) the asymptotic expression

$$I_{dc} \sim \frac{\alpha}{\pi}P\sum_{k=0}^{\infty}\frac{(-\frac{1}{2})_k(-\frac{1}{2})_k}{k!x^k}\,{}_2F_1\left(k-\tfrac{1}{2}, k-\tfrac{1}{2}; 1; \frac{y}{x}\right) \tag{4.2–11}$$

The ${}_2F_1$ may be expressed in terms of the complete elliptic functions E and K of modulus $y^{1/2}x^{-1/2}$. Thus

$$
\begin{aligned}
{}_2F_1\left(-\tfrac{1}{2}, -\tfrac{1}{2}; 1; \frac{y}{x}\right) &= \frac{4}{\pi} E - \frac{2}{\pi}\left(1 - \frac{y}{x}\right) K, \\
{}_2F_1\left(\tfrac{1}{2}, \tfrac{1}{2}; 1; \frac{y}{x}\right) &= \frac{2}{\pi} K
\end{aligned}
\tag{3.10–28}
$$

and the higher terms may be computed from the recurrence relation (3.10–29). The first term, $k = 0$, in (4.2–11) gives I_{dc} when the noise is absent.[44]

The mean square value of $I_{t\ell}$ is

$$\overline{I_{t\ell}^2} = \frac{\alpha^2}{\pi^2} \overline{R^2} = \frac{\alpha^2}{\pi^2} [2\psi_0 + P^2 + Q^2] \tag{4.2–14}$$

From this expression and our expression for I_{dc}, the rms value of the low frequency current, $I_{\ell f}$, excluding the d.c., may be computed. For example, when the noise is small,

$$
\begin{aligned}
\overline{I_{\ell f}^2} \sim \frac{\alpha^2}{\pi^2}\Bigg[P^2 + Q^2 - \left(P\,{}_2F_1\left(-\tfrac{1}{2}, -\tfrac{1}{2}; 1; \frac{y}{x}\right)\right)^2 \\
+ 2\psi_0\left(1 - {}_2F_1\left(-\tfrac{1}{2}, -\tfrac{1}{2}; 1; \frac{y}{x}\right)\frac{K}{\pi}\right)\Bigg]
\end{aligned}
\tag{4.2–15}
$$

The term independent of ψ_0 gives the mean square low frequency current in the absence of noise. As Q goes to zero (4.2–15) approaches the leading term in (4.2–7), as it should. When $P = Q$ our formula breaks down and it appears that we need the asymptotic behavior of[45]

$$I_{dc} = \alpha\left(\frac{\psi_0}{2\pi}\right)^{1/2} \sum_{k=0}^{\infty} \frac{(-\frac{1}{2})_k (2k)!}{[k!]^4} (-x)^k$$

In view of the questionable nature of the derivation given in Section 3.10 of equations (4.2–9) and (4.2–11) it was thought that a numerical check on their equivalence would be worth while. Accordingly, the values $x = 4$, $y = 3$ were used in the second series of (4.2–9). It was found that the largest term (about 130) in the summation occurred at $k = 11$. In all, 24 terms were taken. The result obtained was

$$\frac{\overline{R}}{\sqrt{2\psi_0}} = 2.5502$$

[44] See W. R. Bennett, *B.S.T.J.*, Vol. 12 (1933), 228–243.

[45] This may be done by the method given by W. B. Ford, Asymptotic Developments, Univ. of Mich. Press (1936), Chap. VI.

For the same values of x and y the asymptotic series (4.2–11) gave

$$2.40 + 0.171 + .075 + 0.52 + \cdots.$$

If we stop just before the smallest term we get 2.57 for the sum. If we include the smallest term we get 2.65. This agreement indicates that (4.2–11) is actually the asymptotic expansion of (4.2–9).

When the voltage is of the form

$$V = Q(1 + k \cos pt) \cos qt + V_N$$

we may use

$$\overline{R^n} = (2\psi_0)^{n/2}\Gamma\left(1 + \frac{n}{2}\right)\frac{1}{2\pi}\int_0^{2\pi} {}_1F_1\left[-\frac{n}{2}; 1; -y(1 + k\cos\theta)^2\right] d\theta \tag{4.2–16}$$

where R is the envelope with respect to the frequency $q/2\pi$ and y is given by (4.2–10). The integral may be evaluated by writing ${}_1F_1$ as a power series and integrating termwise using the result

$$\frac{1}{2\pi}\int_0^{2\pi} (1 + k\cos\theta)^{\ell} \cos m\theta \, d\theta = \frac{(-\ell)_m}{2^m m!}(-k)^m \, {}_2F_1\left[\frac{m-\ell}{2}, \frac{m-\ell+1}{2}; m+1; k^2\right] \tag{4.2–17}$$

where m is a non-negative integer, ℓ any number,

$$(\alpha)_m = \alpha(\alpha+1)\cdots(\alpha+m-1), \qquad (\alpha)_0 = 1, \qquad \text{and} \qquad (0)_0 = 1.$$

The integral may also be evaluated in terms of the associated Legendre function.

By applying the methods of Section 3.10 to (4.2–16) we are led to

$$\begin{aligned} \overline{R^2} &= Q^2\left(1 + \frac{k^2}{2}\right) + 2\psi_0 \\ \bar{R} &\sim Q\sum_{s=0}^{\infty} \frac{(-\frac{1}{2})_s(-\frac{1}{2})_s}{s!\,y^s} \, {}_2F_1(s - \tfrac{1}{2}, s; 1; k^2) \end{aligned} \tag{4.2–18}$$

where the asymptotic series holds when y is very large and k is not too close to unity. These expressions give

$$\overline{I_{\ell f}^2} \sim \frac{\alpha^2}{\pi^2}\left(Q^2\frac{k^2}{2} + \psi_0[2 - (1 - k^2)^{-1/2}] + \cdots\right) \tag{4.2–19}$$

The reader might be tempted to associate the coefficient of ψ_0 in (4.2–19) with the continuous portion of the output power spectrum. However, this would not be correct. It appears that the principal contribution of the continuous portion of the power spectrum to $\overline{I_{\ell f}^2}$ is $\alpha^2\psi_0/\pi^2$, just as in (4.2–7) when k is zero. The difference between this and the corresponding term in (4.2–19) seems to arise from the fact that the amplitude of the recovered signal is not exactly $\alpha Qk/\pi$ but is modified by the presence of the noise. This general type of behavior might be expected on physical grounds since changing P, say doubling it, in (4.2–7) does not appreciably affect the $\overline{I_{\ell f}^2}$ in (4.2–7) (which is due entirely to the continuous portion of the noise spectrum). The modulating wave may be regarded as slowly making changes of this sort in P.

4.3 Some Statistical Properties of the Output of a General Non-Linear Device

Our general problem is this: Given a non-linear device whose output I is related to its input V by the relation

$$I = \frac{1}{2\pi}\int_C F(iu)e^{iVu}\,du \tag{4A–1}$$

which is discussed in Appendix 4A. Let the input V contain noise in addition to the signal. Choose some frequency band in the output for study. What are the statistical properties of the current flowing in this band?

It seems to be difficult to handle this general problem. However, it appears that the two following results are true.

1. As the output band is chosen narrower and narrower the statistical properties of the corresponding current approach those of the random noise current discussed in Part III (provided no signal harmonic lies within the band). In particular, the instantaneous current values are distributed normally.

2. When the input V is confined to a relatively narrow band the power spectrum of the output I is clustered around the 0th (d.c.), 1st, 2nd, etc. harmonics of the midband frequency of V. The low frequency output including the d.c. is

$$I_{\ell\ell} = A_0(R) = \frac{1}{2\pi}\int_C F(iu)J_0(uR)\,du \tag{4.3–11}$$

where R is the envelope of V.

The envelope of the nth harmonic of the output, when $n > 0$, is

$$A_n(R) = \frac{1}{\pi}\int_C F(iu)J_n(uR)\,du \tag{4.3–1}$$

The mathematical statement is

$$I = \sum_{n=0}^{\infty} A_n(R) \cos (n\omega_m t + n\theta) \tag{4.3-9}$$

where $f_m = \omega_m/(2\pi)$ is the representative mid-band frequency of V and θ is a relatively slowly varying phase angle. The results of Sections 4.1 and 4.2 are special cases of this.

Middleton's result that the noise power in each of the output bands (in the entire band corresponding to a given harmonic) depends only on $\overline{V_N^2} = \psi_0$ and not on the spectrum of V_N, where V_N is the noise voltage component of V, may also be obtained from (4.3-9). We note that the total power in the nth band depends only on the mean square value of its envelope $A_n(R)$, and that the probability density of the envelope R of the input involves V_N only through ψ_0.

The argument we shall use in discussing the first result is not very satisfactory. It runs as follows. The output current I may be divided into two parts. One consists of sinusoidal terms due to the signal. The other consists of noise. We shall be concerned only with the latter which we shall call I_N. The correlation between two values of I_N separated by an interval of time approaches zero as the interval becomes large. Let τ be an interval long enough to ensure that the two values of I_N are substantially independent. Choose an interval of time T long enough to contain many intervals of length τ. Expand I_N as a Fourier series over this interval. We have

$$\begin{aligned} I_N &= \frac{a_0}{2} + \sum_{n=1}^{\infty} \left[a_n \cos \frac{2\pi nt}{T} + b_n \sin \frac{2\pi nt}{T} \right] \\ a_n - ib_n &= \frac{2}{T} \int_0^T e^{-i2\pi nt/T} I_N(t)\, dt \end{aligned} \tag{4.3-2}$$

Let the band chosen for study be $f_0 - \frac{\beta}{2}$ to $f_0 + \frac{\beta}{2}$ and let

$$T\left(f_0 - \frac{\beta}{2}\right) = n_1, \qquad T\left(f_0 + \frac{\beta}{2}\right) = n_2 \tag{4.3-3}$$

where n_1 and n_2 are integers. The number of components in the band is $(n_2 - n_1)$. We suppose β is such that this is small in comparison with T/τ. The output of the band is

$$J_N = \sum_{n=n_1}^{n_2} \left[a_n \cos \frac{2\pi n}{T} t + b_n \sin \frac{2\pi nt}{T} \right] \tag{4.3-4}$$

where

$$a_n - ib_n = \frac{2}{T}\int_0^T e^{-i2\pi((n/T)-f_0)t}e^{-i2\pi f_0 t} I_N(t)\,dt$$

$$n = \frac{n_1+n_2}{2} + n - \frac{n_1+n_2}{2} = f_0 T + (n - f_0 T) \tag{4.3–5}$$

We choose the band so narrow that

$$n_2 - n_1 \ll T/\tau \quad \text{or} \quad \beta\tau \ll 1 \tag{4.3–6}$$

This enables us to write approximately

$$a_n - ib_n = \sum_{r=1}^{r_1} e^{-i2\pi((n/T)-f_0)r\tau}\frac{2}{T}\int_{(r-1)\tau}^{r\tau} e^{-i2\pi f_0 t} I_N(t)\,dt$$

$r_1 = T/\tau$, T being chosen to make r_1 an integer. Suppose we do this for a large number of intervals of length T. Then $I_N(t)$ will differ from interval to interval. The set of integrals for $r = 1$ gives us an array of values which we regard as defining the distribution of a complex random variable, say x_1. Similarly the set of integrals for $r = 2$ defines the distribution of a second random variable x_2, and so on to x_{r_1}. Because we have chosen τ so large that $I_N(t)$ in any one integral is practically independent of its values in the other integrals we may say that $x_1, x_2, \cdots x_{r_1}$ are independent.

We have

$$a_{n_1} - ib_{n_1} = \sum_{r=1}^{r_1} e^{-i2\pi((n_1/T)-f_0)r\tau} x_r$$

$$a_{n_1+1} - ib_{n_1+1} = \sum_{r=1}^{r_1} e^{-i2\pi((\overline{n_1+1}/T)-f_0)r\tau} x_r$$

$$\vdots \qquad\qquad \vdots$$

$$a_{n_2} - ib_{n_2} = \sum_{r=1}^{r_1} e^{-i2\pi((n_2/T)-f_0)r\tau} x_r$$

and if $n_2 - n_1 \ll r_1$, as was assumed in (4.3–6), we may apply the central limit theorem to show that $a_{n_1}, b_{n_1}, a_{n_1+1}, \cdots a_{n_2}, b_{n_2}$ tend to become independent and normally distributed about zero as we let the band width $\beta \to 0$ and $T \to \infty$ (and hence $r_1 \to \infty$) in such a way as to keep $n_2 - n_1$ fixed. In this work we make use of the fact that $I_N(t)$ is such that the real and imaginary parts of $x_1, x_2, \cdots x_r$ all have the same average and standard deviation. It is convenient to assume f_0T is an integer.

Thus as the band width β approaches zero the band output J_N given by (4.3–4) may be represented in the same way, namely as (2.8–1), as was the random noise current studied in Part III. Hence J_N tends to have the same properties as the random noise current studied thère. For example, the distribution of J_N tends towards a normal law. In our discussion we had to assume that $\beta\tau \ll 1$. If the voltage V applied to the non-linear device is confined to a relatively narrow frequency band, say $f_b - f_a$, it appears that the interval τ (chosen above so that $I(t)$ and $I(t + \tau)$ are substantially independent) may be taken to be of the order of $1/(f_b - f_a)$. In this case J_N tends to behave like a random noise current if $\beta/(f_b - f_a)$ is much smaller than unity.

We now turn our attention to the second statement made at the beginning of this section. Let the applied voltage be confined to a relatively narrow band so that it may be represented by equation (4.1–8) of Section 4.1,

$$V = R \cos (\omega_m t + \theta), \qquad R \geq 0, \qquad (4.1\text{–}8)$$

where $f_m = \omega_m/(2\pi)$ is some representative frequency within the band and R and θ are functions of time which vary slowly in comparison with $\cos \omega_m t$. We call R the envelope of V.

From equation (4A–1)

$$I = \frac{1}{2\pi} \int_C F(iu) e^{iuR \cos (\omega_m t + \theta)} \, du \qquad (4.3\text{–}7)$$

We expand the integrand by means of

$$e^{ix \cos \varphi} = \sum_{n=0}^{\infty} \epsilon_n i^n \cos n\varphi J_n(x) \qquad (4.3\text{–}8)$$

where ϵ_0 is 1 and ϵ_n is 2 when $n > 0$ and $J_n(x)$ is a Bessel function. Thus

$$I = \sum_{n=0}^{\infty} A_n(R) \cos (n\omega_m t + n\theta) \qquad (4.3\text{–}9)$$

where

$$A_n(R) = \epsilon_n \frac{i^n}{2\pi} \int_C F(iu) J_n(uR) \, du \qquad (4.3\text{–}10)$$

Since R is a relatively slowly varying function of time we expect the same to be true of $A_n(R)$, at least for moderately small values of n. Thus from (4.3–9) we see that the power spectrum of I will consist of a succession of bands, the nth band being clustered around the frequency nf_m. If we eliminate all of the bands except the nth by means of a filter we see that the output will have the envelope $A_n(R)$ when $n \geqq 1$. Taking n to be zero, shows that the low frequency output is simply

$$A_0(R) = \frac{1}{2\pi} \int_C F(iu) J_0(uR) \, du \qquad (4.3\text{–}11)$$

Taking n to be one shows that the band around f_m given by

$$\frac{A_1(R)}{R} V \qquad (4.3\text{–}12)$$

The statistical properties of the low frequency output and of the envelopes of the output bands may be obtained from those of R. For example, the probability density of $A_n(R)$ is of the form

$$p(R) \bigg/ \frac{dA_n(R)}{dR} \qquad (4.3\text{–}13)$$

where $p(R)$ is the probability density of R. In this expression R is considered as a function of A_n.

It should be noted that we have been assuming that all of the band surrounding the harmonic frequency nf_m is taken. When we take only a portion of it, presumably the statistical properties will tend to approach those of a random noise current in accordance with the first statement made at the beginning of this section.

When we apply (4.3–11) to the square law device we have

$$F(iu) = \frac{2\alpha}{(iu)^3}$$

$$A_0(R) = -\frac{2\alpha}{2\pi i} \int^{(0+)} \frac{J_0(uR)}{u^3} \, du$$

$$= \frac{\alpha}{2} R^2$$

When we apply (4.3–11) to the linear rectifier:

$$F(iu) = -\frac{\alpha}{u^2}$$

$$A_0(R) = -\frac{\alpha}{2\pi} \int_{-\infty}^{+\infty} \frac{J_0(uR)}{u^2} \, du = \frac{\alpha R}{\pi}$$

where the path of integration passes under the origin. These two results agree with those obtained in Section 4.1 and 4.2 from simple considerations. As a final example we find the low frequency output of a biased linear rectifier in terms of the envelope R of the applied voltage. From the table of $F(iu)$ given in Appendix 4A we see that $F(iu)$ corresponding to

$$I = 0, \qquad V < B$$

$$I = V - B, \qquad V > B$$

is

$$F(iu) = -\frac{e^{-iuB}}{u^2}$$

Consequently, the low frequency output is

$$A_0(R) = -\frac{1}{2\pi}\int_{-\infty}^{+\infty} e^{-iuB}J_0(uR)u^{-2}\,du$$

where the path of integration is indented downwards at the origin. When $B > R$ the value of the integral is zero since then the path of integration may be closed in the lower half plane by an infinite semi-circle This value also follows at once from the physics of the problem. When $-R < B < R$ we may integrate by parts and get

$$\begin{aligned}
A_0(R) &= \frac{1}{2\pi}\int_{-\infty}^{+\infty} e^{-iuB}[iBJ_0(uR) + RJ_1(uR)]u^{-1}\,du \\
&= -\frac{B}{2} + \frac{1}{\pi}\int_0^{\infty} [B \sin uBJ_0(uR) + R \cos uBJ_1(uR)]u^{-1}\,du \\
&= -\frac{B}{2} + \frac{B}{\pi} \text{ arc sin } \frac{B}{R} + \frac{1}{\pi}\sqrt{R^2 - B^2} \\
&= -\frac{B}{2} + \frac{R}{\pi} F\left(-\frac{1}{2}, -\frac{1}{2}; \frac{1}{2}; \frac{B^2}{R^2}\right), \qquad -R < B < R
\end{aligned} \tag{4.3-14}$$

This hypergeometric function turns up again in equation (4.7–6). Also in the range $-R < B < R$,

$$\frac{dA_0}{dR} = \frac{1}{\pi}\sqrt{1 - \frac{B^2}{R^2}}$$

When B is negative and $R < -B$, the path of integration may be closed by an infinite semicircle in the upper half plane and the value of the integral is proportional to the residue of the pole at the origin:

$$\begin{aligned}
A_0(R) &= 2\pi i\left(-\frac{1}{2\pi}\right)(-iB) \\
&= -B
\end{aligned}$$

Thus, to summarize, the low frequency output for our linear rectifier is, for $B > 0$, (R is always positive)

$$A_0(R) = 0, \qquad R < B$$

$$A_0(R) = -\frac{B}{2} + \frac{B}{\pi}\text{ arc sin }\frac{B}{R} + \frac{1}{\pi}\sqrt{R^2 - B^2}, \qquad B < R \tag{4.3-15}$$

and for $B < 0$ it is

$$A_0(R) = |B|, \quad R < |B|$$

$$A_0(R) = +\frac{|B|}{2} + \frac{|B|}{\pi} \arcsin \frac{|B|}{R} + \frac{1}{\pi}\sqrt{R^2 - B^2}, \quad |B| < R \tag{4.3-16}$$

where the arc sines lie between 0 and $\pi/2$. $A_0(R)$ and its first derivative with respect to R are continuous.

From (4.3–15), the d.c. output current is, for $B > 0$,

$$I_{dc} = \int_B^\infty \left[-\frac{B}{2} + \frac{B}{\pi} \arcsin \frac{B}{R} + \frac{1}{\pi}\sqrt{R^2 - B^2} \right] p(R)\, dR \tag{4.3-15}$$

where $p(R)$ is the probability density of the envelope of the input V, e.g., $p(R)$ is of the form (3.7–10) for noise alone, and of the form (3.10–11) for noise plus a sine wave. Similarly, the rms value of the low frequency current $I_{\ell f}$, excluding d.c., may be computed from

$$\overline{I_{\ell f}^2} = \overline{I_{\ell\ell}^2} - I_{dc}^2$$

where, if $B > 0$,

$$\overline{I_{\ell\ell}^2} = \int_B^\infty \left[-\frac{B}{2} + \frac{B}{\pi} \arcsin \frac{B}{R} + \frac{1}{\pi}\sqrt{R^2 - B^2} \right]^2 p(R)\, dR \tag{4.3-16}$$

If V consists of a sine wave of amplitude P plus noise V_N, so it may be represented as (4.1–13), and if $P \gg$ rms V_N, the distribution of R is approximately normal. If, in addition, $P - B \gg$ rms $V_N > 0$, (4.3–15), (4.3–16), and (3.10–19) lead to the approximations

$$\begin{aligned} I_{dc} &\approx -\frac{B}{2} + \frac{B}{\pi} \arcsin \frac{B}{P} + \frac{1}{\pi}\sqrt{P^2 - B^2} + \frac{\psi_0}{2\pi\sqrt{P^2 - B^2}} \\ &\approx -\frac{B}{2} + \frac{P}{\pi} + \frac{B^2 + \psi_0}{2\pi P} \end{aligned} \tag{4.3-17}$$

$$\overline{I_{\ell f}^2} \approx \frac{P^2 - B^2}{\pi^2 P^2}\psi_0$$

The second expression for I_{dc} assumes $P \gg B$. When $B = 0$, these reduce to the first terms of (4.2–5) and (4.2–7). By using a different method Middleton has obtained a more precise form of this result.

Incidentally, for a given applied voltage, $I_{dc}(+)$ for a positive bias $|B|$ is related to $I_{dc}(-)$ for a negative bias $-|B|$ by

$$I_{dc}(-) = |B| + I_{dc}(+) \tag{4.3-18}$$

Also r.m.s. $I_{\ell f}(+)$ is equal to r.m.s. $I_{\ell f}(-)$. Equation (4.3–18) follows from a physical argument based on the areas underneath a curve of I for the two cases. Both of the above relations follow from formulas given by Middleton when V is the sum of a sine wave plus noise. They may also be derived from (4.3–15) and (4.3–16).

4.4 Output Power Spectrum

The remainder of Part IV will be concerned with methods of solving the following problem: Given a non-linear device and an input voltage consisting of noise alone or of a signal plus noise. What is the power spectrum of the output?

In some ways the answer to this problem gives us less information than the methods discussed in the first three sections. For example, beyond giving the rms value, it tells us very little about the probability density of the current corresponding to a given frequency band of the output. On the other hand, this rms value may be found (by integrating the power spectrum) for any band we choose to study. The methods described earlier depended on the input being confined to a relatively narrow band and gave information regarding only the entire band corresponding to a given harmonic (0th, 1st, 2nd, etc.) of the input. There was no way to study the output when part of a band was eliminated by filters except by obtaining the power spectrum of some function of the envelope.

At present there appear to be two general methods available for the determination of the output power spectrum each with its own advantages and disadvantages. First there is the direct method which has been used by W. R. Bennett*, F. C. Williams**, J. R. Ragazzini[46] and others. The noise is represented as the sum of a finite number of sinusoidal components. The typical modulation product is computed and the output power spectrum is obtained by considering the density and amplitude of these products. The chief advantage of this method lies in its close relation to the known theory of modulation in non-linear circuits. Generally, the lower order modulation products are the only ones which contribute significantly to the output power and when they are known, the problem is well along towards solution. The main disadvantage is the labor of counting the modulation products falling in a given interval. However, Bennett has developed a method for doing this.[47]

The fundamental idea of the second method is to obtain the correlation function for the output current. From this the output power spectrum may be obtained by Fourier's transform. The correlation function method and its variations are of more recent origin than the direct method. They have

* Cited in Section 4.0. Also much of this writer's work on interference in broad band communication systems may be carried over to noise theory without any change in the methods used.

** Cited in Section 4.1.

[46] *Proc. I.R.E.* Vol. 30, pp. 277–288 (June 1942), "The Effect of Fluctuation Voltages on the Linear Detector."

[47] *B.S.T.J.*, Vol. 19 (1940), pp. 587–610, Appendix B.

been discovered independently and at about the same time, by several workers. In a paper read before the I.R.E., Jan. 28, 1944, D. O. North described results obtained by using the correlation function. J. H. Van Vleck and D. Middleton have been using the two variations of the method which we shall describe in Sections 4.7 and 4.8, since early in 1943. A primitive form of the method of Section 4.8 had been used by A. D. Fowler and the writer in some unpublished material written in 1942. Recently, I have learned that a method similar to the one used by Fowler and myself had already been used by Kurt Fränz in 1941.[48]

The correlation function method avoids the problem of counting the modulation products. However, in some cases it becomes rather unwieldy. Probably it is best to have both methods in mind when investigating any particular problem. The direct method will be illustrated by applying it to the square law detector. Two approaches to the correlation function method will then be described and applied to examples.

4.5 Noise Through Square Law-Device

Probably the most direct method of obtaining the power spectrum $W(f)$ of I, where

$$I = \alpha V^2, \tag{4.1-1}$$

V being a noise voltage, is to square the expression

$$V = V_N = \sum_{1}^{M} c_m \cos(\omega_m t - \varphi_m) \tag{2.8-6}$$

in which c_m^2 is $2w(f_m)\Delta f$, $\omega_m = 2\pi f_m$, $f_m = m\Delta f$ and $\varphi_1, \varphi_2, \cdots \varphi_M$ are random phase angles.

Considerable simplification of the algebra results when we replace the representation (2.8–6) by

$$V_N = \frac{1}{2}\sum_{-\infty}^{\infty} c_m e^{imat - i\varphi_m} \tag{4.5-1}$$

Here we have added a term $c_0/2$ so as to not have any gaps in the summation and have introduced the definitions

$$\begin{aligned} c_{-m} &= c_m \\ \varphi_{-m} &= -\varphi_m \\ a &= 2\pi\Delta f \end{aligned} \tag{4.5-2}$$

[48] "Die Übertragung von Rauschspannung über den linearen Gleichrichter," *Hochfr. u. Elektroakust.*, June 1941. Other articles by Fränz are (I am indebted to Dr. North for the following references) "Beitrage zur Berechnung des Verhaltnisses von Signal spannung zu Rauschspannung am Ausgang von Empfängern", *E.N.T.*, 17, 215, 1940 and 19, 285, 1942. "Die Amplituden von Geräuschspannungen", *E.N.T.*, 19, 166, 1942. The May 1944 (p. 237), issue of the Wireless Engineer contains an abstract of "The Influence of Carrier Waves on the Noise on the Far Side of Amplitude-Limiters and Linear Rectifiers" by Fränz and Vellat, *E.N.T.*, Vol. 20, pp. 183–189 (Aug. 1943).

Squaring (4.5–1) gives the double series

$$V_N^2 = \frac{1}{4}\sum_{-\infty}^{+\infty}\sum_{-\infty}^{+\infty} c_m c_n e^{i(m+n)at-i\varphi_m-i\varphi_n}$$

$$= \frac{1}{4}\sum_{k=-\infty}^{+\infty}\sum_{n=-\infty}^{+\infty} c_{k-n} c_n e^{ikat-i\varphi_{k-n}-i\varphi_n}$$

Suppose we wish to consider the component of V_N^2 of frequency $f_k = k\Delta f$. It is seen to be

$$A_k \cos(\omega_k t - \psi_k) = \frac{1}{2}\sum_{n=-\infty}^{+\infty} c_{k-n} c_n \cos(kat - \varphi_{k-n} - \varphi_n) \qquad (4.5\text{–}3)$$

The power spectrum $W(f)$ of I at frequency f_k is α^2 times the coefficient of Δf in the mean square value of (4.5–3) where the average is taken over the φ's. Thus

$$W(f_k)\Delta f = \frac{\alpha^2}{4}\sum_{-\infty}^{+\infty}\sum_{-\infty}^{+\infty} c_{k-n} c_n c_{k-m} c_m$$

$$\times \text{ ave. } \cos(kat - \varphi_{k-n} - \varphi_n)\cos(kat - \varphi_{k-m} - \varphi_m)$$

where the summations extend over m and n. Let n be fixed and consider those values of m which give an average different from zero. We see that $m = n$ and $m = k - n$ are two such values. The only other possibilities are $m = -n$ and $m = -k + n$, but these lead to terms containing (except when n or k equal zero) three different angles, φ_n, φ_{k-n}, and φ_{k+n} which average to zero. Using the fact that the average of cosine squared is one-half and that for a given n there are two such terms, we get

$$W(f_k)\Delta f = \frac{\alpha^2}{4}\sum_{n=-\infty}^{+\infty} c_{k-n}^2 c_n^2$$

$$= \alpha^2 \Delta f \sum_{n=-\infty}^{+\infty} w(f_k - f_n) w(f_n)\Delta f \qquad (4.4\text{–}5)$$

where in the last step we have used

$$f_{k-n} = (k - n)\Delta f = f_k - f_n$$

and have implied, from $c_{-n} = c_n$, that

$$w(f_{-n}) = w(-n\Delta f) = w(-f_n)$$

is equal to $w(f_n)$.

Thus, from (4.5–4), we get for the power spectrum of I

$$W(f) = \alpha^2 \int_{-\infty}^{+\infty} w(x) w(f - x)\, dx \qquad (4.5\text{–}5)$$

with the understanding that f is not zero and

$$w(-x) = w(x). \tag{4.5–6}$$

The result which is obtained by using (2.8–6), involving the cosines and only positive values of m, is

$$W(f) = \alpha^2 \int_0^f w(x)w(f-x)\,dx + 2\alpha^2 \int_0^\infty w(x)w(f+x)\,dx \tag{4.5–7}$$

This contains only positive values of frequency. (4.5–5) and (4.5–7) are equivalent and may readily be transformed into each other.

The first integral in (4.5–7) arises from second order modulation products of the sum type and the second integral from products of the difference type. This may be seen by writing the current as

$$I = \text{ъ}V^2 = \alpha \sum_{m=1}^{\infty} \sum_{n=1}^{\infty} c_m c_n \cos(\omega_m t - \varphi_m) \cos(\omega_n t - \varphi_n)$$

$$= \frac{\alpha}{2} \sum_{m=1}^{\infty} \sum_{n=1}^{\infty} c_m c_n \{\cos[(\omega_m - \omega_n)t - \varphi_m + \varphi_n] \tag{4.5–8}$$

$$+ \cos[(\omega_m + \omega_n)t + \varphi_m + \varphi_n]\}$$

The power in the range $f_k, f_k + \Delta f$ is the power due to modulation products of the difference type, $\omega_{k+\ell} - \omega_\ell$, plus the power due to the modulation products of the sum type, $\omega_{k-\ell} + \omega_\ell$. In the first type ℓ runs from 1 to ∞ and in the second type ℓ runs from 1 to $k - 1$.

Consider the difference type first, and for the moment take both k and ℓ to be fixed. The two sets $m = k + \ell$, $n = \ell$ and $m = \ell$, $n = k + \ell$ are the only values of m and n in (4.5–8) leading to $\omega_{k+\ell} - \omega_\ell$. The two corresponding terms in (4.5–8) are equal because $\cos(-x)$ is equal to $\cos x$. The average power contributed by these two terms is

$$\left(\frac{\alpha}{2} c_{k+\ell} c_\ell\right)^2 \times \{\text{Average of } (2\cos[(\omega_{k+\ell} - \omega_\ell)t - \varphi_{k+\ell} + \varphi_\ell])^2\}$$
$$= \tfrac{1}{2}(\alpha c_{k+\ell} c_\ell)^2 \tag{4.5–9}$$

The power contributed to $f_k, f_k + \Delta f$ by the difference modulation products is obtained by summing ℓ from 1 to ∞:

$$\frac{\alpha^2}{2} \sum_{\ell=1}^{\infty} c_{k+\ell}^2 c_\ell^2 = 2\alpha^2 \sum_{\ell=1}^{\infty} w(f_{k+\ell}) w(f_\ell)(\Delta f)^2$$

$$\rightarrow 2\alpha^2 \Delta f \int_0^\infty w(f_k + f) w(f)\,df$$

This leads to the second term in (4.5–7).

Now consider the modulation products of the sum type. The terms of this type in (4.5–8) which give rise to the frequency ω_k are those for which $m + n$ is equal to k. Let n be 1 then $m = k - 1$. The phase of this term is random with respect to all the other terms except the one given by $n = k - 1$, $m = 1$ which has the same phase. The average power contributed by these two terms in (4.5–8) is, as in (4.5–9),

$$\tfrac{1}{2}(\alpha c_1 c_{k-1})^2$$

This disposes of two terms for which $m + n$ is equal to k. Taking n to be 2 and going through the same process gives two more. Thus, assuming for the moment that k is an odd number, the power contributed to the interval $f_k, f_k + \Delta f$ by the sum modulation products is

$$\frac{1}{2}\sum_{n=1}^{(k-1)/2} (\alpha c_n c_{k-n})^2 = \frac{1}{4}\sum_{n=1}^{k-1} (\alpha c_n c_{k-n})^2 \to \alpha^2 \Delta f \int_0^{f_k} w(f)w(f_k - f)\, df$$

and this leads to the second term in (4.5–7).

When the voltage V applied to the square law device is the sum of a noise voltage V_N and a sine wave:

$$V = P \cos pt + V_N, \tag{4.1–13}$$

we have

$$V^2 = P^2 \cos^2 pt + 2PV_N \cos pt + V_N^2 \tag{4.5–10}$$

From the two equations

$$\cos^2 pt = \frac{1}{2} + \frac{1}{2}\cos 2pt$$

$$\text{ave. } V_N^2 = \sum_1^M c_m^2 \frac{1}{2} \to \int_0^\infty w(f)\, df$$

we see that I, or αV^2, has a dc component of

$$\frac{\alpha P^2}{2} + \alpha \int_0^\infty w(f)\, df \tag{4.5–11}$$

which agrees with (4.1–14), and a sinusoidal component

$$\frac{\alpha P^2}{2} \cos 2pt \tag{4.5–12}$$

The continuous power spectrum $W_c(f)$ of the remaining portion of I may be computed from

$$2PV_N \cos pt + V_N^2.$$

Using the representation (2.8–6) we see

$$2PV_N \cos pt = P \sum_1^M c_m[\cos(\omega_m t + pt - \varphi_m) + \cos(\omega_m t - pt - \varphi_m)]$$

For the moment, we take $p = 2\pi r\Delta f$. The terms pertaining to frequency $f_n = n\Delta f$ are those for which

$$\begin{array}{ll} \omega_m + p = 2\pi f_n & |\omega_m - p| = 2\pi f_n \\ m + r = n & |m - r| = n \\ m = n - r & m = r \pm n \end{array}$$

where only positive values of m are to be taken: If $n > r$, then m is $n - r$ or $r + n$. If $n < r$, then m is $r - n$ or $r + n$. In either case the values of m are $|n - r|$ and $n + r$. The terms of frequency f_n in $2PV_N \cos pt$ are therefore

$$Pc_{|n-r|} \cos(2\pi f_n t - \varphi_{|n-r|}) + Pc_{n+r} \cos(2\pi f_n t - \varphi_{n+r})$$

and the mean square value of this expression, the average being taken over the φ's, is

$$\frac{P^2}{2}(c_{|n-r|}^2 + c_{n+r}^2) = P^2 \Delta f[w(f_{|n-r|}) + w(f_{n+r})]$$

$$= P^2 \Delta f[w(|f_n - f_p|) + w(f_n + f_p)]$$

where f_p denotes $p/2\pi$.

By combining this with the expression (4.5–5) which arises from V_N^2 we see that the continuous portion $W_c(f)$ of the power spectrum of I is

$$W_c(f) = \alpha^2 P^2[w(f - f_p) + w(f + f_p)] + \alpha^2 \int_{-\infty}^{+\infty} w(x)w(f - x)\, dx \qquad (4.5\text{–}13)$$

where $w(-f)$ has the same value as $w(f)$.

Equation (4.5–13) has been used to compute $W_c(f)$ as shown in Fig. 8. The input noise is assumed to be uniform over a band of width β centered at f_p, cf. Filter c, Appendix C. By noting the area under the low frequency portion of the spectrum we find

$$\int_0^\beta W_c(f)\, df = \alpha^2 \beta w_0(P^2 + \beta w_0)$$

Since the mean square value of the input V_N is $\psi_0 = \beta w_0$, it is seen that this equation agrees with the expression (4.1–15) for the mean square value of I_{lf}, the low frequency current, excluding the d.c. If audio frequency filters cut out part of the spectrum, $W_c(f)$ may be integrated over the remaining portion to give the mean square value of the corresponding output current. This idea is mentioned in the footnote pertaining to equation (4.1–6).

If V consists of V_N plus two sinusoidal voltages of incommensurable frequencies, say

$$V = P \cos pt + Q \cos qt + V_N ,$$

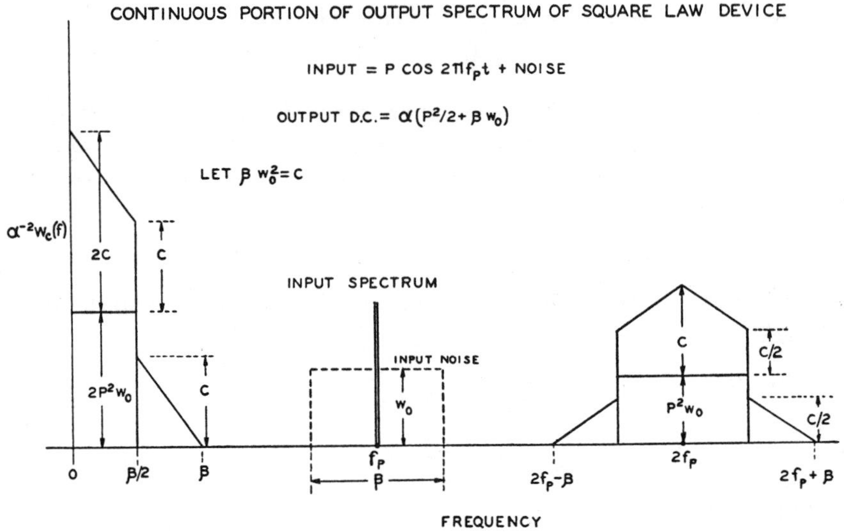

Fig. 8

the continuous portion $W_c(f)$ of the power spectrum of I may be shown to be (4.5–13) plus the additional terms

$$\alpha^2 Q^2[w(f - f_q) + w(f + f_q)] \qquad (4.5\text{–}14)$$

where f_q denotes $q/2\pi$.

When the voltage applied to the square law device (4.1–1) is[49]

$$V(t) = Q(1 + k \cos pt) \cos qt + V_N$$

$$= Q \cos qt + \frac{Qk}{2} \cos (p + q)t + \frac{Qk}{2} \cos (p - q)t + V_N$$

the resulting current contains the dc component

$$\frac{\alpha}{2} Q^2 \left(1 + \frac{k^2}{2}\right) + \alpha \int_0^\infty w(f)\, df \qquad (4.5\text{–}16)$$

[49] A complete discussion of this problem is given by L. A. MacColl in a manuscript being prepared for publication.

The sinusoidal terms of I are obtained by squaring

$$Q(1 + k \cos pt) \cos qt$$

and multiplying by α. The remaining portion of I has a continuous power spectrum given by

$$\begin{aligned} W_c(f) = \alpha^2 Q^2 \Big[& w(f - f_q) + w(f + f_q) \\ & + \frac{k^2}{4} w(f - f_p - f_q) + \frac{k^2}{4} w(f + f_p + f_q) \\ & + \frac{k^2}{4} w(f - f_p + f_q) + \frac{k^2}{4} w(f + f_p - f_q) \Big] \\ & + \alpha^2 \int_{-\infty}^{+\infty} w(x) w(f - x)\, dx \end{aligned} \tag{4.5-17}$$

where f_p denotes $p/2\pi$ and f_q denotes $q/2\pi$.

4.6 Two Correlation Function Methods

As mentioned in Section 4.4 these methods for determining the output power spectrum are based on finding the correlation function $\Psi(\tau)$ for the output current. From this the power spectrum, $W(f)$, of the output current may be obtained from (2.1–5), rewritten as

$$W(f) = 4 \int_0^\infty \Psi(\tau) \cos 2\pi f \tau \, d\tau \tag{4.6-1}$$

It will be recalled that $W(f)\Delta f$ may be regarded as the average power which would be dissipated by those components of I in the band $f, f + \Delta f$ if I were to flow through a resistance of one ohm.

The input of the non-linear device is taken to be a voltage $V(t)$. It may, for example, consist of a noise voltage $V_N(t)$ plus sinusoidal components. The output is taken to be a current $I(t)$. The non-linear device is specified by a relation between $V(t)$ and $I(t)$. In this work $I(t)$ at time t is assumed to be completely determined by the value of $V(t)$ at time t.

Two methods of obtaining $\Psi(\tau)$ will be described.

(a) Integrating the two-dimensional probability density of $V(t)$ and $V(t + \tau)$ over the values allowed by the non-linear device. This method, which is especially direct when applied to noise alone through rectifiers, was discovered independently by Van Vleck and North.

(b) Introducing and using the characteristic function, which for the sake of brevity will be abbreviated to ch. f., of the two-dimensional probability distribution of $V(t)$ and $V(t + \tau)$.

4.7 Linear Detection of Noise—The Van Vleck-North Method

The method due to Van Vleck and North will be illustrated by using it to determine the output power spectrum of a linear detector when the input consists of noise alone.

The linear detector is specified by

$$I(t) = \begin{cases} 0, & V(t) < 0 \\ V(t), & V(t) > 0, \end{cases} \tag{4.7–1}$$

which may be obtained from (4.2–1) by setting α equal to one, and the input voltage is

$$V(t) = V_N(t) \tag{4.7–2}$$

where $V_N(t)$ is a noise voltage whose correlation function is $\psi(\tau)$ and whose power spectrum is $w(f)$.

The correlation function $\Psi(\tau)$ is the average value of $I(t)I(t + \tau)$. This is the same as the average value of the function

$$F(V_1, V_2) = \begin{cases} V_1 V_2, & \text{when both } V_1, V_2 > 0 \\ 0, & \text{all other } V\text{'s}, \end{cases} \tag{4.7–3}$$

where we have set

$$V_1 = V(t)$$

$$V_2 = V(t + \tau)$$

The two-dimensional distribution of V_1 and V_2 is given by (3.2–4), and from this it follows that the average value of any function $F(V_1, V_2)$ is

$$\int_{-\infty}^{+\infty} dV_1 \int_{-\infty}^{+\infty} dV_2 \frac{F(V_1, V_2)}{2\pi |M|^{1/2}} \exp\left[-\frac{1}{2|M|}(\psi_0 V_1^2 + \psi_0 V_2^2 - 2\psi_\tau V_1 V_2)\right] \tag{4.7–4}$$

where

$$|M| = \psi_0^2 - \psi_\tau^2 .$$

For the linear rectifier case, where $F(V_1, V_2)$ is given by (4.7–3), the integral is

$$|M|^{-1/2} \frac{1}{2\pi} \int_0^\infty dV_1 \int_0^\infty dV_2\, V_1 V_2 \exp\left[-\frac{1}{2|M|}(\psi_0 V_1^2 + \psi_0 V_2^2 - 2\psi_\tau V_1 V_2)\right]$$

$$= \frac{1}{2\pi}\left([\psi_0^2 - \psi_\tau^2]^{1/2} + \psi_\tau \cos^{-1}\left[\frac{-\psi_\tau}{\psi_0}\right]\right)$$

where we have used (3.5–4) to evaluate the integral. The arc cosine is taken to be between 0 and π. We therefore have for the correlation function of $I(t)$,

$$\Psi(\tau) = \frac{1}{2\pi}\left([\psi_0^2 - \psi_\tau^2]^{1/2} + \psi_\tau \cos^{-1}\left[\frac{-\psi_\tau}{\psi_0}\right]\right) \tag{4.7–5}$$

The power spectrum $W(f)$ may be obtained from this by use of (4.6–1). For this purpose it is convenient to write (4.7–5) in terms of a hypergeometric function. By expanding and comparing terms it is seen that

$$\begin{aligned}\Psi(\tau) &= \frac{\psi_\tau}{4} + \frac{\psi_0}{2\pi} F\left(-\tfrac{1}{2}, -\tfrac{1}{2}; \tfrac{1}{2}; \frac{\psi_\tau^2}{\psi_0^2}\right) \\ &= \frac{\psi_\tau}{4} + \frac{\psi_0}{2\pi} + \frac{\psi_\tau^2}{4\pi\psi_0} + \text{terms involving } \psi_\tau^4, \psi_\tau^6, \text{etc.}\end{aligned} \tag{4.7–6}$$

As will be discussed more fully in Section 4.8, a constant term A^2 in $\psi(\tau)$ indicates a direct current component of $I(t)$ of A amperes. Thus $I(t)$ has a *dc* component equal to

$$\left[\frac{\psi_0}{2\pi}\right]^{1/2} = \frac{1}{\sqrt{2\pi}} \times \text{rms value of } V(t) \tag{4.7–7}$$

This agrees with (4.2–3) when the P of that equation is set equal to zero.

Integrals of the form

$$G_n(f) = \int_0^\infty \psi_\tau^n \cos 2\pi f\tau \, d\tau$$

which result when (4.7–6) is put in (4.6–1) and integrated termwise are discussed in Appendix 4C. From the results given there it is seen that if we neglect ψ_τ^4 and higher powers we obtain an approximation for the continuous portion $W_c(f)$ of $W(f)$:

$$\begin{aligned}W_c(f) &\approx G_1(f) + \frac{G_2(f)}{\pi\psi_0} \\ &= \frac{w(f)}{4} + \frac{1}{4\pi\psi_0}\cdot\frac{1}{2}\int_{-\infty}^{+\infty} w(x)w(f-x)\,dx\end{aligned} \tag{4.7–8}$$

where $w(-f)$ is defined as $w(f)$.

When $V_N(t)$ is uniform over a relatively narrow band extending from f_a to f_b so that $w(f)$ is equal to w_0 in this band and is zero outside it, we may use the results for Filter c of Appendix 4C. The f_0 and β given there are related to f_a and f_b by

$$f_a = f_0 - \frac{\beta}{2}, \qquad f_b = f_0 + \frac{\beta}{2}$$

and the value of w_0 taken there is the same as here and is ψ_0/β. The value of $G_2(f)$ given there leads to the approximation, for low frequencies:

$$\begin{aligned} W_c(f) &\approx \frac{1}{\pi\psi_0}\frac{\psi_0^2}{4\beta}\left(1 - \frac{f}{\beta}\right) \\ &= \frac{w_0}{4\pi}\left(1 - \frac{f}{f_b - f_a}\right) \end{aligned} \tag{4.7–9}$$

when $0 < f < f_b - f_a$, and to $W_0(f) \approx 0$ for $f_b - f_a < f < f_a$. By setting P equal to zero in the curve given in Fig. 8 for $W_c(f)$ corresponding to the square law detector, we see that the low frequency portion of the power spectrum is triangular in shape and is zero at $f = \beta$. Thus, looking at (4.7–9), we see that to a first approximation the shape of the output power spectrum is the same for a linear detector as for a square law detector when the input consists of a relatively narrow band of noise.

An approximate rms value of the low frequency output current may be obtained by integrating (4.7–9)

$$\begin{aligned} \overline{I_{lf}^2} &= \int_0^{f_b - f_a} W_c(f)\, df \\ &\approx \frac{w_0(f_b - f_a)}{8\pi} = \frac{\psi_0}{8\pi} \end{aligned}$$

$$\text{rms low freq. current} \approx \frac{1}{\sqrt{8\pi}} \times \text{rms applied voltage} \tag{4.7–10}$$

It is seen that this is half of the direct current. It must be kept in mind that (4.7–10) is an approximation because we have neglected ψ_τ^4 and higher powers. The true value may be obtained from (4.2–8). It is seen that the coefficient $(8\pi)^{-1/2} = 0.200$ should be replaced by

$$\frac{1}{\pi}\left(2 - \frac{\pi}{2}\right)^{1/2} = 0.209$$

$W_c(f)$ for other types of band pass filters may be obtained by using the corresponding G's given in appendix 4C. It turns out that (4.7–10) holds for all three types of filters. This is a special case of Middleton's theorem, mentioned several times before, that the total power in any modulation product (it will be shown later in Section 4.9 that the term ψ_τ^n in (4.7–6) corresponds to the nth order modulation products) depends only on the total input power of the applied noise, not on its spectral distribution.

4.8 The Characteristic Function Method

As mentioned in the preceding parts, especially in connection with equation (1.4–3), the ch. f. of a random variable x is the average value of exp

(iux). This is a function of u. The ch. f. of two random variables x and y is the average value of exp $(iux + ivy)$ and is a function of u and v. The ch. f. which we shall use here is the ch. f. of the two random variables $V(t)$ and $V(t + \tau)$ where $V(t)$ is the voltage applied to the non-linear device, and the randomness is introduced by t being selected at random, τ remaining fixed. We may write this characteristic function as

$$g(u, v, \tau) = \operatorname*{Limit}_{T\to\infty} \frac{1}{T} \int_0^T \exp\,[iuV(t) + ivV(t + \tau)]\,dt \quad (4.8\text{–}1)$$

If $V(t)$ contains a noise voltage $V_N(t)$, as it always does in this section, and if we use the representation (2.8–1) or (2.8–6) a large number of random parameters (a_n's and b_n's or φ_n's) will appear in (4.8–1). In accordance with our use of such representations we may average over these parameters without changing the value of (4.8–1) and may thereby simplify the integration.

For example suppose

$$V(t) = V_s(t) + V_N(t) \quad (4.8\text{–}2)$$

where $V_s(t)$ is some regular voltage which may, e.g., consist of one or more sine waves. Substituting this in (4.8–1) and using the result (3.2–7) that the ch. f. of $V_N(t)$ and $V_N(t + \tau)$ is

$$\begin{aligned} g_N(u, v, \tau) &= \text{ave. exp}\,[iuV_N(t) + ivV_N(t + \tau)] \\ &= \exp\left[-\frac{\psi_0}{2}(u^2 + v^2) - \psi_\tau uv\right] \end{aligned} \quad (4.8\text{–}3)$$

$\psi_\tau \equiv \psi(\tau)$ being the correlation function of $V_N(t)$, we obtain for the ch. f. of $V(t)$ and $V(t + \tau)$,

$$\begin{aligned} g(u, v, \tau) &= \exp\left[-\frac{\psi_0}{2}(u^2 + v^2) - \psi_\tau uv\right] \\ &\quad \times \operatorname*{Limit}_{T\to\infty} \frac{1}{T} \int_0^T \exp\,[iuV_s(t) + ivV_s(t + \tau)]\,dt \\ &= g_N(u, v, \tau)g_s(u, v, \tau) \end{aligned} \quad (4.8\text{–}4)$$

In the last line we have used $g_s(u, v, \tau)$ to denote the limit in the line above:

$$g_s(u, v, \tau) = \operatorname*{Limit}_{T\to\infty} \frac{1}{T} \int_0^T \exp\,[iuV_s(t) + ivV_s(t + \tau)]\,dt \quad (4.8\text{–}5)$$

The principal reason we use the ch. f. is because quite a few non-linear devices may be described by the integral

$$I = \frac{1}{2\pi} \int_C F(iu)e^{iVu}\,du \quad (4\text{A–}1)$$

where the function $F(iu)$ and the path of integration C are chosen to fit the device. Examples of such devices are given in Appendix 4A. The correlation function $\Psi(\tau)$ of $I(t)$ is given by

$$\begin{aligned}\Psi(\tau) &= \underset{T\to\infty}{\text{Limit}}\, \frac{1}{T}\int_0^T I(t)I(t+\tau)\,dt \\ &= \underset{T\to\infty}{\text{Limit}}\, \frac{1}{4\pi^2 T}\int_0^T dt \int_C F(iu)e^{iuV(t)}\,du \int_C F(iv)e^{ivV(t+\tau)}\,dv \\ &= \frac{1}{4\pi^2}\int_C F(iu)\,du \int_C F(iv)\,dv \qquad (4.8\text{–}6) \\ &\qquad \underset{T\to\infty}{\text{Limit}}\, \frac{1}{T}\int_0^T \exp\,[iuV(t) + ivV(t+\tau)]\,dt \\ &= \frac{1}{4\pi^2}\int_C F(iu)\,du \int_C F(iv)g(u, v, \tau)\,dv\end{aligned}$$

This is the fundamental formula of the ch. f. method.

When $V(t)$ is the sum of a noise voltage and a regular voltage, as in (4.8–2), (4.8–6) becomes

$$\Psi(\tau) = \frac{1}{4\pi^2}\int_C F(iu)e^{-(\psi_0/2)u^2}\,du \int_C F(iv)e^{-(\psi_0/2)v^2} e^{-\psi_\tau uv} g_s(u, v, \tau)\,dv \qquad (4.8\text{–}7)$$

where $g_s(u, v, \tau)$ is the ch. f. of $V_s(t)$ and $V_s(t+\tau)$ given by (4.8–5). This is a definite expression for $\Psi(\tau)$. All that follows is devoted to the evaluation of this integral and to the evaluation of

$$W(f) = 4\int_0^\infty \Psi(\tau)\cos 2\pi f\tau\,d\tau \qquad (4.6\text{–}1)$$

for the power spectrum of I.

Quite often $I(t)$ will contain *dc* and periodic components. It seems convenient to deal with these separately since they correspond to terms in $\Psi(\tau)$ which cause the integral (4.6–1) for $W(f)$ to diverge. In fact, from Section 2.2 it follows that a correlation function of the form

$$A^2 + \frac{C^2}{2}\cos 2\pi f_0\tau \qquad (2.2\text{–}3)$$

corresponds to a current

$$A + C\cos\,(2\pi f_0 t - \varphi) \qquad (2.2\text{–}2)$$

where the phase angle φ cannot be determined from (2.2–3) since it does not affect the average power.

Consider the correlation function for $V(t) = V_s(t) + V_N(t)$ given by (4.8–2). It is

$$\begin{aligned}\operatorname*{Limit}_{T\to\infty}\frac{1}{T}\Big[\int_0^T V_s(t)V_s(t+\tau)\,dt + \int_0^T V_s(t)V_N(t+\tau)\,dt \\ + \int_0^T V_N(t)V_s(t+\tau)\,dt + \int_0^T V_N(t)V_N(t+\tau)\,dt\Big]\end{aligned} \tag{4.8–8}$$

Since $V_s(t)$ and $V_N(t)$ are unrelated the contributions of the second and third integrals vanish leaving us with the result

$$\begin{aligned}\text{Correlation function of } V(t) &= \text{Correlation function of } V_s(t) \\ &+ \text{Correlation function of } V_N(t).\end{aligned} \tag{4.8–9}$$

Now as $\tau \to \infty$ the correlation function of $V_N(t)$ becomes zero while that of $V_s(t)$ becomes of the type (2.2–3) given above. Hence the correlation function of the regular voltage $V_s(t)$ may be obtained from $V(t)$ by letting $\tau \to \infty$ and picking out the non-vanishing terms. Although we have been speaking of $V(t)$, the same results hold for $I(t)$ and this process may be used to pick out those parts of $\Psi(\tau)$ which correspond to the *dc* and periodic components of $I(t)$. Thus, if we look at (4.8–7) we see that as $\tau \to \infty$, $\psi_\tau \to 0$, while the $g_s(u, v, \tau)$ corresponding to $V_s(t)$ given by (4.8–5) remains unchanged in general magnitude. This last statement may be hard to see, but examination of the cases discussed later show that it is true, at least for these cases. Thus the portion of $\Psi(\tau)$ corresponding to the *dc* and periodic components of $I(t)$ is, setting $\psi_\tau = 0$ in (4.8–7),

$$\Psi_\infty(\tau) = \frac{1}{4\pi^2}\int_C F(iu)e^{-(\psi_0/2)u^2}\,du \int_C F(iv)e^{-(\psi_0/2)v^2} g_s(u, v, \tau)\,dv \tag{4.8–10}$$

where the subscript ∞ indicates that $\Psi_\infty(\tau)$ is that part of $\Psi(\tau)$ which does not vanish as $\tau \to \infty$.

We may write (4.8–9), when applied to $I(t)$, as

$$\Psi(\tau) = \Psi_\infty(\tau) + \Psi_c(\tau) \tag{4.8–11}$$

where $\Psi_c(\tau)$ is the correlation function of the "continuous" portion of the power spectrum of $I(t)$.

Incidentally, the separation of $\Psi(\tau)$ into the two parts shown in (4.8–11) may be avoided if one is willing to use the $\delta(f)$ functions in order to interpret the integral in (4.6–1) as explained in Section 2.2. This method gives the proper *dc* and sinusoidal components even though (4.6–1) does not converge (because of the presence of the terms leading to $\Psi_\infty(\tau)$).

4.9 Noise Plus Sine Wave Applied to Non-Linear Device

In order to illustrate the characteristic function method described in Section 4.8 we shall consider the case of a non-linear device specified by

$$I = \frac{1}{2\pi}\int_C F(iu)e^{iVu}\,du \tag{4A-1}$$

when V consists of a noise voltage plus a sine wave:

$$V(t) = P\cos pt + V_N(t) \tag{4.1-13}$$

As usual, $V_N(t)$ has the power spectrum $w(f)$ and the correlation function $\psi(\tau)$. $\psi(\tau)$ is often written as ψ_τ for the sake of shortness. Comparing (4.1–13) with (4.8–2) gives

$$V_s(t) = P\cos pt \tag{4.9-1}$$

Our first task is to compute the ch. f. $g_s(u, v, \tau)$ for the pair of random variables $V_s(t)$ and $V_s(t+\tau)$. We do this by using the integral (4.8–5):

$$\begin{aligned} g_s(u, v, \tau) &= \underset{T\to\infty}{\text{Limit}}\ \frac{1}{T}\int_0^T \exp\,[iuP\cos pt + ivP\cos p(t+\tau)]\,dt \\ &= J_0(P\sqrt{u^2 + v^2 + 2uv\cos p\tau}) \end{aligned} \tag{4.9-2}$$

where J_0 is a Bessel function. The integration is performed by writing

$$\begin{aligned} u\cos pt + v\cos p\,(t+\tau) &= (u + v\cos p\tau)\cos pt - v\sin p\tau\sin pt \\ &= \sqrt{u^2 + v^2 + 2uv\cos p\tau}\,\cos\,(pt + \text{phase angle}) \end{aligned}$$

and using the integral

$$J_0(z) = \frac{1}{2\pi}\int_0^{2\pi} e^{iz\cos t}\,dt$$

The correlation function for (4.1–13) has also been given in Section 3.10.

The correlation function $\Psi(\tau)$ for $I(t)$ may now be obtained by substituting the above expressions in (4.8–7)

$$\begin{aligned} \Psi(\tau) = \frac{1}{4\pi^2}\int_C du\,F(iu)e^{-(\psi_0/2)u^2}\int_C dv\,F(iv)e^{-(\psi_0/2)v^2} \\ e^{-\psi_\tau uv}J_0(P\sqrt{u^2 + v^2 + 2uv\cos p\tau}). \end{aligned} \tag{4.9-3}$$

$\Psi_\infty(\tau)$, the correlation function for the d.c. and periodic components of I, may, according to (4.8–10), be obtained from this by setting ψ_τ equal to zero.

When we have a particular non-linear device in mind the appropriate $F(iu)$ may often be obtained from Appendix 4*A*. For example, $F(iu)$ for a linear rectifier is $-u^{-2}$. Inserting this value in (4.9–3) gives a definite double integral for $\Psi(\tau)$. If there were some easy way to evaluate this integral then everything would be fine. Unfortunately, no simple method of evaluation has yet been found. However, one method is available which is closely related to the direct method used by Bennett. It is based on the expansion

$$g_s(u, v, \tau) = J_0(P\sqrt{u^2 + v^2 + 2uv \cos p\tau})$$

$$= \sum_{n=0}^{\infty} \epsilon_n(-)^n J_n(Pu)J_n(Pv) \cos np\tau \tag{4.9–4}$$

$$\epsilon_0 = 1, \quad \epsilon_n = 2 \quad \text{for} \quad n \geq 1$$

This expansion enables us to write the troublesome terms in (4.9–3) as

$$e^{-\psi_\tau uv} J_0(P\sqrt{u^2 + v^2 + 2uv \cos p\tau})$$

$$= \sum_{n=0}^{\infty} \sum_{k=0}^{\infty} (-)^{n+k} \epsilon_n \cos np\tau \frac{(\psi_\tau uv)^k}{k!} J_n(Pu)J_n(Pv) \tag{4.9–5}$$

The virtue of this double sum is that it simplifies the integration. Thus, putting it in (4.9–3) and setting

$$h_{nk} = \frac{i^{n+k}}{2\pi} \int_C F(iu)u^k J_n(Pu)e^{-(\psi_0/2)u^2}\, du \tag{4.9–6}$$

gives

$$\Psi(\tau) = \sum_{n=0}^{\infty} \sum_{k=0}^{\infty} \frac{1}{k!} \psi_\tau^k h_{nk}^2 \epsilon_n \cos np\tau \tag{4.9–7}$$

The correlation function $\Psi_\infty(\tau)$ for the dc and periodic components of I are obtained by letting $\tau \to \infty$ where $\psi_\tau \to 0$. Only the terms for which $k = 0$ remain:

$$\Psi_\infty(\tau) = \sum_{n=0}^{\infty} \epsilon_n h_{n0}^2 \cos np\tau \tag{4.9–8}$$

Comparing this with the known fact that the correlation function of

$$A + C \cos (2\pi f_0 t - \varphi) \tag{2.2–2}$$

is

$$A^2 + \frac{C^2}{2} \cos 2\pi f_0 \tau \tag{2.2–3}$$

and remembering that ϵ_0 is one while ϵ_n is two for $n \geq 1$ shows that

$$\text{Amplitude of dc component of } I = h_{00}$$

$$\text{Amplitude of } \frac{np}{2\pi} \text{ component of } I = 2h_{n0} \tag{4.9–9}$$

Incidentally, these expressions for the amplitudes follow almost at once from the direct method of solution. This will be shown in connection with equation (4.9–17).

Since the correlation function $\Psi_c(\tau)$ for the continuous portion $W_c(f)$ of the power spectrum for I is given by

$$\Psi_c(\tau) = \Psi(\tau) - \Psi_\infty(\tau), \tag{4.8–11}$$

we also have

$$\Psi_c(\tau) = \sum_{n=0}^{\infty}\sum_{k=1}^{\infty} \frac{1}{k!} \psi_\tau^k h_{nk}^2 \epsilon_n \cos np\tau \tag{4.9–10}$$

When this is substituted in

$$W_c(f) = 4\int_0^\infty \Psi_c(\tau) \cos 2\pi f\tau \, d\tau \tag{4.9–11}$$

we obtain

$$W_c(f) = \sum_{n=0}^{\infty}\sum_{k=1}^{\infty} \frac{2\epsilon_n}{k!} h_{nk}^2 \left[G_k\left(f - \frac{np}{2\pi}\right) + G_k\left(f + \frac{np}{2\pi}\right)\right] \tag{4.9–12}$$

where

$$G_k(f) = \int_0^\infty \psi_\tau^k \cos 2\pi f\tau \, d\tau \tag{4.9–13}$$

is the function studied in Appendix 4C. $G_k(f)$ is an even function of f. The double series (4.9–12) for W_c looks rather formidable. However, when we are interested in a particular portion of the frequency spectrum often only a few terms of the series are needed.

It has been mentioned above that the direct method of obtaining the output power spectrum is closely related to the equations just derived. We now study this relation.

We start with the following result from modulation theory[50]: Let the voltage

$$\begin{aligned} V &= P_0 \cos x_0 + P_1 \cos x_1 + \cdots + P_N \cos x_N \\ x_k &= p_k t, \qquad k = 0, 1, \cdots N, \end{aligned} \tag{4.9–14}$$

where the p_k's are incommensurable, be applied to the device (4A–1). The output current is

$$I = \sum_{m_0=0}^{\infty} \cdots \sum_{m_N=0}^{\infty} \tfrac{1}{2} A_{m_0 \cdots m_N} \epsilon_{m_0} \cdots \epsilon_{m_N} \cos m_0 x_0 \cos m_1 x_1 \cdots \cos m_N x_N \tag{4.9–15}$$

[50] Bennett and Rice, "Note on Methods of Computing Modulation Products," *Phil. Mag.* S.7, V. 18, pp. 422–424, Sept. 1934, and Bennett's paper cited in Section 4.0.

where $\epsilon_0 = 1$ and $\epsilon_m = 2$ for $m \geq 1$. When the product of the cosines is expressed as a sum of cosines of the angles $m_0 x_0 \pm m_1 x_1 \cdots \pm m_N x_N$, it is seen that the coefficient of the typical term is $A_{m_0 \cdots m_N}$, except when all the m's are zero in which case it is $\frac{1}{2}A_{0 \cdots 0}$. Thus

$$\frac{1}{2}A_{00\cdots 0} = \text{dc component of } I$$

$$| A_{m_0 \cdots m_N} | = \text{amplitude of component of frequency} \qquad (4.9\text{–}16)$$

$$\frac{1}{2\pi} | m_0 p_0 \pm m_1 p_1 \pm \cdots \pm m_N p_N |$$

For all values of the m's,

$$A_{m_0 \cdots m_N} = \frac{i^M}{\pi} \int_C F(iu) \prod_{r=0}^{N} J_{m_r}(P_r u)\, du \qquad (4.9\text{–}17)$$

$$M = m_0 + m_1 + \cdots + m_N$$

Following Bennett's procedure, we identify V as given by (4.9–14), with

$$V = P \cos pt + V_N \qquad (4.1\text{–}13)$$

by setting $P_0 = P$, $p_0 = p$, and representing the noise voltage V_N by the sum of the remaining terms. Since this makes P_1, P_N all very small, Laplace's process indicates that in (4.9–17) we may put

$$\prod_{r=1}^{N} J_0(P_r u) \approx \exp - \frac{u^2}{4} (P_1^2 + \cdots + P_N^2) \qquad (4.9\text{–}18)$$

$$\approx e^{-\psi_0 u^2/2}$$

We have used the fact that ψ_0 is the mean square value of V_N. It follows from these equations that

$$\text{dc component of } I = \frac{1}{2\pi} \int_C F(iu) J_0(Pu) e^{(-\psi_0/2)u^2}\, du$$

$$\text{Component of frequency } \frac{np}{2\pi} = \frac{i^n}{\pi} \int_C F(iu) J_n(Pu) e^{-\psi_0 u^2/2}\, du$$

These results are identical with those of (4.9-9).

The equations just derived show that h_{n0} is to be associated with the nth harmonic of p. In much the same way it may be shown that h_{nk} is to be associated with the modulation products arising from the nth harmonic of p and k of the elementary sinusoidal components representing V_N. We consider only combinations of the form $p_1 \pm p_2 \pm p_3$, taking $k = 3$ for example, and neglect terms of the form $3p_1$ and $2p_1 \pm p_2$. The former type is much more numerous, there being about N^3 of them while there are only about N and N^2, respectively, of the latter type.

We again take $k = 3$ and consider m_1, m_2, m_3 to be one, and $m_4, \cdots m_N$ to be zero, corresponding to the modulation product $np \pm p_1 \pm p_2 \pm p_3$. By making the same sort of approximations as Bennett does we find

$$A_{n,1,1,1,0,0\cdots 0} = \frac{i^{n+3}}{\pi} \frac{P_1 P_2 P_3}{8} \int_C F(iu) J_n(Pu) u^3 e^{(-u^2/2)\psi_0} \, du$$

$$= \frac{P_1 P_2 P_3}{4} h_{n3}$$

When any other modulation product of the form $np \pm p_{r_1} \pm p_{r_2} \pm p_{r_3}$ is considered we get a similar expression in which $P_1 P_2 P_3$ is replaced by $P_{r_1} P_{r_2} P_{r_3}$. This may be done for any value of k. The result indicates that h_{nk}, and consequently also the $(n, k)^{\text{th}}$ terms in the double series (4.9–10) and (4.9–12) for $\Psi_c(\tau)$ and $W_c(f)$, are to be associated with the modulation products of order (n, k), the n referring to the signal and the k to the noise components.

We now may state a theorem due to Middleton regarding the total power in the modulation products of a given order. For a given non-linear device (i.e. $F(iu)$ is given), the total power which would be dissipated by all of the modulation products which are of order (n, k) if I were to flow through a resistance of one ohm is

$$\Psi_{nk}(0) = \frac{\epsilon_n [\psi(0)]^k}{k!} h_{nk}^2 = \frac{\epsilon_n [\overline{V_N^2}]^k h_{nk}^2}{k!} \tag{4.9–19}$$

The important feature of this expression is that it depends only on the r.m.s. value of V_N and on $F(iu)$. It depends not at all upon the spectral distribution of the noise power in the input.

The proof of (4.9–19) is based on the relation

$$\Psi_{nk}(0) = \int_0^\infty W_{nk}(f) \, df$$

between the total power dissipated by all the (n, k) order products and the corresponding correlation function obtained from (4.9–7).

This theorem has been used by Middleton to show that when the input is confined to a relatively narrow frequency band, so that the output spectrum consists of bands, the power in each band depends only on $\overline{V_N^2}$ and not on the spectrum of V_N.

4.10 Miscellaneous Results Obtained by Correlation Function Method

In this section a number of results which may be obtained from the theory given in the sections following 4.6 are given.

When the input to the square law device

$$I = \alpha V^2 \tag{4.1-1}$$

consists of noise only, so that $V = V_N$, the correlation function for I is

$$\Psi(\tau) = \alpha^2[\psi_0^2 + 2\psi_\tau^2] \tag{4.10-1}$$

where ψ_τ is the correlation function of V_N. This may be compared with equation (3.9–7). When V is general,

$$\begin{aligned}\Psi(\tau) &= \text{ave. } I(t)I(t+\tau) \\ &= \text{ave. } \alpha^2 V^2(t)V^2(t+\tau) \\ &= \alpha^2 \times \text{Coefficient of } \frac{(iu)^2}{2!}\frac{(iv)^2}{2!} \text{ in power series expansion} \\ &\qquad\qquad \text{of ch. f. of } V(t),\ V(t+\tau)\end{aligned} \tag{4.10-2}$$

where we have used a known property of the characteristic function. An expression for the ch. f., denoted by $g(u, v, \tau)$, is given by (4.8–4). For example, when V consists of a sine wave plus noise, (4.1–13), the ch. f. is obtainable from (4.9–3). Hence,

$$\begin{aligned}\Psi(\tau) &= \text{Coeff. of } \frac{u^2v^2}{4} \text{ in expansion of} \\ &\qquad \alpha^2 J_0(P\sqrt{u^2+v^2+2uv\cos p\tau}) \\ &\qquad \times \exp\left[-\frac{\psi_0}{2}(u^2+v^2) - \psi_\tau uv\right] \\ &= \alpha^2\left[\left(\frac{P^2}{2}+\psi_0\right)^2 + \frac{P^4}{8}\cos 2p\tau + 2P^2\psi_\tau\cos p\tau + 2\psi_\tau^2\right]\end{aligned} \tag{4.10-3}$$

The first two terms give the dc and second harmonic. The last two terms may be used to compute $W_c(f)$ as given by (4.5–13).

Expressions (4.10–1) and (4.10–3) are special cases of results obtained by Middleton who has studied the general theory of the quadratic rectifier by using the Van Vleck-North method, described in Section 4.7.

As an example to which the theory of Section 4.9 may be applied we consider the sine wave plus noise, (4.1–13), to be applied to the ν-law rectifier

$$\begin{aligned} I &= 0, \qquad V < 0 \\ I &= V^\nu, \qquad V > 0 \end{aligned} \tag{4.10-4}$$

From the table in Appendix 4A it is seen that

$$F(iu) = \Gamma(\nu+1)(iu)^{-\nu-1}$$

and that the path of integration C runs along the real axis from $-\infty$ to ∞ with a downward indentation at the origin. The integral (4.9–6) for h_{nk} becomes

$$h_{nk} = \frac{i^{n+k-\nu-1}}{2\pi} \Gamma(\nu+1) \int_C u^{k-\nu-1} J_n(Pu) e^{-(\psi_0/2)u^2}\, du$$

$$= \frac{\left(\frac{\psi_0}{2}\right)^{(\nu-k)/2} x^{n/2}\,\Gamma(\nu+1)}{2\Gamma\left(\frac{2-k-n+\nu}{2}\right) n!} {}_1F_1\left(\frac{k+n-\nu}{2}; n+1; -x\right) \qquad (4.10\text{–}5)$$

$$x = \frac{P^2}{2\psi_0}$$

where the integration has been performed by expanding $J_n(Pu)$ in powers of u and using

$$\int_C e^{-au^2} u^{2\lambda-1}\, du = ie^{-\lambda i\pi} a^{-\lambda} \sin \lambda\pi\Gamma(\lambda)$$

$$= \frac{a^{-\lambda}}{2}(1 - e^{-2\lambda i\pi})\Gamma(\lambda) \qquad (4.10\text{–}6)$$

$$= \frac{i\pi e^{-\lambda i\pi}}{a^{\lambda}\Gamma(1-\lambda)}$$

it being understood that $\arg u = 0$ on the positive portion of C.

From (4.9–9), the dc component of I is

$$h_{00} = \frac{\Gamma(1+\nu)}{2\Gamma\left(1+\frac{\nu}{2}\right)} \left(\frac{\psi_0}{2}\right)^{\nu/2} {}_1F_1\left(-\frac{\nu}{2}; 1; -x\right) \qquad (4.10\text{–}7)$$

which reduces to the expression (4.2–3) when $\nu = 1$ for the linear rectifier (aside from the factor α).

When the input (sine wave plus noise) is confined to a relatively narrow band, and when we are interested in the low frequency output, consideration of the modulation products suggests that we consider the difference products from the products of order (0, 0), (0, 2), (0, 4), $\cdots$ (1, 1), (1, 3), $\cdots$ (2, 0), (2, 2), $\cdots$ etc. where the typical product is of order (n, k). The orders (0, 0) and (2, 0) give the dc and second harmonic and hence are not considered in the computation of $W_c(f)$. Of the remaining terms, either (0, 2) or (1, 1) gives the greatest contribution to the series (4.9–12) and (4.9–10) for $W_c(f)$ and $\Psi_c(\tau)$. The remaining terms contribute less and less as n and

k increase. The low frequency portion of the continuous portion of the output power spectrum is then, from (4.9–12),

$$W_c(f) = \frac{4}{2!} h_{02}^2 G_2(f) + \frac{4}{4!} h_{04}^2 G_4(f) + \cdots$$
$$+ \frac{4}{1!} h_{11}^2 [G_1(f - f_0) + G_1(f + f_0)] + \frac{4}{3!} h_{13}^2 [G_3(f - f_0) \qquad (4.10\text{–}8)$$
$$+ G_3(f + f_0)] + \frac{4}{2!} h_{22}^2 [G_2(f - 2f_0) + G_2(f + 2f_0)] + \cdots$$

From Table 2 of Appendix 4C we may pick out the low frequency portions of the G's. It must be remembered that $G_m(x)$ is an even function of x and that $0 < f \ll f_0$.

As an example we take the input noise V_N to have the same $w(f)$ and $\psi(\tau)$ as Filter a, the normal law filter, of Appendix 4C, so that

$$w(f) = \frac{\psi_0}{\sigma\sqrt{2\pi}} e^{-(f-f_0)^2/2\sigma^2}$$

and assume that the sine wave signal is at the middle of the band, giving $p = 2\pi f_0$. Thus, from (4.10–8), for low frequencies and the normal law distribution of the input noise power,

$$W_c(f) = \frac{1}{4\sigma\sqrt{\pi}} h_{02}^2 \psi_0^2 e^{-f^2/4\sigma^2} + \frac{1}{64\sigma\sqrt{2\pi}} h_{04}^2 \psi_0^4 e^{-f^2/8\sigma^2}$$
$$+ \frac{2}{\sigma\sqrt{2\pi}} h_{11}^2 \psi_0 e^{-f^2/2\sigma^2} + \frac{1}{4\sigma\sqrt{6\pi}} h_{13}^2 \psi_0^3 e^{-f^2/6\sigma^2} \qquad (4.10\text{–}9)$$
$$+ \frac{1}{4\sigma\sqrt{\pi}} h_{22}^2 \psi_0^2 e^{-f^2/4\sigma^2} + \cdots$$

Although we have been speaking of the ν-law rectifier, equation (4.10–9) gives the low frequency portion of $W_c(f)$, corresponding to a normal law noise power, for any non-linear device provided the proper h_{nk}'s are inserted.

When we set ν equal to one in the expression (4.10–5) for h_{nk} we may obtain the results given by Bennett. Middleton has studied the output of a biased linear rectifier, when the input consists of a sine wave plus noise, and also the special case of the unbiased linear rectifier. He has computed the output for a wide range of the ratios P^2/ψ_0, B^2/ψ_0 where B is the bias. In order to cover the entire range he had to derive two series for the corresponding h_{nk}'s, each series being suitable for its particular portion of the range.

A special case of (4.10–9) occurs when noise alone is applied to a linear rectifier. The low frequency portion of the output power spectrum is

$$W_c(f) = \frac{\psi_0}{\pi} \sum_{m=1}^{\infty} \frac{(-\frac{1}{2})_m(-\frac{1}{2})_m}{m!m!} \frac{1}{\sigma\sqrt{4m\pi}} e^{-f^2/4m\sigma^2}$$

$$= \frac{\psi_0 \pi^{-3/2}}{2\sigma} \left[\tfrac{1}{4} e^{-f^2/4\sigma^2} + \frac{1}{64\sqrt{2}} e^{-f^2/8\sigma^2} + \frac{1}{256\sqrt{3}} e^{-f^2/12\sigma^2} + \cdots \right] \qquad (4.10\text{–}10)$$

where we have used (4.7–6) and Table 2 of Appendix 4C.

The correlation function of

$$V_s = P \cos pt + Q \cos qt,$$

where p and q are incommensurable, is

$$J_0(P\sqrt{u^2 + v^2 + 2uv \cos p\tau}) \times J_0(Q\sqrt{u^2 + v^2 + 2uv \cos q\tau})$$

From equations (4.9–16) and (4.9–17) it is seen immediately that

$$h_{000} = \frac{1}{2\pi} \int_C F(iu) J_0(Pu) J_0(Qu) e^{-(u^2/2)\psi_0}\, du \qquad (4.10\text{–}11)$$

is the d.c. component of I when the applied voltage is

$$P \cos pt + Q \cos qt + V_N . \qquad (4.1\text{–}4)$$

J. R. Ragazzini has obtained an approximate expression for the output power spectrum when the voltage

$$V = V_s + V_N$$
$$V_s = Q(1 + r \cos pt)\cos qt \qquad (4.10\text{–}12)$$

is impressed on a linear rectifier.[46] In terms of our notation his expression for the continuous portion of the power spectrum is (for low frequencies)

$$W_c(f) = \frac{1}{\pi^2 \alpha^2 (Q^2 + 2\psi_0)} \times \left[\begin{matrix} W_c(f) \text{ given by equation} \\ \text{(4.5–17) for square law device} \end{matrix} \right] \qquad (4.10\text{–}13)$$

The α^2 is put in the denominator to cancel the α^2 in the expression (4.5–17). We take the linear rectifier to be

$$I = \begin{cases} 0, & V < 0 \\ V, & 0 < V \end{cases} \qquad (4.10\text{–}14)$$

and replace the index of modulation, k, in (4.5–17) by r.

[46] Equation (12), "The Effect of Fluctuation Voltages on the Linear Detector," *Proc. I.R.E.*, V. 30, pp. 277–288 (June 1942).

Ragazzini's formula is quite accurate when the index of modulation r is small, especially when $y = Q^2/(2\psi_0)$ is large. To show this we put $r = 0$ in (4.10–13) and obtain

$$W_c(f) = \frac{1}{\pi^2(Q^2 + 2\psi_0)} \left[Q^2 w(f_q - f) + Q^2 w(f_q + f) + \int_{-\infty}^{+\infty} w(x) w(f - x)\, dx \right] \qquad (4.10\text{–}15)$$

where $f_q = q/(2\pi)$. This is to be compared with the low frequency portion of $W_c(f)$ obtained by specializing (4.10–8) to obtain the output power spectrum of a linear rectifier when the input consists of a sine wave plus noise. The leading terms in (4.10–8) give

$$W_c(f) = h_{11}^2[w(f_q - f) + w(f_q + f)] + h_{02}^2 \frac{1}{4} \int_{-\infty}^{+\infty} w(x) w(f - x)\, dx \qquad (4.10\text{–}16)$$

The values of the h's appropriate to a linear rectifier are obtained by setting $\nu = 1$ in (4.10–5) and noticing that Q now plays the role of P.

$$\begin{aligned} h_{11} &= \frac{1}{2}\left(\frac{y}{\pi}\right)^{1/2} {}_1F_1(\tfrac{1}{2}; 2; -y) \\ h_{02} &= (2\pi\psi_0)^{-1/2} {}_1F_1(\tfrac{1}{2}; 1; -y) \\ y &= Q^2/(2\psi_0) \end{aligned} \qquad (4.10\text{–}17)$$

Incidentally, the first approximation to the output of a linear rectifier given by (4.10–16) is interesting in its own right. Fig. 9 shows the low frequency portion of $W_c(f)$ as computed from (4.10–16) when the input noise is uniformly distributed over a narrow frequency band of width β, f_q being the mid-band frequency. h_{11} and h_{02} may be obtained from the curves shown in Fig. 10. In these figures P and x replace Q and y of (4.10–17) in order to keep the notation the same as in Fig. 8 for the square law device. These curves may also be obtained from equations (33) to (43) of Bennett's paper.

The following values are useful for our comparison.

When $x = 0$	When x is large
$h_{11} = 0$	$h_{11} = 1/\pi$
$h_{02} = (2\pi\psi_0)^{-1/2}$	$h_{02} = 1/(\pi Q)$.

(4.10–18)

The values for large x are obtained from the asymptotic expansion $(4B - 3)$ given in Appendix $4B$.

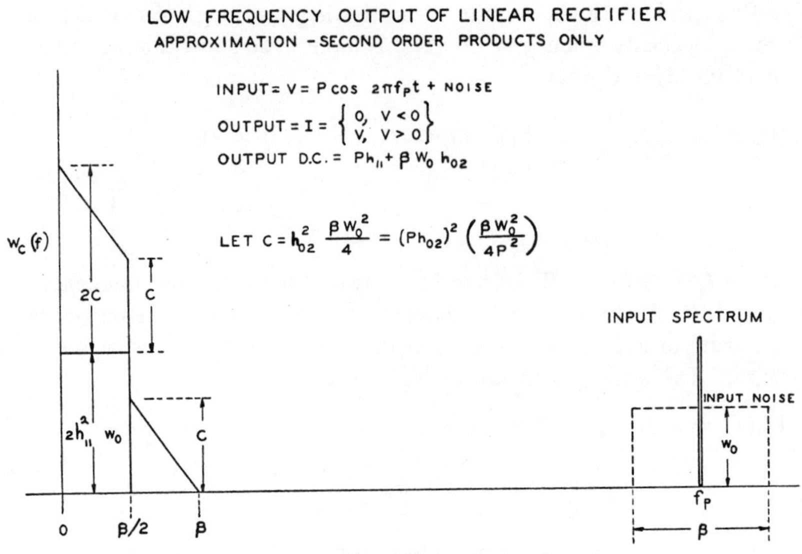

Fig. 9

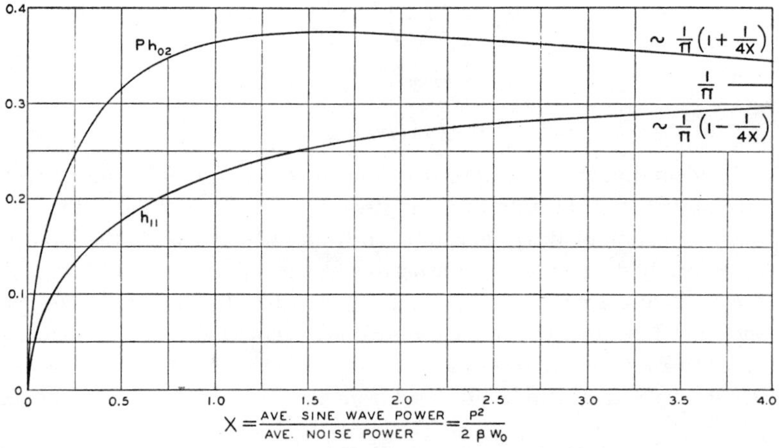

Fig. 10—Coefficients for linear detector output shown on Fig. 9

$$Ph_{02} = \sqrt{\frac{x}{\pi}}\, {}_1F_1(\tfrac{1}{2};\, 1;\, -x) \qquad h_{11} = \frac{1}{2}\sqrt{\frac{x}{\pi}}\, {}_1F_1(\tfrac{1}{2};\, 2;\, -x)$$

We make the first comparison between (4.10–15) and (4.10–16) by letting $Q \to \infty$. It is seen that both reduce to

$$W_c(f) = \frac{1}{\pi^2}\,[w(f_q - f) + w(f_q + f)] \qquad (4.10\text{–}19)$$

which shows that the agreement is perfect in this case. Next we let $Q = 0$. The two expressions then give

$$W_c(f) = \frac{1}{A2\pi\psi_0}\int_{-\infty}^{+\infty} w(x)w(f - x)\,dx$$

where $A = \pi$ for Ragazzini's formula and $A = 4$ for (4.10–16). Thus the agreement is still quite good. The limiting value for (4.10–16) may also be obtained from (4.7–8).

Even if the index of modulation r is not negligibly small it may be shown that when $Q \to \infty$ $W_c(f)$ still approaches the value given by (4.10–19). Ragazzini's formula gives a somewhat larger answer because it includes the additional terms, shown in (4.5–17), which contain $k^2/4$, but this difference does not appear to be serious. If the $Q^2 + 2\psi_0$ in the denominator of (4.10–13) be replaced by $Q^2 + \frac{1}{2}Q^2k^2 + 2\psi_0$ the agreement is improved.

APPENDIX 4A

Table of Non-linear Devices Specified by Integrals

Quite a number of non-linear devices may be specified by integrals of the form

$$I = \frac{1}{2\pi}\int_C F(iu)e^{iVu}\,du \tag{4A–1}$$

where the function $F(iu)$ and the path of integration C are chosen to fit the device.* The table gives examples of such devices. Some important cases cannot be simply represented in this form. An example is the limiter

$$\begin{aligned} I &= -\alpha D, && V < -D \\ I &= \alpha V, && -D < V < D \\ I &= \alpha D, && D < V \end{aligned} \tag{4A–2}$$

which may be represented as

$$\begin{aligned} I &= \frac{2\alpha}{\pi}\int_0^\infty \sin Vu \sin Du \frac{du}{u^2} \\ &= -\alpha D + \frac{2\alpha}{2\pi i}\int_C e^{iVu} \sin Du \frac{du}{u^2} \end{aligned} \tag{4A–3}$$

where C runs from $-\infty$ to $+\infty$ and is indented downward at the origin. This is not of the form assumed in the theory of Part IV. However it appears that it would not be difficult to extend the theory in the particular case of the limiter.

* Reference 50 cited in Section 4.9.

Non-Linear Devices Specified by Integrals

$$I = \frac{1}{2\pi} \int_C F(iu) e^{iVu} \, du$$

I	$F(iu)$	C	Type of Device
$I = \alpha V^n$, n integer	$\dfrac{\alpha\, n!}{(iu)^{n+1}}$	Positive Loop around $u = 0$	nth power device
$I = \alpha(V - B)^n$, n integer	$\dfrac{\alpha\, n!}{(iu)^{n+1}} e^{-iuB}$	Positive Loop around $u = 0$	nth power device with bias
$I = 0, \quad V < 0$ $I = \alpha V, \quad 0 < V$	$\dfrac{\alpha}{(iu)^2} = -\dfrac{\alpha}{u^2}$	Real u axis from $-\infty$ to $+\infty$ with downward indentation at $u = 0$	Linear rectifier cut-off at $V = 0$
$I = 0, \quad V < B$ $I = \alpha(V - B)^\nu$, $V > B$ ν any positive number	$\dfrac{\alpha\Gamma(\nu + 1)}{(iu)^{\nu+1}} e^{-iuB}$	"	νth power rectifier with bias
$I = 0, \quad V < 0$ $I = \alpha V, \quad 0 < V < D$ $I = \alpha D, \quad D < V$	$\dfrac{\alpha(1 - e^{-iuD})}{(iu)^2}$	"	Linear rectifier plus limiter
$I = 0, \quad V < 0$ $I = \varphi(V), \quad V > 0$	$F(p) = \int_0^\infty e^{-pt}\varphi(t)\, dt$	"	

APPENDIX 4B

The Function ${}_1F_1(a; c; x)$

In problems concerning a sine wave plus noise the hypergeometric function

$$ {}_1F_1(a; c; z) = 1 + \frac{az}{c1!} + \frac{a(a+1)}{c(c+1)} \frac{z^2}{2!} + \cdots \tag{4B-1}$$

arises. Here we state some of its properties which are of use in the theory of Part IV. Curves of ${}_1F_1(a; c; z)$ are given for $a = -4, -3.5 \cdots, 3.5, 4.0$ and $c = -1.5, -.5, +.5, 1, 1.5, 2, 3, 4$ in the 1938 edition, page 275, of "Tables of Functions", by Jahnke and Emde. A list of properties of the function and other references are also given. In addition to these references we mention E. T. Copson, "Functions of a Complex Variable" (Oxford, 1935), page 260.

If c is not a negative integer or zero

$$ {}_1F_1(a; c; z) = e^z {}_1F_1(c - a; c; -z). \tag{4B-2}$$

When $R(z) > 0$ we have the asymptotic expansions

$$
\begin{aligned}
{}_1F_1(a; c; z) &\sim \frac{\Gamma(c)e^z}{\Gamma(a)z^{c-a}}\left[1 + \frac{(1-a)(c-a)}{1!z}\right. \\
&\quad \left. + \frac{(1-a)(2-a)(c-a)(c-a+1)}{2!z^2} + \cdots\right] \\
{}_1F_1(a; c; -z) &\sim \frac{\Gamma(c)}{\Gamma(c-a)z^a}\left[1 + \frac{a(1+a-c)}{1!z}\right. \\
&\quad \left. + \frac{a(a+1)(1+a-c)(2+a-c)}{2!z^2} + \cdots\right]
\end{aligned}
\tag{4B-3}
$$

Many of the hypergeometric functions encountered may be expressed in terms of Bessel functions of the first kind for imaginary argument. The connection may be made by means of the relation[51]

$$
{}_1F_1\left(\nu + \frac{1}{2}; 2\nu + 1; z\right) = 2^{2\nu}\Gamma(\nu + 1)z^{-\nu}e^{z/2}I_\nu\left(\frac{z}{2}\right) \tag{4B-4}
$$

together with the recurrence relations

	F_{a+}	F_{a-}	F_{c+}	F_{c-}	F
1.	a	$(a - c)$			$c - 2a - z$
2.	ac		$(c - a)z$		$-c(a + z)$
3.	a			$1 - c$	$c - a - 1$
4.		$-c$	$-z$		c
5.		$a - c$		$c - 1$	$1 - a - z$
6.			$(c - a)z$	$c(c - 1)$	$c(1 - c - z)$

For example, the first recurrence relation is obtained from line 1 as follows

$$
\begin{aligned}
aF(a+1; c; z) &+ (a - c)F(a - 1; c; z) \\
&+ (c - 2a - z)F(a; c; z) = 0
\end{aligned}
\tag{4B-5}
$$

These six relations between the contiguous ${}_1F_1$ functions are analogous to the 15 relations, given by Gauss, between the contiguous ${}_2F_1$ hypergeometric functions and may be derived from these by using

$$
{}_1F_1(a; c; z) = \operatorname*{Limit}_{b\to\infty} {}_2F_1\left(a, b; c; \frac{z}{b}\right) \tag{4B-6}
$$

A recurrence relation involving two ${}_1F_1$'s of the type (4B-4) may be obtained by replacing a by $a + 1$ in the relation given by row four of the table

[51] G. N. Watson, "Theory of Bessel Functions" (Cambridge, 1922), p. 191.

and then eliminating ${}_1F_1(a+1;c;z)$ from this relation and the one obtained from row 3 of the table. There results

$$ {}_1F_1(a;c;z) = {}_1F_1(a;c-1;z) + \frac{za}{c(1-c)} F(a+1;c+1;z) \qquad \text{(4B–7)} $$

Setting ν equal to zero and one in (4B–4) and a equal to $\frac{1}{2}$, c equal to 2 in (4B–7) gives

$$ \begin{aligned} {}_1F_1\left(\frac{1}{2};1;z\right) &= e^{z/2} I_0\left(\frac{z}{2}\right) \\ {}_1F_1\left(\frac{3}{2};3;z\right) &= 4z^{-1}e^{z/2} I_1\left(\frac{z}{2}\right) \\ {}_1F_1\left(\frac{1}{2};2;z\right) &= e^{z/2}\left[I_0\left(\frac{z}{2}\right) - I_1\left(\frac{z}{2}\right)\right] \end{aligned} \qquad \text{(4B–8)} $$

Starting with these relations the relations in the table enable us to find an expression for ${}_1F_1(n+\frac{1}{2};m;z)$ where n and m are integers. A number of these are given in Bennett's paper. In particular, using (4B–2),

$$ {}_1F_1\left(-\frac{1}{2};1;-z\right) = e^{-z/2}\left[(1+z)I_0\left(\frac{z}{2}\right) + zI_1\left(\frac{z}{2}\right)\right]. \qquad \text{(4B–9)} $$

APPENDIX 4C

The Power Spectrum Corresponding to ψ_τ^n

Quite often we encounter the integral

$$ G_n(f) = \int_0^\infty [\psi(\tau)]^n \cos 2\pi f\tau \, d\tau \qquad \text{(4C–1)} $$

where $\psi(\tau)$ is the correlation function corresponding to the power spectrum $w(f)$. From the fundamental relation between $w(f)$ and $\psi(\tau)$ given by (2.1–5),

$$ G_1(f) = \frac{w(f)}{4} \qquad \text{(4C–2)} $$

The expression for the spectrum of the product of two functions enables us to write $G_n(f)$ in terms of $w(f)$. We shall use the following form of this expression: Let $F_r(f)$ be the spectrum of the function $\varphi_r(\tau)$ so that

$$ \varphi_r(\tau) = \int_{-\infty}^{+\infty} F_r(f) e^{2\pi i f\tau} \, df, \quad r = 1, 2 $$

$$ F_r(f) = \int_{-\infty}^{+\infty} \varphi_r(\tau) e^{-2\pi i f\tau} \, dt $$

Then

$$\int_{-\infty}^{+\infty} \varphi_1(\tau)\varphi_2(\tau)e^{-2\pi i f\tau}\,d\tau = \int_{-\infty}^{+\infty} F_1(x)F_2(f-x)\,dx \tag{4C-3}$$

i.e., the spectrum of the product $\varphi_1(\tau)\varphi_2(\tau)$ is the integral on the right. If $\varphi_1(\tau)$ and $\varphi_2(\tau)$ are real even functions of τ, (4C–3) may be written as

$$\int_0^{\infty} \varphi_1(\tau)\varphi_2(\tau)\cos 2\pi f\tau\,d\tau = \frac{1}{2}\int_{-\infty}^{+\infty} F_1(x)F_2(f-x)\,dx \tag{4C-4}$$

In order to obtain $G_2(f)$ we set $\varphi_1(\tau)$ and $\varphi_2(\tau)$ equal to $\psi(\tau)$. We may then use (4C-4) since $\psi(\tau)$ is an even real function of τ. When $\varphi_r(\tau)$ is an even real function of τ we see, from the Fourier integral for $F_r(f)$, that $F_r(f)$ must be an even real function of f. We therefore set

$$2F_r(f) = w(f), \qquad r = 1, 2$$

and define $w(f)$ for negative f by

$$w(-f) = w(f) \tag{4C-5}$$

Equation (4C–4) then gives

$$\begin{aligned} G_2(f) &= \frac{1}{8}\int_{-\infty}^{+\infty} w(x)w(f-x)\,dx \\ &= \frac{1}{8}\int_0^{f} w(x)w(f-x)\,dx \\ &\quad + \frac{1}{4}\int_0^{\infty} w(x)w(f+x)\,dx \end{aligned} \tag{4C-6}$$

where in the second equation only positive values of the argument of $w(f)$ appear.

In order to get $G_3(f)$ we set $\varphi_1(\tau)$ equal to $\psi(\tau)$, $2F_1(f)$ equal to $w(f)$, and $\varphi_2(\tau)$ equal to $\psi^2(\tau)$. Then

$$\begin{aligned} F_2(f) &= 2\int_0^{\infty} \varphi_2(\tau)\cos 2\pi f\tau\,d\tau \\ &= 2G_2(f) \end{aligned}$$

and from (4C–4) we obtain

$$\begin{aligned} G_3(f) &= \frac{1}{2}\int_{-\infty}^{+\infty} w(x)G_2(f-x)\,dx \\ &= \frac{1}{16}\int_{-\infty}^{+\infty} w(x)\,dx \int_{-\infty}^{+\infty} w(y)w(f-y)\,dy \end{aligned} \tag{4C-7}$$

Equation (4C–7) suggests that we may write the expression for $G_2(f)$ as

$$G_2(f) = \frac{1}{2}\int_{-\infty}^{+\infty} w(x)G_1(f - x)\,dx \qquad (4C\text{–}8)$$

This is seen to be true from (4C–2) and (4C–6). In fact it appears that

$$G_n(f) = \frac{1}{2}\int_{-\infty}^{+\infty} w(f - x)G_{n-1}(x)\,dx \qquad (4C\text{–}9)$$

might be used for a step by step computation of $G_n(f)$.

We now consider $G_n(f)$ for the case of relatively narrow band pass filters. As examples we take filters whose characteristics give the following $w(f)$'s and $\psi(\tau)$'s

TABLE 1

Filter	$w(f)$ for $f > 0$	$\psi(\tau)$
a	$\frac{\psi_0}{\sigma\sqrt{2\pi}} e^{-(f-f_0)^2/2\sigma^2}$	$\psi_0 e^{-2(\pi\sigma\tau)^2} \cos 2\pi f_0 \tau$
b	$\frac{\psi_0 \alpha}{\pi} \frac{1}{\alpha^2 + (f - f_0)^2}$	$\psi_0 e^{-2\pi\alpha\lvert\tau\rvert} \cos 2\pi f_0 \tau$
c	$w(f) = w_0 = \psi_0/\beta$ for $f_0 - \frac{\beta}{2} < f < f_0 + \frac{\beta}{2}$; $w(f) = 0$ elsewhere	$\psi_0 \frac{\sin \pi\beta\tau}{\pi\beta\tau} \cos 2\pi f_0 \tau$

We shall refer to these filters as Filter a, Filter b, and Filter c, respectively. All have f_0 as the mid-frequency of the pass band. The constants have been chosen so that they all pass the same average power when a wide band voltage is applied:

$$\psi_0 = \int_0^{\infty} w(f)\,df = \text{mean square value of } I(t) \text{ or } V(t)$$

and it is assumed that $f_0 \gg \sigma$, $f_0 \gg \alpha$, $f_0 \gg \beta$ so that the pass bands are relatively narrow.

Expressions for $G_n(f)$ corresponding to several values of n are given in Table 2. When $n = 1$, $G_1(f)$ is simply $w(f)/4$. $G_2(f)$ is obtained by setting $n = 2$ in the definition (4C–1) for $G_n(f)$, squaring the $\psi(\tau)$'s of Table 1, and using

$$\cos^2 2\pi f_0\tau = \tfrac{1}{2} + \tfrac{1}{2}\cos 4\pi f_0\tau$$

TABLE 2

$G_n(f)$	Filter a	Filter b
$G_1(f)$	$\frac{\psi_0}{4\sigma\sqrt{2\pi}} e^{-(f-f_0)^2/2\sigma^2}$	$\frac{\alpha\psi_0}{4\pi} \frac{1}{\alpha^2 + (f - f_0)^2}$
$G_2(f)$	$\frac{\psi_0^2}{8\sigma\sqrt{4\pi}} [2e^{-f^2/4\sigma^2} + e^{-(f-2f_0)^2/4\sigma^2}]$	$\frac{2\alpha\psi_0^2}{8\pi} \left[\frac{2}{4\alpha^2 + f^2} + \frac{1}{4\alpha^2 + (f - 2f_0)^2} \right]$
$G_3(f)$	$\frac{\psi_0^3}{16\sigma\sqrt{6\pi}} [3e^{-(f-f_0)^2/6\sigma^2} + e^{-(f-3f_0)^2/6\sigma^2}]$	$\frac{3\alpha\psi_0^3}{16\pi} \left[\frac{3}{9\alpha^2 + (f - f_0)^2} + \frac{1}{9\alpha^2 + (f - 3f_0)^2} \right]$
$G_4(f)$	$\frac{\psi_0^4}{32\sigma\sqrt{8\pi}} [6e^{-f^2/8\sigma^2} + 4e^{-(f-2f_0)^2/8\sigma^2} + e^{-(f-4f_0)^2/8\sigma^2}]$	$\frac{4\alpha\psi_0^4}{32\pi} \left[\frac{6}{16\alpha^2 + f^2} + \frac{4}{16\alpha^2 + (f - 2f_0)^2} + \frac{1}{16\alpha^2 + (f - 4f_0)^2} \right]$
$G_n(f)$ n odd f small	0	0
$G_n(f)$ n even f small	$\frac{\psi_0^n n!}{\left(\frac{n}{2}\right)! \left(\frac{n}{2}\right)!} \frac{e^{-f^2/2n\sigma^2}}{2^{n+1} \sigma \sqrt{2n\pi}}$	$\frac{\psi_0^n n!}{\left(\frac{n}{2}\right)! \left(\frac{n}{2}\right)! \, 2^n} \frac{1}{2^{n+1} \pi n\alpha} \frac{1}{1 + \left(\frac{f}{n\alpha}\right)^2}$
$G_n(f)$ n even n large f small	$\frac{1}{2\pi\sigma n} e^{-f^2/2n\sigma^2}$	$\frac{2}{\alpha(2\pi n)^{3/2}} \frac{1}{1 + \left(\frac{f}{n\alpha}\right)^2}$
Filter c $G_1(f)$	$\frac{\psi_0}{4\beta}$ when $f_0 - \frac{\beta}{2} < f < f_0 + \frac{\beta}{2}$ 0 elsewhere	Filter c $G_2(f)$: $\frac{\psi_0^2}{4\beta}\left(1 - \frac{f}{\beta}\right)$ when $0 \leq f \leq \beta$ $\frac{\psi_0^2}{8\beta^2} (f - 2f_0 + \beta)$ " $2f_0 - \beta \leq f \leq 2f_0$ $\frac{\psi_0^2}{8\beta^2} (2f_0 + \beta - f)$ " $2f_0 \leq f \leq 2f_0 + \beta$

The expression for $G_2(f)$ given in Table 2 corresponding to Filter c is exact. The expressions for Filters a and b give good approximations around $f = 0$ and $f = 2f_0$ where $G_2(f)$ is large. However, they are not exact because terms involving $f + 2f_0$ have been omitted. It is seen that all three G_2's behave in the same manner. Each has a peak symmetrical about $2f_0$ whose width is twice that of the original $w(f)$, is almost zero between 0 and $2f_0$, and rises to a peak at 0 whose height is twice that at $2f_0$.

$G_3(f)$ is obtained by cubing the $\psi(\tau)$ given in Table 1 and using

$$\cos^3 2\pi f_0\tau = \tfrac{3}{4} \cos 2\pi f_0\tau + \tfrac{1}{4} \cos 6\pi f_0\tau.$$

From the way in which the cosine terms combine with $\cos 2\pi f\tau$ in (4C–1) we see that $G_3(f)$, for our relatively narrow band pass filters, has peaks at f_0 and $3f_0$, the first peak being three times as high as the second. The expressions given for $G_3(f)$ and $G_4(f)$ are approximate in the same sense as are those for $G_2(f)$. It will be observed that the coefficients within the brackets, for Filters a and b, are the binomial coefficients for the value of n concerned. Thus for $n = 2$, they are 2 and 1, for $n = 3$ they are 3 and 1, and for $n = 4$ they are 6, 4, and 1.

The higher $G_n(f)$'s for Filters a and b may be computed in the same way. The integrals to be used are

$$\int_0^\infty e^{-2n(\pi\sigma\tau)^2} \cos 2\pi f\tau \, d\tau = \frac{e^{-f^2/2n\sigma^2}}{2\sigma\sqrt{2n\pi}}$$

$$\int_0^\infty e^{-2n\pi\alpha\tau} \cos 2\pi f\tau \, d\tau = \frac{1}{2\pi} \frac{n\alpha}{n^2\alpha^2 + f^2}$$

In many of our examples we are interested only in the values $G_n(f)$ for f near zero, i.e., only in that peak which is at zero. It is seen that $G_n(f)$ has such a peak only when n is even, this peak arising from the constant term in the expansion

$$\cos^{2k} x = \frac{1}{2^{2k-1}} \left[\cos 2kx + 2k \cos 2(k-1)x + \frac{(2k)(2k-1)}{2!} \cos 2(k-2)x + \cdots + \frac{(2k)!}{(k-1)!(k+1)!} \cos 2x + \frac{(2k)!}{k!k!2} \right]$$

RANDOM WALK AND THE THEORY OF BROWNIAN MOTION*

MARK KAC,† Cornell University

1. Introduction. In 1827 an English botanist, Robert Brown, noticed that small particles suspended in fluids perform peculiarly erratic movements. This phenomenon, which can also be observed in gases, is referred to as Brownian motion. Although it soon became clear that Brownian motion is an outward manifestation of the molecular motion postulated by the kinetic theory of matter, it was not until 1905 that Albert Einstein first advanced a satisfactory theory.

The theory was then considerably generalized and extended by the Polish physicist Marjan Smoluchowski, and further important contributions were made by Fokker, Planck, Burger, Fürth, Ornstein, Uhlenbeck, Chandrasekhar, Kramers and others [1]. On the purely mathematical side various aspects of the theory were analyzed by Wiener, Kolomgoroff, Feller, Lévy, Doob, and Fortet [2]. Einstein considered the case of the *free* particle, that is, a particle on which no forces other than those due to the molecules of the surrounding medium are acting. His results can be briefly summarized as follows.

Consider the motion of the projection of the free particle‡ on a straight line which we shall call the x-axis. What one wants is the probability

$$\int_{x_1}^{x_2} P(x_0 \mid x; t)dx$$

that at time t the particle will be between x_1 and x_2 if it was at x_0 at time $t=0$. Einstein was then able to show that the "probability density" $P(x_0 \mid x; t)$§ must satisfy the partial differential equation

$$\frac{\partial P}{\partial t} = D \frac{\partial^2 P}{\partial x^2}, \tag{1}$$

where D is a certain physical constant. The conditions imposed on P are

$$(2) \qquad \begin{aligned} &(a) \qquad P \geqq 0 \\ &(b) \qquad \int_{-\infty}^{\infty} P(x_0 \mid x; t)dx = 1 \\ &(c) \qquad \lim_{t\to 0} P(x_0 \mid x; t) = 0, \qquad \text{for } x \neq x_0. \end{aligned}$$

---

* This is an extended version of an address delivered at the annual meeting of the Association at Swarthmore, Pennsylvania, December 26–27, 1946.

† John Simon Guggenheim Memorial Fellow.

‡ In what follows we shall identify this projection with the particle itself and hence consider the so-called one-dimensional Brownian motion.

§ The notation $P(x_0 \mid x; t)$ and $P(n \mid m; s)$ for conditional probabilities is that used by Wang and Uhlenbeck [1]. It does not conform with the notation adopted in the statistical literature. Had we adopted the latter notation we would write $P(x; t \mid x_0)$ and $P(m; s \mid n)$.

Conditions (*a*) and (*b*) are the usual ones imposed upon a probability density and condition (*c*) expresses the *certainty* that at $t=0$ the particle was at x_0. It is well known that (1) and (2) imply that

$$P(x_0 \mid x;\, t) = \frac{1}{2\sqrt{\pi D t}}\, e^{-(x-x_0)^2/4Dt} \tag{3}$$

and that the solution (3) is unique.

The greatness of Einstein's contribution was not, however, solely due to the derivation of (1), and hence (3). From the point of view of physical applications it was equally, or perhaps even more, important that he was able to show that

$$D = \frac{2RT}{Nf}, \tag{4}$$

where R is the universal gas constant, T the absolute temperature, N the Avogadro number, and f the friction coefficient. The friction coefficient f, in the case the medium is a liquid or a gas at ordinary pressure, can in turn be expressed in terms of viscosity and the size of the particle [3].

It was relation (4) that made possible the determination of the Avogadro number from Brownian motion experiments, an achievement for which Perrin was awarded the Nobel prize in 1926. However, the derivation of (4) belongs to physics proper, and presents no particular mathematical interest; we shall therefore not be concerned with it in the sequel.

As soon as the theory for the free particle was established, a natural question arose as to how it should be modified in order to take into account outside forces as, for example, gravity. Assuming that the outside force acts in the direction of the x-axis and is given by an expression $F(x)$, Smoluchowski has shown that (1) should in this case be replaced by

$$\frac{\partial P}{\partial t} = -\frac{1}{f}\frac{\partial}{\partial x}(PF) + D\frac{\partial^2 P}{\partial x^2}. \tag{5}$$

Two cases of special interest and importance are:

$F(x) = -a$; field of constant force (for example, gravity).

$F(x) = -bx$; elastically bound particle (for example, small pendulum).

At this point it must be strongly emphasized that theories based on (1) and (5) are only approximate. They are valid only for relatively large t and, in the case of an elastically bound particle, only in the overdamped case, that is, when the friction coefficient is sufficiently large. These limitations of the theory were already recognized by Einstein and Smoluchowski but are often disregarded by writers who stress that in Brownian motion the velocity of the particle is infinite. This paradoxical conclusion is a result of stretching the theory beyond the bounds of its applicability. An improved theory (known as "exact") was advanced by Uhlenbeck and Ornstein and by Kramers. The Uhlenbeck-Ornstein approach was further elaborated by Chandrasekhar and Doob.

In what follows we shall be concerned with a discrete approach to the Einstein-Smoluchowski (approximate) theory. This approach was first suggested by Smoluchowski himself; it consists in treating Brownian motion as a discrete random walk. Smoluchowski used this approach only in connection with a free particle but we shall also treat other classical cases. Moreover, a re-interpretation of one of the discrete models will allow us to discuss the important question of recurrence and irreversibility in thermodynamics.

The main advantages of a discrete approach are pedagogical, inasmuch as one is able to circumvent various conceptual difficulties inherent to the continuous approach. It is also not without a purely scientific interest and it is hoped that it may suggest various generalizations which will contribute to the development of the Calculus of Probability.

2. **The free particle.** Imagine a particle which moves along the x-axis in such a way that in each step it can move either Δ to the right or Δ to the left, the duration of each step being τ. The fact that we are dealing with a free particle is interpreted by assuming that the probabilities of moving to the right or to the left are equal, and hence each equal $\frac{1}{2}$. Instead of $P(x_0 | x; t)$ we now consider $P(n\Delta | m\Delta; s\tau) = P(n | m; s)$ which is the probability that the particle is at $m\Delta$ at time $s\tau$, if at the beginning it was at $n\Delta$. Noticing that $P(n | m; s)$ is also the probability that after s games of "heads or tails" the gain of a player is $\nu = m - n$, we can write

$$(6)\quad P(n | m; s) = \begin{cases} \dfrac{1}{2^s} \dfrac{s!}{\left(\dfrac{s + |\nu|}{2}\right)! \left(\dfrac{s - |\nu|}{2}\right)!} & \text{if } |\nu| \leqq s \text{ and } |\nu| + s \text{ is even,} \\ 0 & \text{otherwise.} \end{cases}$$

Suppose now that Δ and τ approach 0 in such a way that

$$(7)\quad \frac{\Delta^2}{2\tau} = D, \qquad n\Delta \to x_0, \qquad s\tau = t.$$

It then follows from the classical Laplace-De Moivre theorem [4] that

$$(8)\quad \lim \sum_{x_1 < m\Delta < x_2} P(n | m; s) = \frac{1}{2\sqrt{\pi D t}} \int_{x_1}^{x_2} e^{-(x - x_0)^2/4Dt} dx,$$

and hence the fundamental result of Einstein emerges as a consequence of what in probability theory we call a "limit theorem."

It is both important and instructive to point out a striking formal connection between the discrete (random walk) and the continuous (Einstein) approaches. We notice that $P(n | m; s)$ satisfies the difference equation

$$(9)\quad P(n | m; s + 1) = \tfrac{1}{2}P(n | m - 1; s) + \tfrac{1}{2}P(n | m + 1; s),$$

which we write in the equivalent form

$$\frac{P(n\Delta \mid m\Delta; (s+1)\tau) - P(n\Delta \mid m\Delta; s\tau)}{\tau}$$

(10)

$$= \frac{\Delta^2}{2\tau}\left\{\frac{P(n\Delta \mid (m+1)\Delta; s\tau) - 2P(n\Delta \mid m\Delta; s\tau) + P(n\Delta \mid (m-1)\Delta; s\tau)}{\Delta^2}\right\}.$$

In the limit (7) this difference equation goes over formally into the differential equation

$$\frac{\partial P}{\partial t} = D\frac{\partial^2 P}{\partial x^2}, \tag{11}$$

which as noted before was the basis of Einstein's theory. This formal connection between the two approaches can be made rigorous, but we shall not go into this. However, we shall use it as a guiding heuristic principle in constructing models of Brownian motion when outside forces are taken into account.

Finally, let us mention that it is the relation

$$\frac{\Delta^2}{2\tau} = D,$$

which is responsible for the conclusion that the velocity of a Brownian particle is infinite. In fact, in our model, the ratio Δ/τ plays the role of the instantaneous velocity and it obviously approaches infinity as $\Delta \to 0$.

3. **Particle in a field of constant force and in the presence of a reflecting barrier.** We again consider random walk along the x-axis in which a particle can move Δ to the right or Δ to the left, the duration of each step being τ. We now introduce the following new assumptions:

(a) The probability of a move to the right is $q = \frac{1}{2} - \beta\Delta$, and consequently the probability of a move to the left is $p = \frac{1}{2} + \beta\Delta$. Here β is a certain physical constant, and Δ must be chosen sufficiently small so that $q > 0$.

(b) When the particle reaches the point $x = 0$ (*reflecting barrier*) it must, in the next step, move Δ to the right.

Without the assumption (b) the problem would be quite simple and of no great physical interest. In actual experiments with heavy Brownian particles, like those of Perrin, the bottom of the container acts as a reflecting barrier and the elucidation of its influence on the Brownian motion is of considerable theoretical interest.

This problem has been solved by Smoluchowski, on the basis of his equation (5) but we shall show that one can solve the discrete problem and obtain Smoluchowski's result by passing to a limit.

Assuming that the particle starts from $n\Delta \geq 0$ (n an integer) we seek an explicit expression for $P(n \mid m; s)$. We first notice that $P(n \mid m; s)$ satisfies, for

$m \geqq 2$, the difference equation

$$P(n \mid m; s+1) = qP(n \mid m-1; s) + pP(n \mid m+1; s), \tag{12}$$

and that for $m=1$ and $m=0$ we have

$$P(n \mid 1; s+1) = P(n \mid 0; s) + pP(n \mid 2; s), \tag{12a}$$

$$P(n \mid 0; s+1) = pP(n \mid 1; s). \tag{12b}$$

We also have the initial condition

$$P(n \mid m; 0) = \delta(m, n), \tag{13}$$

where $\delta(m, n)$ denotes, as usual, the Kronecker delta.

The difference equation (12) when rewritten in the form analogous to (10) can be shown to go over formally (in the limit $\Delta \to 0$, $\tau \to 0$, $\Delta^2/2\tau = D$, $n\Delta \to x_0$, $m\Delta \to x$, $s\tau = t$) into the differential equation

$$\frac{\partial P}{\partial t} = D \frac{\partial^2 P}{\partial x^2} + 4\beta D \frac{\partial P}{\partial x}, \tag{14}$$

which is of the form (5) with $F(x) = -a = -4\beta Df$.

To find $P(n \mid m; s)$ we use a method which is basic in the study of the so-called Markoff chains, of which our problem is but a particular example, and which in its essentials goes back to Poincaré [5]. Let $(p)_s$ be the (infinite) vector

$$(p)_s = \begin{bmatrix} P(n \mid 0; s) \\ P(n \mid 1; s) \\ P(n \mid 2; s) \\ \vdots \end{bmatrix} \tag{15}$$

and A the infinite matrix

$$A = \begin{Vmatrix} 0 & p & 0 & 0 & 0 \cdots \\ 1 & 0 & p & 0 & 0 \cdots \\ 0 & q & 0 & p & 0 \cdots \\ 0 & 0 & q & 0 & p \cdots \\ \cdots & \cdots & \cdots & \cdots & \cdots \end{Vmatrix}. \tag{16}$$

Then, the difference equation (12), (12*a*) and (12*b*), can be written in the matrix form as

$$(p)_{s+1} = A(p)_s. \tag{17}$$

Thus it follows immediately that

$$(p)_s = A^s(p)_0, \tag{18}$$

where $(p)_0$ is the vector

$$(p)_0 = \begin{pmatrix} 0 \\ \cdot \\ \cdot \\ \cdot \\ 0 \\ 1 \\ 0 \\ \cdot \\ \cdot \\ \cdot \end{pmatrix},$$

1 being the nth component, the components being numbered from zero on. Interpreting (18), we see that

(19) $$P(n \mid m; s) = \text{the } (m, n) \text{ element of } A^s.$$

To make use of (19), we notice that if $R > n+s$ and we consider the finite matrix A_R, which is the upper left R by R submatrix of A, then for $m < R$

(20) $$\text{the } (m, n) \text{ element of } A^s = \text{the } (m, n) \text{ element of } A_R^s,$$

or equivalently,

(21) $$\text{the } (m, n) \text{ element of } A^s = \lim_{R \to \infty} \text{ of the } (m, n) \text{ element of } A_R^s.$$

For each R there exist matrices P_R and Q_R such that

(22) $$P_R Q_R = I$$

and

(23) $$A_R = P_R \begin{bmatrix} \lambda_0(R) & & & 0 \\ & \lambda_1(R) & & \\ & & \ddots & \\ 0 & & & \lambda_{R-1}(R) \end{bmatrix} Q_R,$$

$\lambda_0(R), \lambda_1(R), \cdots, \lambda_{R-1}(R)$ being the eigenvalues of the matrix A_R. To simplify the notation we write λ_j for $\lambda_j(R)$.

Multiplying the matrix A_R s times by itself and making use of (22), we obtain

(24) $$A_R^s = P_R \begin{bmatrix} \lambda_0^s & & & 0 \\ & \lambda_1^s & & \\ & & \ddots & \\ 0 & & & \lambda_{R-1}^s \end{bmatrix} Q_R$$

and one can calculate the (m, n) element of A_R^s explicitly provided the diagonalization (23) can be performed explicitly. This indeed is the case.

Let $(x)_0, (x)_1, \cdots, (x)_{R-1}$ be the "right" and $(y)_0, (y)_1, \cdots, (y)_{R-1}$ the "left" eigenvectors belonging respectively to the eigenvalues $\lambda_0, \lambda_1, \cdots, \lambda_{R-1}$. In other words, for $k=0, 1, \cdots, R-1$, we have

$$A_R(x)_k = \lambda_k(x)_k,$$
$$A_R'(y)_k = \lambda_k(y)_k,$$

where A_R' is the transpose of A_R.

Suppose furthermore that the vectors can be so normalized that

$$(x)_k \cdot (y)_k = 1, \qquad k = 0, 1, \cdots, R-1, \tag{25}$$

where $(x)_k \cdot (y)_k$ is the inner (dot) product of the vectors. Since it is well known that

$$(x)_j \cdot (y)_k = 0 \qquad \text{for } \lambda_j \neq \lambda_k$$

we see that, in the case when all the eigenvalues are distinct, we can take as P_R the matrix whose columns are the vectors $(x)_k$ and for Q_R the matrix whose rows are the vectors $(y)_k$.

In order to determine the eigenvalues and the right eigenvectors we consider the system of linear equations

$$\begin{aligned} px_1 &= \lambda x_0 \\ x_0 + px_2 &= \lambda x_1 \\ qx_1 + px_3 &= \lambda x_2 \\ &\cdots\cdots\cdots \\ qx_{R-2} \qquad &= \lambda x_{R-1}, \end{aligned} \tag{26}$$

and the extended infinite system

$$\begin{aligned} px_1 &= \lambda x_0 \\ x_0 + px_2 &= \lambda x_1 \\ &\cdots\cdots\cdots\cdots \\ qx_{R-1} + px_{R+1} &= \lambda x_R \\ &\cdots\cdots\cdots\cdots . \end{aligned} \tag{27}$$

If we can find non-trivial solutions of (27), for which

$$x_R = 0, \tag{28}$$

we will have found solutions of (26).

It turns out that (28) will yield an equation in λ which has R distinct roots and thus our procedure gives us all eigenvalues, and consequently all right eigenvectors. Multiplying the members of the equations of (27) by 1,

$z, z^2, \ldots$ „ and adding, we obtain formally

$$x_0 z + q\sum_1^\infty x_k z^{k+1} + p\sum_1^\infty x_k z^{k-1} = \lambda \sum_0^\infty x_k z^k$$

or, upon introducing the abbreviation,

$$f(z) = \sum_0^\infty x_k z^k, \tag{29}$$

we have

$$x_0 z + qz[f(z) - x_0] + \frac{p}{z}[f(z) - x_0] = \lambda f(z). \tag{30}$$

From (30) we obtain

$$f(z) = \frac{p}{q} x_0 \left\{ -1 + \frac{1 - \lambda z}{qz^2 - \lambda z + p} \right\}, \tag{31}$$

and since this function is analytic in the neighborhood of zero the formal procedure used above can be justified.

Let ρ_1 and ρ_2 be the *reciprocals* of the roots of

$$qz^2 - \lambda z + p = 0. \tag{32}$$

We have then

$$f(z) = \frac{p}{q} x_0 \left\{ -1 + \frac{1 - \lambda z}{p(1 - \rho_1 z)(1 - \rho_2 z)} \right\}, \tag{33}$$

and introducing partial fractions,

$$\frac{1 - \lambda z}{p(1 - \rho_1 z)(1 - \rho_2 z)} = \frac{1}{p} \frac{\lambda - \rho_1}{\rho_2 - \rho_1} \frac{1}{1 - \rho_1 z} + \frac{1}{p} \frac{\rho_2 - \lambda}{\rho_2 - \rho_1} \frac{1}{1 - \rho_2 z}. \tag{34}$$

Thus

$$x_k = \frac{x_0}{q}\left(\frac{\lambda - \rho_1}{\rho_2 - \rho_1} \rho_1^k + \frac{\rho_2 - \lambda}{\rho_2 - \rho_1} \rho_2^k \right) \qquad \text{for } k \geqq 1, \tag{35}$$

and, in particular, the equation $x_R = 0$ assumes the form

$$\frac{\lambda - \rho_1}{\rho_2 - \rho_1} \rho_1^R + \frac{\rho_2 - \lambda}{\rho_2 - \rho_1} \rho_2^R = 0. \tag{36}$$

Equation (36) must now be solved for λ. Assuming R to be even, and seeking solutions in the form

$$\lambda = 2\sqrt{pq} \cos \Theta, \qquad 0 \leqq \Theta \leqq \pi,$$

we are led to the equation

$$\frac{\tan R\Theta}{\tan \Theta} = \frac{1}{2p - 1}.$$

For $R > (2p-1)^{-1}$ this equation is seen to have $R-2$ distinct roots, $\Theta_1, \Theta_2, \cdots, \Theta_{R-2}$, which lie in the subintervals

$$\left(\frac{j\pi}{R} - \frac{\pi}{2R}, \frac{j\pi}{R} + \frac{\pi}{2R}\right),$$

where j ranges through the integers from 1 to $R-1$ with the exception of $j = R/2$. Corresponding to $\Theta_1, \Theta_2, \cdots, \Theta_{R-2}$ we have $R-2$ distinct eigenvalues,

$$\lambda_k = 2\sqrt{pq} \cos \Theta_k, \qquad k = 1, 2, \cdots, R-2,$$

and the components of the right eigenvector belonging to λ_k can be written in the form

$$x_k^{(m)} = a_k \left(\frac{q}{p}\right)_*^{m/2} \left(\cos m\Theta_k - 2\beta\Delta \frac{\sin m\Theta_k}{\sin \Theta_k} \cos \Theta_k\right),$$

where

$$\left(\frac{q}{p}\right)_*^{\mu} = \begin{cases} \left(\dfrac{q}{p}\right)^{\mu} & \text{if} \quad \mu > 0, \\ q & \text{if} \quad \mu = 0, \end{cases}$$

and a_k is a normalizing constant which will be fixed later. For sufficiently large R the remaining eigenvalues λ_0 and λ_{R-1} can be shown to be given by the formulas

$$\lambda_0 = 2\sqrt{pq} \cosh \theta_0, \qquad \lambda_{R-1} = -\lambda_0,$$

where θ_0 is the only positive root of the equation

$$\frac{\tanh R\theta}{\tanh \theta} = \frac{1}{2p - 1}.$$

The components of the corresponding right eigenvectors are given by the expressions

$$x_0^{(m)} = a_0 \left(\frac{q}{p}\right)_*^{m/2} \left(\cosh m\theta_0 - 2\beta\Delta \frac{\sinh m\theta_0}{\sinh \theta_0} \cosh \theta_0\right)$$

$$x_{R-1}^{(m)} = a_{R-1}(-1)^m \left(\frac{q}{p}\right)_*^{m/2} \left(\cosh m\theta_0 - 2\beta\Delta \frac{\sinh m\theta_0}{\sinh \theta_0} \cosh \theta_0\right).$$

It remains now to find the left eigenvectors. This can be accomplished in exactly the same manner and we merely quote the results. We obtain

$$y_k^{(m)} = b_k\left(\frac{p}{q}\right)^{m/2}\left(\cos m\Theta_k - 2\beta\Delta\frac{\sin m\Theta_k}{\sin \Theta_k}\cos\Theta_k\right)$$

for $m=0, 1,\ldots, R-1$; $k=1, 2, \cdots, R-2$, and

$$y_0^{(m)} = b_0\left(\frac{p}{q}\right)^{m/2}\left(\cosh m\theta_0 - 2\beta\Delta\frac{\sinh m\theta_0}{\sinh \theta_0}\cosh\theta_0\right)$$

$$y_{R-1}^{(m)} = b_{R-1}(-1)^m\left(\frac{p}{q}\right)^{m/2}\left(\cosh m\theta_0 - 2\beta\Delta\frac{\sinh m\theta_0}{\sinh \theta_0}\cosh\theta_0\right).$$

To satisfy the normalization conditions (25) we must have

(37) $$a_k b_k\left(q + \sum_{m=1}^{R-1} f_m^2(\Theta_k)\right) = 1, \qquad k = 1, 2, \cdots, R-2,$$

(38) $$a_k b_k\left(q + \sum_{m=1}^{R-1} F_m^2(\theta_0)\right) = 1, \qquad k = 0, R-1,$$

where

$$f_m(\Theta) = \cos m\Theta - 2\beta\Delta\frac{\sin m\Theta}{\sin\Theta}\cos\Theta_k$$

and

$$F_m(\theta) = \cosh m\theta - 2\beta\Delta\frac{\sinh m\theta}{\sinh\theta}\cosh\theta_0 .$$

We can, of course, put $a_0=a_1= \cdots =a_{R-1}=1$, and determine the b's from (37) and (38). Referring back to (19), (20), and (24), and recalling that columns of P_R are the right eigenvectors $(x)_k$, and the rows of Q_R are the left eigenvectors $(y)_k$, we obtain

(39) $$P(n \mid m; s) = \sum_{k=0}^{R-1} \lambda_k^s x_k^{(m)} y_k^{(n)},$$

or, more explicitly,

(40) $$\begin{aligned} P(n \mid m; s) = {} & b_0(2\sqrt{pq}\cosh\theta_0)^s\left(\frac{p}{q}\right)^{n/2}\left(\frac{q}{p}\right)_*^{m/2} F_m(\theta_0)F_n(\theta_0)\left[1 + (-1)^{m+n+s}\right] \\ & + \left(\frac{p}{q}\right)^{n/2}\left(\frac{q}{p}\right)_*^{m/2}(2\sqrt{pq})^s\sum_{k=1}^{R-2} b_k\cos^s\Theta_k f_m(\Theta_k)f_n(\Theta_k). \end{aligned}$$

Making use of (21), we can achieve considerable simplification by letting $R\to\infty$. In fact, it can be shown that

(41) $$P(n \mid m; s) = \frac{p-q}{2pq}\left(\frac{q}{p}\right)_*^m\left[1 + (-1)^{m+n+s}\right]$$

$$+\frac{2}{\pi}\left(\frac{p}{q}\right)^{n/2}\left(\frac{q}{p}\right)_*^{m/2}(2\sqrt{pq})^s\int_0^\pi \cos^s\Theta\frac{\tan^2\Theta}{(p-q)^2+\tan^2\Theta}f_n(\Theta)f_m(\Theta)d\Theta.$$

Although in various places we have tacitly assumed that p and q are different from $\frac{1}{2}$, the final formula (41) can easily be seen to be valid also for the case $p=q=\frac{1}{2}$. In this case (free particle in the presence of a reflecting barrier) the formula assumes the remarkably simple form

$$P(n \mid m;\, s) = \frac{2}{\pi}(1)_*^{m/2}\int_0^\pi \cos^s\Theta\cos m\Theta\cos n\Theta d\Theta, \tag{42}$$

and the right member can be expressed in terms of binomial coefficients. This formula can also be derived in a much simpler way using, for instance, the classical method of images.

In the limit

$$\Delta \to 0, \qquad \tau \to 0, \qquad \frac{\Delta^2}{2\tau} = D, \qquad n\Delta \to x_0, \qquad s\tau = t,$$

one can show that

$$\lim \sum_{x_1 < m\Delta < x_2} P(n \mid m;\, s) = \int_{x_1}^{x_2} P(x_0 \mid x;\, t)dx,$$

where

$$P(x_0 \mid x;\, t) = 4\beta e^{-4\beta x} + e^{-2\beta(x-x_0)}e^{-4\beta^2 Dt}\frac{2}{\pi}\int_0^\infty e^{-Dy^2t}\frac{y^2}{4\beta^2+y^2}g(x,y)g(x_0,y)dy, \tag{43}$$

and

$$g(x,\, y) = \cos xy - 2\beta(\sin xy)/y.$$

The proof of this theorem is not elementary inasmuch as it utilizes the so called "continuity theorem for Fourier-Stieltjes transforms" [6]. Formula (43) can be shown to be equivalent with Smoluchowski's formula given in [1].

4. An elastically bound particle. Again the particle can move either Δ to the right or Δ to the left, and the duration of each step is τ. However, the probability of moving in either direction depends on the position of the particle. More precisely, if the particle is at $k\Delta$ the probabilities of moving right or left are

$$\frac{1}{2}\left(1-\frac{k}{R}\right) \quad \text{or} \quad \frac{1}{2}\left(1+\frac{k}{R}\right),$$

respectively. R is a certain integer, and possible positions of the particle are limited by the condition $-R \leq k \leq R$. The basic probabilities $P(n \mid m;\, s)$ now satisfy the difference equation

$$(44)\quad P(n\,|\,m; s+1) = \frac{R+m+1}{2R}P(n\,|\,m+1; s) + \frac{R-m+1}{2R}P(n\,|\,m-1; s),$$

which must be solved with the initial condition

$$(45)\qquad P(n\,|\,m; 0) = \delta(m, n).$$

In the limit

$$\Delta \to 0, \qquad \tau \to 0, \qquad R \to \infty, \qquad \frac{\Delta^2}{2\tau} = D, \qquad \frac{1}{R\tau} \to \gamma,$$

$$s\tau = t, \qquad n\Delta \to x_0, \qquad m\Delta \to x,$$

the difference equation (44) is seen to go over formally into the differential equation

$$(46)\qquad \frac{\partial P}{\partial t} = \gamma \frac{\partial(xP)}{\partial x} + D\frac{\partial^2 P}{\partial x^2}$$

which is Smoluchowski's equation (5) with $F(x) = -x\gamma f$.

The discrete problem in a different form and in a different connection was first proposed and discussed by P. and T. Ehrenfest in 1907 [7]. In the next section we shall discuss their original formulation. A fairly detailed treatment was given by Schrödinger and Kohlrausch in 1926 [8] and a brief exposition can be found in the review article of Wang and Uhlenbeck [1]. It seems that Schrödinger and Kohlrausch were the first to point out the connection between the Ehrenfest model and Brownian motion of an elastically bound particle. However, an explicit solution of (44) with the initial condition (45) was apparently not known. I have recently found such a solution using the matrix method described in Section 3 [9]. Instead of the infinite matrix of that section we must now consider the finite matrix

$$(47)\qquad B = \begin{pmatrix} 0 & \frac{1}{2R} & 0 & 0 & 0 \cdots 0 \\ 1 & 0 & \frac{2}{2R} & 0 & 0 \cdots 0 \\ 0 & 1-\frac{1}{2R} & 0 & \frac{3}{2R} & 0 \cdots 0 \\ \cdot & \cdot & \cdot & \cdot & \cdot \\ 0 & 0 & \cdots & \frac{1}{2R} & 0 \end{pmatrix}$$

and the problem is again reduced to finding the eigenvalues $\lambda_{-R}, \lambda_{-R+1}, \cdots, \lambda_0, \cdots, \lambda_{R-1}, \lambda_R$ of B and matrices P and Q such that

$$PQ = I \tag{48}$$

and

$$B = P\begin{pmatrix} \lambda_{-R} & & & & \\ & \lambda_{-R+1} & & 0 & \\ & & \cdot & & \\ & 0 & & \lambda_{R-1} & \\ & & & & \lambda_R \end{pmatrix}Q. \tag{49}$$

As before, $P(n|m; s)$ is the (m, n) element of B^s, where

$$B^s = P\begin{pmatrix} \lambda_{-R}^s & & & & \\ & \lambda_{-R+1}^s & & 0 & \\ & & \cdot & & \\ & 0 & & \lambda_{R-1}^s & \\ & & & & \lambda_R^s \end{pmatrix}Q. \tag{50}$$

In order to perform the diagonalization (49) explicitly we start (following the procedure of Section 3) by trying to find the eigenvalues and the right eigenvectors of B. For this purpose we consider the system of linear equations

$$\begin{aligned} \frac{1}{2R}\, x_1 &= \lambda x_0 \\ x_0 + \frac{2}{2R}\, x_2 &= \lambda x_1 \\ \left(1 - \frac{1}{2R}\right) x_1 + \frac{3}{2R}\, x_3 &= \lambda x_2 \\ \cdots\cdots\cdots\cdots & \\ \frac{1}{2R}\, x_{2R-1} &= \lambda x_{2R}, \end{aligned} \tag{51}$$

and the auxiliary infinite system

$$\begin{aligned} \frac{1}{2R}\, x_1 &= \lambda x_0 \\ x_0 + \frac{2}{2R}\, x_2 &= \lambda x_1 \\ \left(1 - \frac{1}{2R}\right) x_1 + \frac{3}{2R}\, x_3 &= \lambda x_2 \\ \cdots\cdots\cdots\cdots & \end{aligned} \tag{52}$$

$$\frac{1}{2R} x_{2R-1} + \frac{2R+1}{2R} x_{2R+1} = \lambda x_{2R}$$

$$\frac{2R+2}{2R} x_{2R+2} = \lambda x_{2R+1}$$

$$\cdots\cdots\cdots\cdots\cdots\cdots\cdots\cdots$$

If we can find non-trivial solutions of (52) for which

(53) $$x_{2R+1} = 0$$

we will have found solutions of (51). It will turn out that this procedure will again yield all eigenvalues and right eigenvectors. Multiplying the members of the equations of (52) by $1, z, z^2, \cdots$, and adding, we obtain formally

$$\sum_{k=0}^{\infty}\left(1 - \frac{k}{2R}\right) x_k z^{k+1} + \sum_{k=0}^{\infty} \frac{k}{2R} x_k z^{k-1} = \lambda \sum_{k=0}^{\infty} x_k z^k,$$

or, introducing the abbreviation

$$f(z) = \sum_{k=0}^{\infty} x_k z^k,$$

$$zf(z) - \frac{z^2}{2R} f'(z) + \frac{1}{2R} f'(z) = \lambda f(z).$$

We thus get the differential equation

(54) $$f'(z) = 2R \frac{\lambda - z}{1 - z^2} f(z),$$

whose solution satisfying $f(0) = x_0$ is easily found to be

(55) $$f(z) = x_0(1 - z)^{R(1-\lambda)}(1 + z)^{R(1+\lambda)}.$$

Since $f(z)$ is analytic in the neighborhood of $z = 0$ the formal procedure can be justified.

We now notice that if

(56) $$\lambda = \frac{j}{R}, \qquad j = -R, -R+1, \cdots, 0, \cdots, R-1, R,$$

$f(z)$ is a polynomial of degree $2R$, and hence $x_{2R+1} = 0$. The numbers (56) are thus seen to be eigenvalues of B and, since there are $2R+1$ of them, we see that we have found *all* the eigenvalues. It also follows that the components of the right eigenvector belonging to the eigenvalue $\lambda_j = j/R$ can be taken as

$$C_0^{(j)} = 1,\ C_1^{(j)},\ C_2^{(j)},\ \cdots,\ C_{2R}^{(j)},$$

where the C's are defined by the identity

$$(1 - z)^{R-j}(1 + z)^{R+j} \equiv \sum_{k=0}^{2R} C_k^{(j)} z^k. \tag{57}$$

So far we have followed very closely the procedure described in Section 3. Surprisingly enough, we encounter unexpected difficulties in trying to carry out the analogy still further and determine by similar means the left eigenvectors.

To find the matrix Q we resort to a different method. Let us first recall that P can be taken as the matrix whose jth column (for convenience columns and rows are numbered from $-R$ to R) is

$$\begin{matrix} 1 \\ C_1^{(j)} \\ C_2^{(j)} \\ \cdot \\ \cdot \\ \cdot \\ C_{2R}^{(j)}. \end{matrix}$$

Matrix Q must satisfy the equation

$$P'Q' = I,$$

which is an immediate consequence of the equation $PQ=I$, and hence denoting by $\alpha_{-R}, \cdots, \alpha_0, \cdots, \alpha_R$, the consecutive elements of the jth column of Q', we must have

$$\sum_{k=-R}^{R} C_{R+r}^{(k)}\alpha_k = \delta(j, r), \qquad r = -R, \cdots, R. \tag{58}$$

From (58) it follows that

$$z^{R+j} = \sum_{r=-R}^{R} \delta(j, r) z^{R+r} = \sum_{r=-R}^{R} z^{R+r} \sum_{k=-R}^{R} C_{R+r}^{(k)}\alpha_k = \sum_{k=-R}^{R} \alpha_k \sum_{r=-R}^{R} C_{R+r}^{(k)} z^{R+r}$$

$$= \sum_{k=-R}^{R} \alpha_k \sum_{s=0}^{2R} C_s^{(k)} z^s,$$

or, by virtue of (57),

$$z^{R+j} = \sum_{k=-R}^{R} \alpha_k(1 - z)^{R-k}(1 + z)^{R+k} = (1 - z)^{2R} \sum_{l=0}^{2R} \alpha_{l-R}\left(\frac{1 + z}{1 - z}\right)^l.$$

Thus

$$\frac{z^{R+j}}{(1 - z)^{2R}} = \sum_{l=0}^{2R} \alpha_{l-R}\left(\frac{1 + z}{1 - z}\right)^l. \tag{59}$$

Let

$$\zeta = \frac{1+z}{1-z},$$

so that

$$z = -\frac{1-\zeta}{1+\zeta} \quad \text{and} \quad 1 - z = \frac{2}{1+\zeta}.$$

In terms of ζ (59) assumes the form

$$\frac{(-1)^{R+j}}{2^{2R}}(1-\zeta)^{R+j}(1+\zeta)^{R-j} = \sum_{l=0}^{2R} \alpha_{l-R}\zeta^l, \tag{60}$$

and since by (57)

$$(1-\zeta)^{R+j}(1+\zeta)^{R-j} = \sum_{l=0}^{2R} C_l^{(-j)}\zeta^l,$$

we obtain, by comparing coefficients of corresponding powers of ζ,

$$\alpha_{l-R} = \frac{(-1)^{R+j}}{2^{2R}} C_l^{(-j)},$$

or finally,

$$\alpha_s = \frac{(-1)^{R+j}}{2^{2R}} C_{R+s}^{(-j)}. \tag{61}$$

Formula (61) determines explicitly the elements of Q' (and hence of Q), and it is now possible to write an explicit expression for $P(n \mid m; s)$. In fact, making use of (50), we obtain

$$P(n \mid m; s) = \frac{(-1)^{R+n}}{2^{2R}} \sum_{j=-R}^{R} \left(\frac{j}{R}\right)^s C_{R+j}^{(-n)} C_{R+m}^{(j)}. \tag{62}$$

In the limit

$$\Delta \to 0, \qquad \tau \to 0, \qquad \frac{\Delta^2}{2\tau} = D, \qquad \frac{1}{R\tau} \to \gamma, \qquad s\tau = t, \qquad n\Delta \to x_0,$$

we have

$$\lim \sum_{x_1 < m\Delta < x_2} P(n \mid m; s) = \int_{x_1}^{x_2} P(x_0 \mid x; t)\,dx,$$

where

$$P(x_0 \mid x; t) = \frac{\sqrt{\gamma}}{\sqrt{2\pi D(1-e^{-2\gamma t})}} e^{-\gamma(x - x_0 e^{-\gamma t})^2/2\gamma(1-e^{-2\gamma t})}. \tag{63}$$

The proof is again made to depend on the continuity theorm for Fourier-Stieltjes transforms.

The frequency function (63) was first discovered by Lord Rayleigh [10]. Its connection with Brownian motion of an elastically bound particle, in the strongly overdamped case, was established by Smoluchowski who arrived at it quite independently.

5. **The Ehrenfest model. Irreversibility and recurrence.** Imagine $2R$ balls numbered consecutively from 1 to $2R$, distributed in two boxes (I and II) so that at the beginning there are $R+n$, $-R \leqq n \leqq R$, balls in box I. We chose at random an integer between 1 and $2R$ (all these integers are assumed to be equiprobable) and move the ball, whose number has been drawn from the box in which it is, to the other box. This process is then repeated s times and we ask for the probability $Q(R+n|R+m; s)$ that after s drawings there should be $R+m$ balls in box I.

A moment's reflection will persuade one that this formulation (originally proposed by P. and T. Ehrenfest) [7] is equivalent to the random walk formulation of Section 4, if one interprets the excess over R of balls in box I as the displacement of the particle ($\Delta = 1$). Thus

$$Q(R + n \mid R + m; \ ) = P(n \mid m; s),$$

where $P(n|m; s)$ has the meaning of Section 4.

In the present formulation we have a simple and convenient model of heat exchange between two isolated bodies of unequal temperatures. The temperatures are symbolized by the numbers of balls in the boxes and the heat exchange is not an orderly process, as in classical thermodynamics, but a random one like in the kinetic theory of matter. The realistic value of the model is greatly enhanced by the fact that the average excess over R of the number of balls in box I, namely, the quantity

$$\sum_{m=-R}^{R} mP(n \mid m; s)$$

can easily be shown to be equal to

$$n\left(1 - \frac{1}{R}\right)^s \tag{64}$$

which in the limit $R \to \infty$, $1/R\tau \to \gamma$, $s\tau = t$, gives

$$ne^{-\gamma t},$$

or the Newton law of cooling.

There are several proofs of (64) [11]. The most straightforward one, which is not however the simplest, is based on formula (62).

The Ehrenfest model is also particularly suited for the discussion of a famous paradox which at the turn of this century nearly wrecked Boltzmann's inspired efforts to explain thermodynamics on the basis of kinetic theory. In classical thermodynamics the process of heat exchange of two isolated bodies of unequal temperatures is irreversible. On the other hand, if the bodies are treated as a dynamical system the famed "Wiederkehrsatz" of Poincaré asserts that "almost every" state (except for a set of states which, when interpreted as points in phase space, form a set of Lebesgue measure 0) of the system will be, to an arbitrarily prescribed degree of accuracy, again approximately achieved. Thus, argued Zermelo, the irreversibility postulated in thermodynamics and the "recurrence" properties of dynamical systems are irreconcilable. Boltzmann then replied that the "Poincaré cycles" (time intervals after which states "nearly recur" for the first time,—the word "nearly" requiring further specification) are so long compared to time intervals involved in ordinary experiences that predictions based on classical thermodynamics can be fully trusted. This explanation, though correct in principle, was set forth in a manner which was not quite convincing and the controversy raged on. It was mainly through the efforts of Ehrenfest and Smoluchowski that the situation became completely clarified, and the irreversibility interpreted in a proper statistical manner.

It will now be easy to discuss this explanation by appealing to the Ehrenfest model. Let $P'(n|m; s)$ denote the probability that after s drawings (the duration of each drawing is τ) $R+m$ balls will be observed *for the first time* in box I if there were $R+n$ balls in that box at the beginning. In particular, $P'(n|n; s)$ is the probability that the recurrence time of the state "n" (defined by the presence of $R+n$ balls in box I) is $s\tau$. One can then show that

$$\sum_{s=1}^{\infty} P'(n \mid n; s) = 1, \tag{65}$$

or, in other words: *each state is bound to recur with probability 1.* This is the statistical analogue of the "Wiederkehrsatz." One can show furthermore that the mean recurrence time, namely, the quantity

$$\theta_n = \sum_{s=1}^{\infty} s\tau P'(n \mid n; s)$$

is equal to

$$\tau \frac{(R+n)!(R-n)!}{(2R)!} 2^{2R}. \tag{66}$$

This is the statistical analogue of a "Poincaré cycle," and it tells us, roughly speaking, how long, on the average, one will have to wait for the state "n" to recur.

If $R+n$ and $R-n$ differ considerably, θ_n is enormous. For example, if $R=10000$, $n=10000$, $\tau=1$ second, we get

$$\theta = 2^{20000} \text{ seconds (of the order of } 10^{6000} \text{ years!)}.$$

If on the other hand, $R+n$ and $R-n$ are nearly equal, θ_n is quite short. If in the above example we set $n=0$ we get (using Stirling's formula)

$$\theta \sim 100\sqrt{\pi} \text{ seconds} \sim 175 \text{ seconds}.$$

It was Smoluchowski who advanced the rule [12] that if one starts in a state with a long recurrence time the process will appear as irreversible. In our example if one starts with 20000 balls in one box and none in the other, one should observe, for a long time, an essentially irreversible flow of balls. On the other hand, if the mean recurrence time is short, there is no sense to speak about irreversibility.

We now give the proofs of (65) and (66). We shall base our considerations on a formula which Professor Uhlenbeck used for similar purposes in some of his unpublished notes. The formula in question is:

$$P(n \mid m; s) = P'(n \mid m; s) + \sum_{k=1}^{s-1} P'(n \mid m; k) P(m \mid m; s-k). \tag{67}$$

To convince oneself of the validity of this formula we divide all possible ways of reaching "m" from "n" in s steps into classes according to when "m" has been reached for the first time. We then observe that starting from "n" one can reach "m" in s steps in the following s mutually exclusive ways:

(1) "m" is reached for the first time after s steps.

(2) "m" is reached for the first time in 1 step and then, starting from "m" it is again reached in $s-1$ steps.

(3) "m" is reached for the first time in 2 steps and then, starting from "m", it is reached again in $s-2$ steps, and so forth. We note furthermore that the probability that "m" will be reached for the first time in k steps and then, starting from "m," it will be reached again in $s-k$ steps, is

$$P'(n \mid m; k) P(m \mid m; s-k). \tag{68}$$

This completes the proof of (67).

It should be emphasized that the justification of using the product of probabilities in (68) rests upon the fact that in our process the past is independent of the future. In other words, once we know that the system starts, say, from "m," its subsequent behavior is independent of the way in which "m" was reached in the first place.

We introduce now the generating functions

$$h(n \mid m; z) = \sum_{s=1}^{\infty} P(n \mid m; s) z^s \tag{69}$$

$$g(n \mid m; z) = \sum_{s=1}^{\infty} P'(n \mid m; s) z^s, \tag{70}$$

and note that (67) is equivalent to

$$h(n \mid m; z) = g(n \mid m; z) + h(m \mid m; z) g(n \mid m; z),$$

or

(71) $$g(n \mid m; z) = \frac{h(n \mid m; z)}{1 + h(m \mid m; z)}.$$

In particular,

(72) $$g(n \mid n; z) = \frac{h(n \mid n; z)}{1 + h(n \mid n; z)} = 1 - \frac{1}{1 + h(n \mid n; z)},$$

and we also note that

(73) $$\frac{dg(n \mid n; z)}{dz} = \frac{\dfrac{dh(n \mid n; z)}{dz}}{(1 + h(n \mid n; z))^2}.$$

From the definition of $g(n|n; z)$, we obtain

(74) $$\lim_{z\to 1} g(n \mid n; z) = \sum_{s=1}^{\infty} P'(n \mid n; s)$$

(75) $$\tau \lim_{z\to 1} \frac{dg(n \mid n; z)}{dz} = \sum_{s=1}^{\infty} s\tau P'(n \mid n; s).$$

It is from these formulas that we shall derive (65) and (66). We have, using (62)

$$h(n \mid n; z) = \frac{(-1)^{R+n}}{2^{2R}} \sum_{j=-R}^{R} \sum_{s=1}^{\infty} \left(\frac{jz}{R}\right)^s C_{R+j}^{(-n)} C_{R+n}^{(j)},$$

and since

$$1 = \frac{(-1)^{R+n}}{2^{2R}} \sum_{j=-R}^{R} C_{R+j}^{(-n)} C_{R+n}^{(j)},$$

we obtain

(76) $$1 + h(n \mid n; z) = \frac{(-1)^{R+n}}{2^{2R}} \sum_{j=-R}^{R} \frac{1}{1 - \dfrac{j}{R} z} C_{R+j}^{(-n)} C_{R+n}^{(j)}.$$

All terms in the sum on the right hand side of (76) are regular at $z=1$ except the term corresponding to $j=R$, which has a simple pole at that point. Thus we can write

$$1 + h(n \mid n; z) = p(z) + \frac{(-1)^{R+n}}{2^{2R}} C_{2R}^{(-n)} C_{R+n}^{(R)} \frac{1}{1 - z},$$

where $p(z)$ is regular at $z=1$. We see that

$$\lim_{z\to 1} (1 + h(n \mid n; z)) = \infty$$

and hence, using (72) and (74)

$$\sum_{s=1}^{\infty} P'(n \mid n; s) = 1.$$

It is easy to see that

$$\frac{(-1)^{R+n}}{2^{2R}} C_{2R}^{(-n)} C_{R+n}^{(R)} = \frac{1}{2^{2R}} \frac{(2R)!}{(R+n)!(R-n)!},$$

and, denoting this expression by ω, we have (for $|z| < 1$)

$$\frac{dg(n \mid n; z)}{dz} = \frac{(1-z)^2 p'(z) + \omega}{[(1-z)p(z) + \omega]^2},$$

and hence

$$\lim_{z \to 1} \frac{dg(n \mid n; z)}{dz} = \frac{1}{\omega} \cdot$$

This together with (75) yields (66).

The above considerations can be extended to more general processes. However, Markoffian processes (*i.e.*, processes for which (68) is valid) are still the only ones for which one can also calculate the "fluctuation" of the recurrence time, namely, the quantity

$$(77) \qquad \sum_{s=1}^{\infty} s^2 \tau^2 P'(n \mid n; s) - \theta_n^2.$$

Without going into the details, let us mention that (77) can be calculated in terms of

$$\lim_{z \to 1} \frac{d^2 g(n \mid n; z)}{dz^2} \cdot$$

The fluctuation (77) gives us a measure of stability of the mean recurrence time inasmuch as it permits us to estimate how likely (or unlikely) it is to get a specified deviation of the actual recurrence time from the mean. It may seem that since the generating function $g(n|n; z)$ is known explictly it should be easy to get an explicit expression for $P'(n|n; s)$. This, however, is not the case. We have not succeeded in finding such an expression, except for $P'(0|0; s)$, and even then we had to use a different method. We shall give a brief description of this method. Let

$$P(n \mid m; 1) = p_{nm}.$$

Then,

$$P'(n \mid n; s) = \sum_{m_1, \cdots, m_{s-1}}' p_{nm_1} p_{m_1 m_2} \cdots p_{m_{s-1} n},$$

where the accent on the summation sign indicates that $m_j \neq n, j=1, 2, \cdots, s-1$. Now let

$$\epsilon_i = \begin{cases} 0 & \text{if } i = n \\ 1 & \text{if } i \neq n. \end{cases}$$

Noticing that

$$\epsilon_i^2 = \epsilon_i,$$

we can write

$$P'(n \mid n; s) = \sum_{m_1, \cdots, m_{s-1}} p_{nm_1}\epsilon_{m_1}p_{m_1m_2}\epsilon_{m_2}p_{m_2m_3} \cdots \epsilon_{m_{s-2}}p_{m_{s-2}m_{s-1}}\epsilon_{m_{s-1}}p_{m_{s-1}n},$$

where the summation is now extended over all m_j. If B is the matrix

$$((p_{nm})),$$

and B_1 the matrix

$$((\epsilon_n p_{nm} \epsilon_m)),$$

we see that

$$P'(n \mid n; s) = (n, n) \text{ element of } BB_1^{s-2}B. \tag{78}$$

We may note that B_1 is obtained from B by crossing out the nth row and the nth column of the latter, and replacing them by a row and column consisting entirely of zeros. If B_1 can be explicitly diagonalized, that is, written in the form

$$B_1 = P_1 \begin{pmatrix} \mu_1 & & 0 \\ & \mu_2 & \\ & & \ddots \\ 0 & & & \cdot \end{pmatrix} Q_1,$$

where

$$P_1Q_1 = I$$

one can calculate $P'(n \mid n; s)$, explicitly using (78).

We have applied this method to the Ehrenfest model, but only in the case when the middle (zeroth) row and column of B are replaced by a row and column consisting entirely of zeros have we been able to diagonalize explicitly the resulting matrix B_1. The diagonalization proceeds very much as in Section 4, but it has proved necessary to distinguish between the cases when R is even or odd. In case R is even, we were able to derive the formula

$$P'(0 \mid 0; s) = -\frac{1}{2^{2R-1}} \frac{R+1}{2R} \sum \left(\frac{j}{R}\right)^{s-2} C_{R-j}^{(-1)} C_{R-1}^{(j)}, \qquad s \geq 2, \tag{79}$$

where the summation is extended over all odd integers j between $-R$ and R. The details of the derivation are somewhat tedious and will not be reproduced here. Formula (79) furnishes a partial solution to a question left open by Wang and Uhlenbeck [1].

References

1. An extensive list of references to the physical literature can be found in the following articles: G. E. Uhlenbeck and L. S. Ornstein, On the theory of Brownian motion, Phys. Rev. 36 (1930) pp. 823–841, M. C. Wang and G. E. Uhlenbeck, On the theory of Brownian motion II, Rev. Mod. Phys. 17 (1945) pp. 323–342. For a very complete summary of earlier results see the important paper of Smoluchowski, Drei Vortrage über Diffusion, Brownsche Molekularbewegung und Koagulation von Kolloidteilchen, Phys. Zeit. 17 (1916) pp. 557–571 and 585–599.

2. References to the work of Wiener, Kolmogoroff, Doob and Feller are given in the second article in [1]. See also P. Lévy, Sur certains processus stochastiques homogènes, Comp. Math. 7 (1939) pp. 283–339, R. Fortet, Les fonctions aléatoires du type de Markoff associées à certain équations paraboliques, Jour. de Math. 22 (1943) pp. 177–243.

3. For a brief discussion of the nature of the friction coefficient f see the first article in [1].

4. For proofs of this theorem the reader is referred to H. Cramér, Mathematical Methods of Statistics, Princeton University Press (1946), in particular, pp. 198–203.

5. For a complete presentation of the matrix method as applied to Markoff chains see M. Fréchet, Traité du Calcul des Probabilités, Tome I, Fasc. III Second Livre, Paris, Gauthier-Villars (1936).

6. For the proof of the continuity theorem in its most general form see Cramér's book [4] pp. 96–100.

7. Über zwei bekannte Einwände gegen das Boltzmannsche H-Theorem, Phys. Zeit. 8 (1907) pp. 311–314.

8. Das Ehrenfestsche Model der H-Kurve, Phys. Zeit. 27 (1926), pp. 306–313.

9. M. Kac, Bull. Am. Math. Soc. 52 (1946) p. 621 (abstract).

10. See footnote 27 in the second paper of [1].

11. See the paper [8] and for a derivation based on an entirely different principle H. Steinhaus, La théorie and les applications des fonctions indépendantes au sens stochastique, Actualités Scientifiques et Industrielles 738 (1938) Paris, Hermann et Cie, pp. 57–73, in particular, pp. 61–64.

12. See the third article in [1] p. 568.

Annals of Mathematics
Vol. 43, No. 2, April, 1942

THE BROWNIAN MOVEMENT AND STOCHASTIC EQUATIONS

By J. L. Doob

(Received January 14, 1942)

The irregular movements of small particles immersed in a liquid, caused by the impacts of the molecules of the liquid, were described by Brown in 1828.[1] Since 1905 the Brownian movement has been treated statistically, on the basis of the fundamental work of Einstein and Smoluchowski. Let $x(t)$ be the x-coordinate of a particle at time t. Einstein and Smoluchowski treated $x(t)$ as a chance variable. They found the distribution of $x(t) - x(0)$ to be Gaussian, with mean 0 and variance $\alpha \mid t \mid$, where α is a positive constant which can be calculated from the physical characteristics of the moving particles and the given liquid. More exactly, such a family of chance variables $\{x(t)\}$ is now described as the family of chance variables determining a temporally homogeneous differential stochastic process: the distribution of $x(s + t) - x(t)$ is Gaussian, with mean 0, variance $\alpha \mid t \mid$, and if $t_1 < \cdots < t_n$,

$$x(t_2) - x(t_1), \cdots, x(t_n) - x(t_{n-1})$$

are mutually independent chance variables. Wiener, who was the first to discuss this stochastic process rigorously, proved in 1923 that the functions $x(t)$ of this stochastic process are continuous, with probability 1.[2] This is of course a desirable result, which makes the stochastic process somewhat more acceptable as the mathematical idealization of the Brownian movement. It was not expected, however, that the above distribution of $x(s + t) - x(s)$ would prove correct for small t. Even if the derivation did not break down for small t, the mathematical fact that $x(s + t) - x(s)$ has standard deviation $\alpha \mid t \mid$ so that $x(s + t) - x(s)$ is of the order of magnitude of $\mid t \mid^{\frac{1}{2}}$, implying that $dx(s)/ds$ cannot be finite, would suggest the desirability of modifications of the Einstein-Smoluchowski distributions. In fact it is easily seen that (with probability 1) $x(t)$ is not even of bounded variation, so that the path curves of the Einstein-Smoluchowski process have infinite length!

A different stochastic process describing the $x(t)$ was in fact derived in 1930 by Ornstein and Uhlenbeck (15),[3] and later by S. Bernstein (1), (2) and Krutkow (11), all using different methods. This new distribution of $x(s + t) - x(s)$ is

[1] For a historical account of the subject up to 1913, see Haas-Lorentz (6). (The numbered references will refer to the bibliography at the end of the paper.)

[2] Wiener (18, pp. 148–151) has since given a more simple proof. For a discussion of the exact meaning of such a statement concerning the continuity of paths, cf. Doob (3) and (5), §2. The result means that $x(t)$ can be treated as representing one of a multiplicity of continuous functions of t, and integrated, etc. Probability here is formally the study of measures on certain spaces of functions.

[3] Cf. also Ornstein and Wijk (16) and Wijk (17). References to work since 1913 are given in Ornstein and Uhlenbeck (15).

Gaussian, with mean 0 and variance $(\alpha/\beta)(e^{-\beta|t|} - 1 + \beta | t |)$, approximately $\alpha | t |$ for large t, but $\alpha\beta t^2/2$ for small t. (Here β is a second physically determined constant.)

The purpose of the present paper is to apply the methods and results of modern probability theory to the analysis of the Ornstein-Uhlenbeck distribution, its properties and its derivation. It will be seen that the use of rigorous methods actually simplifies some of the formal work, besides clarifying the hypotheses. A stochastic differential equation will be introduced in a rigorous way to give a precise meaning to the Langevin differential equation for the velocity function $dx(s)/ds$. This will avoid the usual embarrassing situation in which the Langevin equation, involving the second derivative of $x(s)$ is used to find a solution $x(s)$ not having a second derivative.

1. The velocity distribution

The displacement function $x(t)$, as discussed by Ornstein and Uhlenbeck, has a derivative $u(t)$, and all the probability relations needed can be derived from those of $u(t)$, as will be seen below. The distribution of $u(t)$ can be described as follows: the conditional distribution of $u(s + t)$ $(t > 0)$ for given $u(s) = u_0$, is Gaussian, with mean $u_0 e^{-\beta t}$ and variance $\sigma_0^2(1 - e^{-2\beta t})$. Here σ_0^2, β are physically determined constants. When $t \to \infty$, this distribution becomes the Maxwell distribution of velocities, furnishing stationary absolute (unconditioned) probabilities for the process, if these are desired. Using these absolute probabilities, which make the distribution easier to describe, the full description of the $u(t)$ distribution can then be stated as follows: for each t, $u(t)$ is a chance variable with a Gaussian distribution, having mean 0, variance σ_0^2; the transition probabilities are as just described; the process is a Markoff process.[4] This last fact means that the Maxwell distribution of $u(t_0)$ for each fixed t_0, and the transition probabilities just described determine the full set of probability relations of the process. Under these conditions, if $t_1 < t_2$, the pair $u(t_1)$, $u(t_2)$ has a bivariate Gaussian distribution, with zero means, equal variances σ_0^2, and correlation coefficient $e^{-\beta(t_2 - t_1)}$. This stochastic process goes back at least to Smoluchowski, although it was first derived by Ornstein and Uhlenbeck as the process describing the velocity of a particle in Brownian motion. Ornstein and Uhlenbeck were only interested in the transition probabilities. The formal manipulations made below will show that there are technical advantages in defining (unconditioned) probabilities for the $u(t)$ also. The above described process will be called the O. U. process below.

The following theorem shows that such a process is essentially determined by three fundamental properties, of which at least the first two have simple physical

---

[4] A process is called a Markoff process if whenever $t_1 < \cdots < t_n$, the conditional distribution of $u(t_n)$ for given values of $u(t_1), \cdots, u(t_{n-1})$ actually depends only on $u(t_{n-1})$. It is in this case, and only in this case, that the Smoluchowski equation between the transition probabilities, and the Fokker-Planck differential equations for the transitional probabilities are valid.

significance. (We can exclude Case A of the theorem, since it obviously does not fit the physical picture.)

THEOREM 1.1. *Let $u(t)$ $(-\infty < t < +\infty)$ be a one-parameter family of chance variables, determining a stochastic process with the following properties.*

1. *The process is temporally homogeneous.*[5]
2. *The process is a Markoff process.*
3. *If s, t are arbitrary distinct numbers, $u(s)$, $u(t)$ have a (non-singular) bivariate Gaussian distribution.*

Define m, σ_0^2 by

$$m = E\{u(t)\}, \qquad \sigma_0^2 = E\{[u(t) - m]^2\}.\text{[6]} \tag{1.1.1}$$

Then the given process is one of the following two types.

(A) *If $t_1 < \cdots < t_n$, $u(t_1), \cdots, u(t_n)$ are mutually independent Gaussian chance variables, with mean m and variance σ_0^2.*

(B) *(O. U. process) There is a constant $\beta > 0$ such that if $t_1 < \cdots < t_n$, $u(t_1), \cdots, u(t_n)$ have an n-variate Gaussian distribution, with common mean m and variance σ_0^2, and correlation coefficients determined by the equation $E\{[u(t) - m][u(s) - m]\} = \sigma_0^2 e^{-\beta|t-s|}$.*

Instead of considering $u(t)$, we can consider $(1/\sigma_0)[u(t) - m]$, which has mean 0 and variance 1. Then we shall assume in the following that $u(t)$ itself has these properties: $m = 0$, $\sigma_0^2 = 1$. Let $\rho(t)$ be the correlation function: $\rho(t) = E\{u(s + t)u(s)\}$, independent of s by Property 1. If $s < t$, the conditional distribution of $u(t)$ for given $u(s)$ has density

$$\frac{1}{(2\pi)^{\frac{1}{2}}(1-\rho^2)^{\frac{1}{2}}} \exp\left(-\frac{1}{2}\frac{[u(t) - \rho u(s)]^2}{1-\rho^2}\right), \qquad \rho = \rho(t-s), \tag{1.1.2}$$

(Property 3). If $t_1 < \cdots < t_n$, $u(t_1), \cdots, u(t_n)$ then have an n-variate Gaussian distribution with density

$$\frac{1}{(2\pi)^{\frac{n}{2}} \prod_1^{n-1} (1-\rho_j^2)^{\frac{1}{2}}} \exp\left(-\frac{1}{2}u_1^2 - \frac{1}{2}\sum_1^{n-1}\frac{(u_{j+1} - \rho_j u_j)^2}{1-\rho_j^2}\right), \tag{1.1.3}$$

$$\rho_j = \rho(t_{j+1} - t_j), \; u_j = u(t_j)$$

using Property 2. Now if $u_1, \cdots, u_n$ have an n-variate Gaussian distribution with density

$$\frac{1}{\Delta}\exp\left(-\frac{1}{2}\sum_{i,j} a_{ij}u_i u_j\right), \tag{1.1.4}$$

$\Delta = \det(E\{u_i u_j\})$ is the determinant of the matrix of variances and covariances, and (a_{ij}) is the inverse of this matrix. Using these facts we can calculate $\rho(t_3 - t_1) = E\{u_1 u_3\}$ in (1.1.3) with $n = 3$, and find that $\rho(t_3 - t_1) = \rho_1\rho_3$, that is

---

[5] That is, the probability distributions are unaffected by translations of the t-axis.

[6] The expectation of the chance variable v will be denoted by $E\{v\}$.

$$(1.1.5) \qquad \rho(t_3 - t_1) = \rho(t_2 - t_1)\rho(t_3 - t_2).$$

Then $\rho(t)$ is an even function; $| \rho(t) | \leqq 1$ (Schwarz's inequality); and according to (1.1.5) $\rho(s + t) = \rho(s)\rho(t)$ for all positive s, t. Under these conditions either $\rho(t) \equiv 0$ or there is a constant $\beta \geqq 0$ such that

$$(1.1.6) \qquad \rho(t) = e^{-\beta|t|}.$$

In the present case, $\beta > 0$, by Property 3 (non-singularity of the given bivariate distributions). Evidently $\rho(t) \equiv 0$ furnishes Case A of the theorem, which certainly has the three given properties. If $\rho(t)$ is given by (1.1.6) with $\beta > 0$, we show first that the matrix (a_{ij}), the inverse of $(\rho(t_j - t_i))$ actually determines a Gaussian density distribution (1.1.4). To see this we consider the density function (1.1.3) with $\rho_j = e^{-\beta(t_{j+1}-t_j)}$. The coefficients of the quadratic form in the exponent of (1.1.3) are easily evaluated and the matrix of the form is found to be the inverse of the matrix $(e^{-\beta|t_j-t_i|})$. Thus (1.1.3) actually is the required probability density. Moreover the probability densities obtained in this way (as the t_i vary) are mutually consistent, because integrating out any variable leaves a quadratic form of the same type, without the integrated variable, but with the same rule determining the coefficients. The correlation function (1.1.6) therefore determines a stochastic process. The process obviously is a Markoff process because of the form of the probability density (1.1.3): an initial factor involving u_1 only, followed by the product of functions of pairs of adjacent variables. The proof of the theorem is now complete.

According to a theorem of Khintchine ((9) p. 608), $\rho(t)$ is the correlation function of a temporally homogeneous stochastic process if and only if it can be put in the form

$$(1.1.7) \qquad \rho(t) = \int_0^\infty \cos \lambda t \, dF(\lambda),$$

where $F(\lambda)$ is monotone non-decreasing and bounded. In Case B of the theorem, (1.1.7) is true when $F(\lambda)$ is given by

$$(1.1.8) \qquad F(\lambda) = \frac{2\beta\sigma_0^2}{\pi} \int_0^\lambda \frac{d\lambda}{\beta^2 + \lambda^2}.$$

In the stochastic process of Case B, the variance of $u(s + t) - u(s)$ is $2\sigma_0^2\beta \, | t |$ for small t:

$$(1.1.9) \qquad E\{[u(s + t) - u(s)]^2\} = 2\sigma_0^2(1 - e^{-\beta|t|}) \sim 2\sigma_0^2\beta \, | t |.$$

Thus $u(s + t) - u(s)$ is of the order of magnitude of $| t |^{\frac{1}{2}}$, and du/dt cannot exist. Physically this means that the particles in question do not have a finite acceleration (if the given stochastic process represents the Brownian movement that closely).

THEOREM 1.2. *If $u(t)$ is the representative function of the stochastic process of Theorem* 1.1 *Case B, $u(t)$ is a continuous function of t, with probability* 1.

Let $v(t)$ be determined by the equation

$$v(t) = t^{\frac{1}{2}} u\left(\frac{1}{2\beta} \log t\right), \qquad t > 0. \tag{1.2.1}$$

Then $v(t)$ has the property that if $t_1 < \cdots < t_n$, $v(t_1), \cdots, v(t_n)$ have an n-variate Gaussian distribution. We find by direct calculation (taking $m = 0$):

$$\begin{aligned} E\{v(s+t) - v(s)\} &= 0, \\ E\{[v(s+t) - v(s)]^2\} &= \sigma_0^2 t, \\ E\{[v(s_2) - v(s_1)][v(t_2) - v(t_1)]\} &= 0, \qquad (s_1 < s_2 \leqq t_1 < t_2). \end{aligned} \tag{1.2.2}$$

Then $v(t)$ determines a differential process—in fact precisely the original Einstein-Smoluchowski process. Since Wiener has proved continuity of the path functions in this case, the theorem follows.

The transition from $u(t)$ to $v(t)$ just used reduces every property of the Ornstein-Uhlenbeck stochastic process to a corresponding property of the Einstein-Smoluchowski process, and vice versa. Many properties of the individual functions of the latter process, that is, properties possessed by almost all the individual functions, in other words possessed "with probability 1," have been proved in recent years, besides the continuity property we have just used. The following theorem gives the counterparts of two of these for the O. U. process.

THEOREM 1.3. *If $u(t)$ is the representative function of the O. U. process of Theorem* 1.1 *Case B,*

$$\limsup_{t\to 0} \frac{u(t) - u(0)}{(4\sigma_0^2 \beta t \log \log (1/t))^{\frac{1}{2}}} = 1, \qquad \limsup_{t\to 0} \frac{u(t)}{(2\sigma_0^2 \log t)^{\frac{1}{2}}} = 1, \tag{1.3.1}$$

with probability 1.

Let $v(t)$ be defined by (1.2.1). Then Khintchine ((10) pp. 68–75) has proved

$$\limsup_{t\to 0} \frac{v(1+t) - v(1)}{(2\sigma_0^2 t \log \log (1/t))^{\frac{1}{2}}} = 1, \qquad \limsup_{t\to \infty} \frac{v(t) - v(0)}{(2\sigma_0^2 t \log \log t)^{\frac{1}{2}}} = 1, \tag{1.3.2}$$

and (1.3.2) becomes (1.3.1) when $v(t)$ is expressed in terms of $u(t)$.

2. The distribution of displacements

It does not seem to have been realized by earlier writers that the distribution of displacements in the O. U. process can be obtained directly from that of the velocities. In fact, we have seen that as t varies, $u(t)$ considered as one of a multiplicity of continuous functions of t. Integration of $u(t)$ is therefore admissible, and will give the displacement function. If $x(t)$ is the x-coordinate of a particle at time t,

$$x(t) - x(0) = \int_0^t u(s)\, ds \tag{2.1}$$

with probability 1 (that is, neglecting the discontinuous $u(t)$ functions which have total probability 0). The main advantages of the rigorous approach to stochastic processes depending on a continuous parameter is precisely that the $u(t)$ of the process, as t varies, can be regarded as an individual function or rather, as one of many functions with whatever regularity properties the given probability distributions imply. Theorem 1.3 limits the actual upper bounds of the velocity functions $u(t)$. The following result takes advantage of the oscillations in sign.

THEOREM 2.1. *If $u(t)$ is the representative function of the O. U. process of Theorem* 1.1 *Case B, with* $m = 0$,

$$\lim_{t\to\infty} \frac{1}{t}\int_0^t u(s)\,ds = \lim_{t\to\infty} \frac{x(t)-x(0)}{t} = 0, \tag{2.1.1}$$

with probability 1.

This theorem is simply the ergodic theorem applied to the $u(t)$ process to give the strong law of large numbers, (cf. Doob (4) p. 294). From (2.2.3) below, it is quite obvious that the expectation of the square of the left side of (2.1.1) goes to 0 as $t \to \infty$, so that the left side goes to 0 in the mean. The strength of (2.1.1) is that it is a statement about the path of the individual path functions, or physically, a statement about the path of a single particle. The same was true in Theorems 1.2 and 1.3.

In order to find the distribution of $x(t) - x(0)$ we proceed as follows. Riemann integrability of $u(t)$ implies that (with probability 1)

$$x(t) - x(0) = \lim_{n\to\infty} \sum_{j=1}^{n} u(tj/n)t/n. \tag{2.2.1}$$

Now the n-variate distribution of the variables summed is Gaussian. Then the sum is Gaussian, so the distribution of $x(t) - x(0)$ is also Gaussian, if it can be shown that the variance of $x(t) - x(0)$ is positive. The distribution of $x(t) - x(0)$ is thus completely determined by its first two moments, which we proceed to calculate. We shall suppose, that $E\{u(t)\} = 0$, $E\{u(t)^2\} = \sigma_0^2$. Then we find

$$E\{x(t) - x(0)\} = \int_0^t E\{u(s)\}\,ds = 0,^7 \tag{2.2.2}$$

and, if $t > 0$,

$$\begin{aligned} E\{[x(t) - x(0)]^2\} &= \int_0^t \int_0^t E\{u(s)u(s')\}\,ds\,ds' \\ &= \sigma_0^2 \int_0^t \int_0^t e^{-\beta|s-s'|}\,ds\,ds' = \frac{2\sigma_0^2}{\beta^2}(e^{-\beta t} - 1 + \beta t). \end{aligned} \tag{2.2.3}$$

[7] By Fubini's integration theorem, we can find the expectations under the integral sign, before integrating with respect to s.

The same sort of argument shows that if $t_1, \cdots, t_n$ are any distinct numbers, the chance variables

$$\{x(t_j) - x(0), u(t_j), \quad j = 1, \cdots, n\}$$

have a $2n$-variate Gaussian distribution, which can then be evaluated explicitly by finding the first and second moments. For example, the following equations determine the bivariate distribution of $x(t) - x(0)$, $u(t)$, $(t > 0)$:

$$E\{[x(t) - x(0)]u(t)\} = \int_0^t E\{u(t)u(s)\}\, ds = \frac{\sigma_0^2}{\beta}(1 - e^{-\beta t}),$$

$$E\{x(t) - x(0)\} = 0, \tag{2.2.4}$$

$$E\{[x(t) - x(0)]^2\} = \frac{2\sigma_0^2}{\beta^2}(e^{-\beta t} - 1 + \beta t), \quad E\{u(t)\} = 0, \quad E\{u(t)^2\} = \sigma_0^2.$$

Thus the bivariate density of $x(t) - x(0)$, $u(t)$ is Gaussian, with common mean 0, and variances $(2\sigma_0^2/\beta^2)(e^{-\beta t} - 1 + \beta t)$, σ_0^2, respectively, and correlation coefficient

$$\frac{1 - e^{-\beta t}}{2^{\frac{1}{2}}(e^{-\beta t} - 1 + \beta t)^{\frac{1}{2}}}. \tag{2.2.5}$$

It is to be expected that if $s_1 < s_2 \leqq t_1 < t_2$, $x(s_2) - x(s_1)$ and $x(t_2) - x(t_1)$ become independent as $t_1 \to \infty$. In fact, these two normally distributed variables have correlation coefficient.

$$\frac{(e^{\beta s_2} - e^{\beta s_1})(e^{-\beta t_1} - e^{-\beta t_2})}{2(e^{-\beta(s_2 - s_1)} - 1 + \beta(s_2 - s_1))^{\frac{1}{2}}(e^{-\beta(t_2 - t_1)} - 1 + \beta(t_2 - t_1))^{\frac{1}{2}}}; \tag{2.2.6}$$

which goes to 0 when t_1 and t_2 become infinite.

If in this discussion only the conditional distribution functions are wanted, for $u(0) = u_0$, for example, two procedures are possible. Setting $u(0) \equiv u_0$ instead of using the initial distribution we have used above, carrying out the same type calculations as above, now would give the desired conditional probabilities. Or the conditional distributions could be calculated from the distributions just derived, since the conditional distributions of a multivariate Gaussian distribution are easily found. Theorems 1.2, 1.3 and 2.1 hold no matter what initial distribution is assigned to $u(0)$.

Finally, there is one more fact which we shall need in the next section. Define $B(t)$ by

$$B(t) = \beta[x(t) - x(0)] + u(t) - u(0). \tag{2.2.7}$$

Then $B(t)$ has for each t a Gaussian distribution, with mean 0. Evidently the distribution of $B(s + t) - B(s)$ is independent of s. It is Gaussian, with mean 0, and the variance is easily calculated to be $2\sigma_0^2\beta \mid t \mid$. Moreover, if $s_1 < s_2 \leqq t_1 < t_2$.

$$E\{[B(t_2) - B(t_1)][B(s_2) - B(s_1)]\} = 0. \tag{2.2.8}$$

Thus the $B(t)$-process is again the Einstein-Smoluchowski process.

3. Derivation of the velocity distribution using the Langevin equation

Ornstein and Uhlenbeck base their investigation on the Langevin equation

$$\frac{du}{dt} = -\beta u(t) + A(t), \tag{3.1}$$

which is simply Newton's law of motion applied to a particle, after dividing through by the mass. The first term on the right is due to the frictional resistance or its analogue, which is supposed proportional to the velocity. The second term represents the random forces (molecular impacts). Probability hypotheses are imposed on the $A(t)$, including relations between $A(t)$ and $u(t)$, to determine the $u(t)$ distribution. Unfortunately this $u(t)$ distribution (Case B of Theorem 1.1), as we have seen, has the property that the velocity function has no time derivative. Then the solution can hardly satisfy (3.1).

Bernstein ((2) p. 361) replaces (3.1) by a finite difference equation:

$$\Delta\left(\frac{\Delta\xi_n}{\Delta t}\right) = -\beta\Delta\xi_n + \alpha_n, \qquad n = 1, 2, \cdots. \tag{3.2}$$

Here $\xi_1, \xi_2, \cdots$ is a sequence of chance variables, $\Delta\xi_n = \xi_{n+1} - \xi_n$ etc., and $\alpha_1, \alpha_2, \cdots$ is a given sequence of mutually independent chance variables. If we think of ξ_j as the analogue of $x(j\Delta t)$, the correspondence between (3.2) and (3.1) is clear. The equations of (3.2) determine definite distributions for the ξ_j in terms of those of the α_j. Bernstein shows that as $\Delta t \to 0$ the distribution of $\Delta\xi_n/\Delta t$ ($\sim \Delta x/\Delta t$) becomes the $u(t)$ distribution we have been discussing, if suitable hypotheses are made on the α_j. This approach is essentially different from that of Ornstein and Uhlenbeck in that Bernstein, as he states explicitly ((1) pp. 5, 6) is not writing a difference equation in the displacement functions $x(t)$ themselves: (3.2) determines distributions only, and these are approximated by the limiting distributions described in Theorem 1.1 Case B.

In our treatment, we shall replace the Langevin equation by a formalized differential equation for the velocity function $u(t)$. This equation is to be exact, not merely asymptotically true. The equation will be perfectly proper mathematically, so that solution by ordinary methods will provide all the information relevant to the desired distributions, and solution of more general problems, involving external forces, will require no special methods.

The problem is to find a proper stochastic analogue of the Langevin equation, remembering that we do not expect $u'(t)$ to exist. We write the equation in the following form:

$$du(t) = -\beta u(t)\, dt + dB(t), \tag{3.3}$$

and try to give these differentials a suitable interpretation. We shall suppose that the $B(t)$-process is a differential process: that is, if $t_1 < \cdots < t_n$, we suppose that

$$B(t_2) - B(t_1), \cdots, B(t_n) - B(t_{n-1})$$

are mutually independent chance variables. We also suppose temporal homogeneity, that is that the distribution of $B(s + t) - B(s)$ is independent of s. The physical meaning of these hypotheses is clear, and they will be justified further below. Equation (3.3) can be interpreted roughly in terms of small changes in momentum. An important particular case is that in which the second moments of the $B(t)$-process are finite:

$$\sigma^2(t) = E\{[B(s + t) - B(s)]^2\} < \infty. \tag{3.4}$$

The first moment $E\{B(s + t) - B(s)\}$ then exists. If this first moment vanishes, $\sigma^2(t)$ satisfies the functional equation

$$\sigma^2(s + t) = \sigma^2(s) + \sigma^2(t).$$

Then $\sigma^2(t)$ must be proportional to t: $\sigma^2(t) = t\sigma^2$. If $f(t)$ is continuous,

$$\int_a^b f(t)\, dB(t) \tag{3.5}$$

has been defined under these hypotheses (Wiener (18), pp. 151–157, Doob (3), pp. 131–134), even though the functions $B(t)$ are known not to be of bounded variation. The definition makes all the formal processes correct. For example, if $f'(t)$ exists and is continuous,

$$\int_a^b f(t)\, dB(t) = f(t)[B(t) - B(0)]\Big|_a^b - \int_a^b [B(t) - B(0)]f'(t)\, dt \tag{3.6}$$ [8]

with probability 1. The usual Riemann-Stieltjes sums converge to (3.5) in the mean. Moreover

$$\begin{aligned} &E\left\{\int_a^b f(t)\, dB(t)\right\} = 0, \\ &E\left\{\left[\int_a^b f(t)\, dB(t)\right]\left[\int_a^b g(t)\, dB(t)\right]\right\} = \sigma^2 \int_a^b f(t)g(t)\, dt. \end{aligned} \tag{3.7}$$

Now it can be shown even without the hypothesis of the finiteness of the second moment in (3.4) that the formal integral in (3.5) can be defined, and will satisfy (3.6). The form of the characteristic function of $B(s + t) - B(s)$ has been derived by Lévy ((14) Chapter VII) and using this it is easy to prove that the

[8] We never write $B(t)$ alone in an equation, since strictly speaking only differences like $B(t) - B(0)$ are defined. It is unnecessary to define $B(0)$ itself, although for convenience it can be taken identically 0, without affecting any of the equations to be used. Differential processes have been discussed in detail by Lévy ((12), (13), (14) Chapter VII) and Doob ((3) §3).

usual Riemann-Stieltjes sums for the integral (3.5) converge in probability. The integral is defined as the limit, and (3.6) is readily verified. On the other hand, (3.7) cannot be expected to hold, since if $f(t) = 1$ the integral becomes $B(b) - B(a)$, and we have not supposed that the expectation of this difference is finite. The special case in which the second moment is finite is the only important one for the purposes of this section, but less restrictive conditions will be needed in §5. We shall justify later the assumption that the $B(t)$ process is a differential process.

We shall interpret an equation in differentials like (3.3) to mean the truth (with probability 1, that is for almost all functions $u(t)$) of

$$\int_a^b f(t)\,du(t) = -\beta \int_a^b f(t)u(t)\,dt + \int_a^b f(t)\,dB(t) \tag{3.8}$$

for all a, b, whenever f is a continuous function. Here the first two integrals are to be defined as the limits (in probability) of the usual Riemann or Riemann-Stieltjes sums. Equation (2.2.7) implies

$$\begin{aligned}\int_a^b f(t)\,du(t) &= -\beta \int_a^b f(t)\,dx(t) + \int_a^b f(t)\,dB(t)\\ &= -\beta \int_a^b f(t)u(t)\,dt + \int_a^b f(t)\,dB(t).\end{aligned} \tag{3.9}$$

Thus (3.3) holds for the $u(t)$ of the O. U. distribution if the $B(t)$ is defined by (2.2.7). Moreover (2.2.7) with $B(t)$ replaced by $B(t) - B(0)$ is an immediate consequence of (3.3). In this case, $B(t)$ has the property that the differences $B(s + t) - B(s)$ have finite second moments and even Gaussian distributions, but we are not making either assumption in solving (3.3).

If (3.3) is true, then (with probability 1)

$$\int_0^t e^{\beta\tau}\,du(\tau) = -\beta \int_0^t e^{\beta\tau}\,u(\tau)\,d\tau + \int_0^t e^{\beta\tau}\,dB(\tau), \tag{3.10}$$

which implies, since integration by parts is applicable,

$$u(t) = u(0)e^{-\beta t} + e^{-\beta t} \int_0^t e^{\beta\tau}\,dB(\tau) \tag{3.11}$$

for all t, with probability 1. Conversely suppose that $u(t)$ is defined by (3.11). Since $B(t)$ is known to be continuous in t except for non-oscillatory discontinuities (jumps) (Lévy (12) pp. 359–364, (13); Doob (3), pp. 134–138), the same must be true of the right side of (3.11), and therefore of $u(t)$. Then $u(t)$ is Riemann integrable with probability 1. Moreover

$$\int_a^b f(t)\,e^{-\beta t}\,d_t \int_0^t e^{\beta\tau}\,dB(\tau) = \int_a^b f(t)\,dB(t), \tag{3.12}$$

so that from (3.11)

$$\int_a^b f(t)e^{-\beta t}\, d[e^{\beta t}u(t) - u(0)] = \int_a^b f(t)\, dB(t), \tag{3.13}$$

proving incidentally that the left side exists. The left side can be simplified to

$$\beta \int_a^b f(t)u(t)\, dt + \int_a^b f(t)\, du(t), \tag{3.14}$$

and putting this into (3.13) we find that (3.8) is satisfied. Then (3.11) furnishes the complete solution of (3.3) under the stated conditions. We stress again that although (3.11) implies strong connections between the $u(t)$ and $B(t)$ processes, we have made no such hypothesis in the derivation not implicit in (3.3). Lévy ((14) pp. 166–167) has shown that the only differential processes whose path functions $B(t) - B(0)$ do not have jumps have the property that the distribution of $B(t) - B(0)$ is Gaussian. Then it is only in this case, which will lead to the O. U. process, that $u(t)$ will not have jumps.

The term $\beta u(t)$ in the Langevin equation is supposed to account for the total frictional effect, including the Doppler friction, caused by the fact that more impacts decelerate than accelerate the motion of a moving particle. The term $A(t)$ in (3.1) or $dB(t)$ in (3.3) represents the "purely random" impulses, that is, the residual effect after the frictional effect has been subtracted out. One idea running through any treatment of the Langevin equation is that this term or, sometimes, $x(t)$ itself, is independent of the given velocity at any time. This hypothesis goes back to Langevin, and has caused much controversy. We shall make the hypothesis only to the following extent. The chance variable $u(0)$ will be given various initial distributions, but will always be made independent of the $B(t)$-process for $t \geqq 0$. This means that if $0 \leqq t_1 < \cdots < t_n$ the chance variable $u(0)$ is supposed independent of the set of chance variables

$$\{B(t_{j+1}) - B(t_j), \qquad j = 1, \cdots, n-1\}.$$

We shall describe the above hypothesis in the following physical terms: *the initial velocity $u(0)$ is independent of later residual random impacts.* It would be a serious drawback to the whole treatment if when $u(0)$ is so chosen $u(t_0)$ for each $t_0 > 0$ were not independent of the $B(t)$-process for $t \geqq t_0$, that is if $u(t_0)$ were not independent of later residual random impacts for all t_0. We can prove, however, the following statement, which incidentally justifies our hypothesis that the $B(t)$-process is a differential process. *Let the $B(t)$ process be a differential process, and define $u(t)$ by* (3.11). *If the chance variable $u(0)$ is independent of the $B(t)$-process for $t \geqq 0$, then $u(t_0)$ will be independent of the $B(t)$-process for $t \geqq t_0$, for all $t_0 > 0$. Conversely suppose only that the $B(t)$-process is regular enough that the integral* (3.5) *can be defined as the limit in probability of the usual sums, and that* (3.6) *is true. Then if $u(t)$ is defined by* (3.11), *and if choosing $u(0)$ independent of the $B(t)$ process for $t \geqq 0$ implies that $u(t_0)$ will be independent of the $B(t)$-process for $t \geqq t_0$, for all $t_0 > 0$, then the $B(t)$-process is a differential process.*

PROOF. Let the $B(t)$-process be a differential process, define $u(t)$ by (3.11) and let $u(0)$ be independent of the $B(t)$-process for $t \geqq 0$. Then from (3.11) with $t = t_0$, $u(t_0)$ involves only $u(0)$ and the $B(t)$-process for $t \leqq t_0$. Then $u(t_0)$ is independent of the $B(t)$-process for $t \geqq t_0$ because the $B(t)$-process is a differential one, with differences involving t-values beyond t_0 independent of those involving t-values before t_0. Conversely suppose that choosing $u(0)$ independent of the $B(t)$-process for $t \geqq 0$ implies that $u(t_0)$ will be independent of the $B(t)$-process for $t \geqq t_0$, for all $t_0 > 0$. Then if $u(0)$ is so chosen,

$$u(0) + \int_0^{t_0} e^{\beta\tau}\, dB(\tau)$$

and therefore

$$\int_0^{t_0} e^{\beta\tau}\, dB(\tau)$$

are independent of the $B(t)$-process for $t \geqq t_0$. This fact implies that the preceding integral determines a differential process, that is, if $t_1 < \cdots < t_n$, the integrals

$$\int_{t_j}^{t_{j+1}} e^{\beta\tau}\, dB(\tau)$$

are mutually independent. Then (applying this fact to subintervals of the intervals (t_j, t_{j+1}))

$$\int_{t_j}^{t_{j+1}} e^{-\beta\tau}\, d_t \int_{t_j}^{t} e^{\beta\tau}\, dB(\tau), \qquad j = 1, \cdots, n$$

are mutually independent, and these repeated integrals are simply

$$B(t_{j+1}) - B(t_j) \qquad j = 1, \cdots, n - 1.$$

The latter differences are therefore mutually independent, as was to be proved.

We shall need the following lemma.

LEMMA 3. *Suppose that $a < 1$, and let $x_0, x_1, \cdots$ be mutually independent chance variables with a common distribution function. If there is a chance variable y with a Gaussian distribution such that the distribution function of $\sum_{j=0}^{n-1} a^{n-j}x_j$ approaches that of y as $n \to \infty$, then the x_j have Gaussian distributions.*

Many of the hypotheses of the lemma are unnecessary, but its statement is general enough for our purposes, and the proof will apply to a situation to be discussed in §5, where the distribution of y will not be Gaussian. The hypotheses imply that the distribution of $\sum_m^n a^j x_j$ approaches that of $a^{m-1}y$ as $n \to \infty$. If $\varphi(t)$ is the characteristic function of x_1 and $\psi(t)$ that of y, writing $\sum_1^n a^j x_j$ in the form $ax_1 + \sum_2^n a^j x_j$ shows that

$$\psi(t) = \varphi(at)\cdot\psi(at).$$

Solving for φ we find that it is the characteristic function of a Gaussian distribution, as was to be proved.

In the physical picture under discussion, further conditions on the solution of (3.3) are known. In fact the Brownian movement is simply a visible example of molecular or near molecular movement. The general principles of such movements are therefore applicable, and the principle of equipartition of energy leads to the Maxwell distribution of velocities. Let k be the Boltzmann constant, and T the absolute temperature. We can formulate the significance of the Maxwell distribution (as much as we shall need it) as follows.

M_1. *Tendency towards the Maxwell distribution.* Whatever the initial distribution of $u(0)$, the transition probabilities have the property that when $t \to \infty$ the distribution function of $u(t)$ converges to the Gaussian distribution function with mean 0 and variance kT/m. (Here m is the mass of the moving particle.)

M_2. *Stability of the Maxwell distribution.* If $u(0)$ is independent of later residual random impacts, and if it has the Gaussian distribution described in M_1, $u(t)$ will have this same distribution for every positive t.

These two statements are closely related, but neither apparently can be deduced from the other without further assumptions. Since these principles act the part of a *deus ex machina* in a discussion of the Langevin equation, we shall use them as little as possible. It will usually be sufficient to use a weakened form of M_1:

M_1'. There is an initial distribution of $u(0)$, such that the transition probabilities have the property that when $t \to \infty$ the distribution function of $u(t)$ converges to the Gaussian distribution function with mean 0 and variance kT/m. It is understood here as before that $u(0)$ is to be independent of later residual random impacts.

Conditions M_1 and M_2 restrict the possibilities for the $B(t)$-process. In fact suppose that condition M_1' is satisfied. Then (3.11) shows that

$$e^{-\beta t} \int_0^t e^{\beta\tau} \, dB(\tau)$$

is nearly Gaussian for large t, with mean 0 and variance kT/m. We write this integral as a sum, replacing t by nt:

$$e^{-\beta n t} \int_0^{nt} e^{\beta\tau} \, dB(\tau) = \sum_0^{n-1} e^{-\beta t(n-j)} x_j, \tag{3.15}$$

where

$$x_j = \int_{jt}^{(j+1)t} e^{\beta(\tau - jt)} \, dB(\tau). \tag{3.16}$$

Since the $B(t)$-process is a differential process, and is temporally homogeneous, the x_j are mutually independent, with identical distributions. According to the lemma, the right side of (3.15) cannot become Gaussian for large t unless the distribution of x_1 is Gaussian. Thus, since t is arbitrary in the above discussion,

$$\int_s^t e^{\beta\tau} \, dB(\tau)$$

has a Gaussian distribution for all s, t. Since the chance variables

$$\int_{jt/n}^{(j+1)t/n} e^{\beta\tau}\,dB(\tau), \qquad j = 1, \cdots, n \tag{3.17}$$

are mutually independent and Gaussian, the chance variable

$$\sum_{0}^{n-1} e^{-\beta jt/n} \int_{jt/n}^{(j+1)t/n} e^{\beta\tau}\,dB(\tau) \tag{3.18}$$

also has a Gaussian distribution. When n becomes infinite, (3.18) becomes $B(t) - B(0)$, with probability 1. The latter difference thus has a Gaussian distribution, with mean 0. The $B(t)$-process therefore has finite second moments $\sigma^2(t) = t\sigma^2$ as defined in (3.4). According to (3.7) the last term in (3.11), which we now know has a Gaussian distribution, has mean 0 and variance $\sigma^2(1 - e^{-2\beta t})/2\beta$. Then $u(t) - e^{-\beta t}u(0)$ has this same distribution. The variance becomes $\sigma^2/2\beta$ when $t \to \infty$, and therefore, according to M_1', $\sigma^2 = 2\beta kT/m$. Thus condition M_1' completely determines the $B(t)$-process. We show next that condition M_2 determines this same $B(t)$-process. In fact suppose condition M_2 is true, and assign to $u(0)$ the distribution of that condition. Then $u(0)$ is independent of the integral in (3.11), and in (3.11), $u(t)$ (which has a Gaussian distribution, according to condition M_2) is expressed as the sum of two independent chance variables, of which the first is Gaussian. The characteristic function of the second is the quotient of the characteristic functions of two Gaussian distributions, and is therefore the characteristic function of a Gaussian distribution. Thus the expression

$$e^{-\beta t} \int_0^t e^{\beta\tau}\,dB(\tau) \tag{3.19}$$

has a Gaussian distribution for all t, and this implies, as above, that $B(t) - B(0)$ has a Gaussian distribution, with variance $\sigma^2 t$. The variances on the right side of (3.11) add up to that on the left, giving an equation for σ^2:

$$\frac{kT}{m} = e^{-2\beta t}\frac{kT}{m} + \frac{1 - e^{-2\beta t}}{2\beta}\sigma^2. \tag{3.20}$$

Then $\sigma^2 = 2\beta kT/m$ as above.

We can now finally derive the O. U. velocity process as the solution of the Langevin equation. Suppose the $B(t)$-process is the one derived in the preceding paragraphs, and choose the chance variable $u(0)$ to be independent of the $B(t)$-process for $t \geqq 0$. Then $u(0)$ is independent of the integral in (3.11), and this means that the conditional distribution of $u(t)$ for $u(0) = u_0$ is Gaussian, with mean 0 and variance $kT(1 - e^{-2\beta t})/m$. Moreover, (3.11) implies

$$u(s + t) = u(s)e^{-\beta t} + e^{-\beta(s+t)} \int_s^{s+t} e^{\beta\tau}\,dB(\tau). \tag{3.21}$$

As we have seen, $u(s)$ is independent of the $B(t)$-process as far as it appears in (3.21) and therefore is independent of the integral. Thus the transition probabilities from s to $s + t$ are the same as those from 0 to t, which are precisely those of the O. U. process. Incidentally it follows that the full condition M_1 is satisfied. Finally, if $u(0)$ is not only supposed independent of the $B(t)$-process, for $t \geqq 0$, but also is supposed to have a Gaussian distribution with mean 0 and variance kT/m, the same will be true of $u(t)$ (as can be calculated from (3.11)) and condition M_2 is thus satisfied. We can summarize all our results as follows.

* THEOREM 3. *Let the $B(t)$-process be a temporally homogeneous differential process. Then* (3.11) *furnishes the solution of* (3.3). *The following conditions on the solution are equivalent.*

(i) *The solution satisfies condition* M_1'.

(ii) *The solution satisfies condition* M_1.

(iii) *The solution satisfies condition* M_2.

(iv) $B(t) - B(0)$ *has a Gaussian distribution, with mean* 0 *and variance* $\sigma^2 t = t2\beta kT/m$.

If the above conditions are satisfied, $u(t) - e^{-\beta t}u(0)$ *will have a Gaussian distribution with mean* 0 *and variance* $kT(1 - e^{-2\beta t})/m$; *if* $u(0)$ *is independent of the* $B(t)$*-process for* $t \geqq 0$, $u(s)$ *is independent of the* $B(t)$*-process for* $t \geqq s$ *for all* $s > 0$, *and the transition probabilities of the* $u(t)$*-process are those of the O. U. velocity process. If in addition* $u(0)$ *has the Gaussian distribution with mean* 0 *and variance* kT/m, *the* $u(t)$*-process becomes the O. U. process, with* $m = 0$, $\sigma_0^2 = kT/m$.

The Langevin equation gives a physical interpretation to every property of the O. U. process. It is interesting to verify that as $h \to 0$ the correlation coefficient of the pair $B(s + h) - B(s)$, $u(t)$ (any s, t) goes to 0. In this sense then, $dB(s)$, the effect of the residual random impacts at time s, is independent of the velocity at any particular time t. Since in (3.11) $u(t)$ is written in terms of the $B(t)$-process, $u(t)$ is of course not independent of this process.

We have written $u(t)$ in terms of the $B(t)$-process. It is easy to write $x(t)$ in terms of the $B(t)$ process by combining (2.1) with (3.11):

$$x(t) = x(0) + \frac{1 - e^{-\beta t}}{\beta} u(0) + \frac{1}{\beta} \int_0^t [1 - e^{-\beta(t-\tau)}] \, dB(\tau). \tag{3.22}$$

Instead of finding the distributions of the displacement and velocity processes, and their correlations, as at the beginning of the paper, we could easily derive the desired results using (3.11) and (3.22). The various expectations can be calculated using (3.7).

In physical applications, the correlation function $E\{u(s)u(s + t)\}$ is sometimes wanted as a time average. Now the transformation S_h taking $B(t) - B(0)$ into $B(t + h) - B(h)$ preserves the $B(t)$ probability relations (temporal homogeneity), and the family of transformations $\{S_h\}$ is well known to be metrically transitive.[9] Then applying the ergodic theorem to the function $u(0)u(h)$, considered as a function of the $B(t)$, we find that

---

[9] Cf. for example Doob, (3) p. 125.

* See Note on page 369.

$$\lim_{t\to\infty}\frac{1}{t}\int_0^t u(s)u(s+h)\,ds = E\{u(0)u(h)\} = \frac{kT}{m}e^{-\beta|h|}, \tag{3.23}$$

with probability 1, that is for almost all functions $u(t)$. The ergodic theorem was applied to the $B(t)$-process in essentially this way by Wiener ((18) p. 169) who has been interested in the harmonic analysis of functions like the $u(t)$ discussed here. The work of this paper verifies in this particular case the importance Wiener gave to the functions of the $B(t)$-process of the type (3.5).

There is no difficulty in extending the above results to bound particles. For example, the Langevin equation of the harmonically bound particle is

$$\frac{du}{dt} = -\beta u - \omega^2 x + A(t), \tag{3.24}$$

which in our treatment becomes

$$du = -\beta u\,dt - \omega^2 x\,dt + dB. \tag{3.25}$$

The usual methods of solving the differential equation (3.24) are still applicable to (3.25) and again the distribution of u turns out to be Gaussian.[10] The distribution of displacements is then obtained as above.

4. The $B(t)$-impact process

When the $B(t)$-process and the initial conditions on $u(0)$ are given, the solution $u(t)$ is determined by (3.11). Conversely if the solution $u(t)$ is known, $B(t)$ is determined by the equation

$$B(t) - B(0) = \beta\int_0^t u(s)\,ds + u(t) - u(0) \tag{4.1}$$

which is derived immediately from (3.3). The O. U. velocity distribution for the $u(t)$-process can therefore be given only by the $B(t)$-process described in §3. We shall investigate the possibility that a different choice of the $B(t)$-process might have led to a different velocity process compatible with the known physical conditions like M_1 and M_2. If we suppose that $u(0)$ can be chosen so that the velocity at each moment is independent of subsequent residual random impacts, then we have seen that the $B(t)$-process must be differential, and is then uniquely determined by conditions M_1' or M_2. Any velocity process other than the O. U. process satisfying the Langevin equation and M_1' or M_2 would therefore imply dependence between velocity and later residual impacts. This is really another way of saying that the frictional resistance cannot be considered as proportional to the velocity. Before going further we put a condition going back to Maxwell in its modern setting. We formulate a hypothesis M_3 as follows.

M_3. In two or more dimensions (using any orthogonal axes) the velocity components are mutually independent.

---

[10] Cf. Ornstein and Wijk (16) and Wijk (17). The $B(t, \Delta)$ used in these papers corresponds formally to our $dB(t)$. The difference is that it is possible to give a precise description of the $B(t)$-distribution.

In conjunction with the following lemma, due to Kaç ((8) p. 278), hypothesis M_3 implies that all quantities linear in the displacement or velocity functions have Gaussian distributions.

LEMMA.[11] *Let $(x_1, y_1), \cdots, (x_n, y_n)$ be $2n$ chance variables with the property that the sets of chance variables*

$$\{x_j \cos\theta + y_j \sin\theta, j = 1, \cdots, n\}\{-x_j \sin\theta + y_j \cos\theta, j = 1, \cdots, n\}$$

are mutually independent for each value of θ. Then $(x_1, \cdots, x_n)$ have an n-variate Gaussian distribution or a singular Gaussian distribution.

We can combine the Maxwell hypotheses to obtain another justification of the O. U. velocity process.

THEOREM 4. *Let the $B(t)$-process be any process such that the distribution of $B(t_2) - B(t_1)$ or of any quantity depending on such differences is unaffected by translations of the t-axis, and that the integral* (3.5) *can be defined as the limit in probability of the usual sums, with* (3.6) *valid. Then if $u(t)$ is defined by* (3.11), *and if conditions M_2 and M_3 are satisfied, the $B(t)$-process must be precisely that finally obtained in* §3, *leading to the O. U. velocity process.*

Suppose that condition M_2 is satisfied, and let $u(0)$ be fixed as in that condition. Just as in §3, (3.11) then implies that the integral

$$B^*(t) = \int_0^t e^{\beta\tau}\, dB(\tau)$$

has a Gaussian distribution with mean 0 and variance $(kT/m)(e^{2\beta t} - 1)$. If condition M_3 is also satisfied, $B^*(t_2) - B^*(t_1)$, and more generally any finite set of such differences, has a one or more dimensional Gaussian distribution. Using the fact that the distribution of $e^{-\beta s}[B^*(s + t) - B^*(s)]$ is the same as that of $B^*(t)$, in evaluating the expectations in the following equation

$$E\{B^*(s + t)^2\} = E\{[B^*(s) + [B^*(s + t) - B^*(s)]]^2\}, \tag{4.2}$$

we find that $B^*(s) = B^*(s) - B^*(0)$ and $B^*(s + t) - B^*(s)$ are uncorrelated. These two variables are therefore independent. Going further, similar calculations show that any differences $B^*(t_2) - B^*(t_1)$, $B^*(s_2) - B^*(s_1)$ with $0 \leqq s_1 < s_2 \leqq t_1 < t_2$ are independent. Using the fact (derived from condition M_3) that any finite set of differences has a multivariate Gaussian distribution, the $B^*(t)$-process is thus a differential process. This means, by a method we have used above, that the $B(t)$-process is a differential process, leading to the O. U. velocity distribution, because condition M_2 is satisfied.

It is easily seen from counterexamples that Theorem 4 is no longer correct if condition M_1 is supposed instead of condition M_2.

5. Velocity processes not subject to Maxwell's laws

In all the above work the role of the Maxwell velocity distribution has been fundamental. In certain studies, however, other distributions play a somewhat

[11] The result is stated slightly incorrectly by Kaç.

analogous role.[12] It is interesting to note that the Langevin equation can be solved to give a distribution whose transition probabilities are asymptotically any of the symmetric stable distributions classified by Lévy ((14) §30, §56, §57). Such a distribution has characteristic function

$$e^{-\sigma_0^2|z|^\gamma}$$

where σ_0^2 is a positive parameter and $0 < \gamma \leqq 2$. The Gaussian distribution is obtained when $\gamma = 2$. The parameter σ_0^2 plays the role of the variance, although the second moment is never finite when $\gamma < 2$. The velocity process we shall derive will be called the O. U. (γ) process. It is the O. U. process when $\gamma = 2$. The O. U. (γ) process can be described as follows.

1. The process is temporally homogeneous, that is translations of the t-axis do not affect the probability distributions.

2. The process is a Markoff process.

3. For each fixed t, $u(t)$ has a symmetric stable distribution with parameter value σ_0^2, exponent γ. The conditional distribution of $u(s + t)$ for $u(s) = u_0$ is the stable distribution symmetric about $u_0e^{-\beta t}$, with parameter value $\sigma_0^2(1 - e^{-\gamma\beta|t|})$ and exponent γ.

We can obtain this process as a solution of the Langevin equation by choosing the $B(t)$-process properly. In fact, let the $B(t)$-process be the temporally homogeneous differential process in which $B(s + t) - B(s)$ has a symmetric stable distribution with exponent γ and parameter value $\sigma^2 t$. Let $u(t)$ be the corresponding solution of the Langevin equation, given by (3.11). If y is the sum of two independent chance variables with stable symmetric distributions, having parameter values σ_1^2, σ_2^2, and with the same exponent γ then y also has a symmetric stable distribution, with the same exponent, γ, and with parameter value $\sigma_1^2 + \sigma_2^2$. From this fact it is simple to check that the integral (3.5) in the present case has a symmetric stable distribution with exponent γ and parameter value.

$$\int_a^b |f(t)|^\gamma \, dt.$$

If $u(0)$ is given a symmetric stable distribution independent of the $B(t)$-process for $t \geqq 0$, with parameter value $\sigma^2/\gamma\beta$, the distribution of $u(t)$ can be calculated, using characteristic functions, and is found to be symmetric and stable, with exponent γ and parameter value $\sigma^2/\gamma\beta$. The $u(t)$ thus defined determines an O. U. (γ) process, with the above three properties, setting $\sigma_0^2 = \sigma^2/\gamma\beta$.

We shall not spend any time on the details of the analysis of the O. U. (γ) process, since the work runs parallel to that for the case $\gamma = 2$, already discussed. There are, however, a few essential differences. If $v(t)$ is determined by the equation

$$v(t) = t^{1/\gamma} u\left(\frac{1}{\gamma\beta} \log t\right), \qquad t > 0, \tag{1.2.1}$$

[12] Cf. Holtzmark (7).

the $v(t)$ process can be analyzed using (3.11). The $v(t)$-process has the same distribution as the $B(t)$-process just described. The continuity properties of the velocity process can now be derived from those of the $v(t)$-process, which are known. When $\gamma < 2$, the velocity function $u(t)$ is no longer a continuous function of t with probability 1, but is certain to have discontinuities. These discontinuities are however non-oscillatory (jumps).[13] We omit the details of the analogue of Theorem 1.3. Theorem 2.1 is still true if $\gamma \geqq 1$. The considerations of §3 have their obvious counterparts here. Lemma 3 played an essential role, but its statement and proof are correct if the variable y of the lemma is supposed to have a symmetric stable distribution and if the conclusion is that the x_j have a symmetric stable distribution with the same exponent as y.

University of Illinois
and
Institute for Advanced Study

Bibliography

1. S. Bernstein, Comptes Rendus de l'Académie des Sciences de l'URSS, N.S. (1934) pp. 1–9.
2. S. Bernstein, Comptes Rendus de l'Académie des Sciences de l'URSS, N.S. (1934), pp. 361–364.
3. J. L. Doob, Transactions of the American Mathematical Society 42 (1937), pp. 107–140.
4. J. L. Doob, Duke Mathematical Journal 6 (1940), pp. 290–306.
5. J. L. Doob, Transactions of the American Mathematical Society 47 (1940), pp. 455–486.
6. G. L. de Haas-Lorentz, *Die Brownsche Bewegung und einige verwandte Erscheinungen*, Braunschweig (1913).
7. J. Holtzmark, Annalen der Physik 58 (1919), pp. 577–630.
8. M. Kaç, American Journal of Mathematics 61 (1939), pp. 726–728.
9. A. Khintchine, Mathematische Annalen 109 (1934), pp. 604–615.
10. A. Khintchine, Ergebnisse der Mathematik 4 No. 3.
11. G. Krutkow, Physikalische Zeitschrift der Sowjet-Union 5 (1934), pp. 287–300
12. P. Lévy, Pisa Annali Series 2 vol. 3 (1934), pp. 337–366.
13. P. Lévy, Pisa Annali Series 2 vol. 4 (1935), pp. 217–218.
14. P. Lévy, *Théorie de l'addition des variables aléatoires*, Paris 1937.
15. L. S. Ornstein and G. E. Uhlenbeck, Physical Review 36 (1930), pp. 823–841.
16. L. S. Ornstein and W. R. van Wijk, Physica 1 (1934), pp. 235–254, errata p. 966.
17. W. R. van Wijk, Physica 3 (1936), pp. 1111–1119.
18. N. Wiener and R. E. A. C. Paley, *Fourier Transforms in the Complex Domain*, American Mathematical Society Colloquium Publications Vol. XIX.

---

[13] For further details, cf. Lévy (14) Chapter VII.

NOTE:

m is unfortunately used in two senses in the last two lines of Theorem 3. The m which is to vanish is of course the mean value in the O. U. process. Every other m in the statement of the theorem refers to the mass of a Brownian particle.

A CATALOGUE OF SELECTED DOVER BOOKS IN ALL FIELDS OF INTEREST

A CATALOGUE OF SELECTED DOVER BOOKS IN ALL FIELDS OF INTEREST

WHAT IS SCIENCE?, *N. Campbell*
The role of experiment and measurement, the function of mathematics, the nature of scientific laws, the difference between laws and theories, the limitations of science, and many similarly provocative topics are treated clearly and without technicalities by an eminent scientist. "Still an excellent introduction to scientific philosophy," H. Margenau in *Physics Today*. "A first-rate primer . . . deserves a wide audience," *Scientific American*. 192pp. 5⅜ x 8.
60043-2 Paperbound $1.25

THE NATURE OF LIGHT AND COLOUR IN THE OPEN AIR, *M. Minnaert*
Why are shadows sometimes blue, sometimes green, or other colors depending on the light and surroundings? What causes mirages? Why do multiple suns and moons appear in the sky? Professor Minnaert explains these unusual phenomena and hundreds of others in simple, easy-to-understand terms based on optical laws and the properties of light and color. No mathematics is required but artists, scientists, students, and everyone fascinated by these "tricks" of nature will find thousands of useful and amazing pieces of information. Hundreds of observational experiments are suggested which require no special equipment. 200 illustrations; 42 photos. xvi + 362pp. 5⅜ x 8.
20196-1 Paperbound $2.00

THE STRANGE STORY OF THE QUANTUM, AN ACCOUNT FOR THE GENERAL READER OF THE GROWTH OF IDEAS UNDERLYING OUR PRESENT ATOMIC KNOWLEDGE, *B. Hoffmann*
Presents lucidly and expertly, with barest amount of mathematics, the problems and theories which led to modern quantum physics. Dr. Hoffmann begins with the closing years of the 19th century, when certain trifling discrepancies were noticed, and with illuminating analogies and examples takes you through the brilliant concepts of Planck, Einstein, Pauli, Broglie, Bohr, Schroedinger, Heisenberg, Dirac, Sommerfeld, Feynman, etc. This edition includes a new, long postscript carrying the story through 1958. "Of the books attempting an account of the history and contents of our modern atomic physics which have come to my attention, this is the best," H. Margenau, Yale University, in *American Journal of Physics*. 32 tables and line illustrations. Index. 275pp. 5⅜ x 8. 20518-5 Paperbound $2.00

GREAT IDEAS OF MODERN MATHEMATICS: THEIR NATURE AND USE, *Jagjit Singh*
Reader with only high school math will understand main mathematical ideas of modern physics, astronomy, genetics, psychology, evolution, etc. better than many who use them as tools, but comprehend little of their basic structure. Author uses his wide knowledge of non-mathematical fields in brilliant exposition of differential equations, matrices, group theory, logic, statistics, problems of mathematical foundations, imaginary numbers, vectors, etc. Original publication. 2 appendixes. 2 indexes. 65 ills. 322pp. 5⅜ x 8.
20587-8 Paperbound $2.25

PRINCIPLES OF ART HISTORY,
H. Wölfflin
Analyzing such terms as "baroque," "classic," "neoclassic," "primitive," "picturesque," and 164 different works by artists like Botticelli, van Cleve, Dürer, Hobbema, Holbein, Hals, Rembrandt, Titian, Brueghel, Vermeer, and many others, the author establishes the classifications of art history and style on a firm, concrete basis. This classic of art criticism shows what really occurred between the 14th-century primitives and the sophistication of the 18th century in terms of basic attitudes and philosophies. "A remarkable lesson in the art of seeing," *Sat. Rev. of Literature*. Translated from the 7th German edition. 150 illustrations. 254pp. 6⅛ x 9¼. 20276-3 Paperbound $2.25

PRIMITIVE ART,
Franz Boas
This authoritative and exhaustive work by a great American anthropologist covers the entire gamut of primitive art. Pottery, leatherwork, metal work, stone work, wood, basketry, are treated in detail. Theories of primitive art, historical depth in art history, technical virtuosity, unconscious levels of patterning, symbolism, styles, literature, music, dance, etc. A must book for the interested layman, the anthropologist, artist, handicrafter (hundreds of unusual motifs), and the historian. Over 900 illustrations (50 ceramic vessels, 12 totem poles, etc.). 376pp. 5⅜ x 8. 20025-6 Paperbound $2.50

THE GENTLEMAN AND CABINET MAKER'S DIRECTOR,
Thomas Chippendale
A reprint of the 1762 catalogue of furniture designs that went on to influence generations of English and Colonial and Early Republic American furniture makers. The 200 plates, most of them full-page sized, show Chippendale's designs for French (Louis XV), Gothic, and Chinese-manner chairs, sofas, canopy and dome beds, cornices, chamber organs, cabinets, shaving tables, commodes, picture frames, frets, candle stands, chimney pieces, decorations, etc. The drawings are all elegant and highly detailed; many include construction diagrams and elevations. A supplement of 24 photographs shows surviving pieces of original and Chippendale-style pieces of furniture. Brief biography of Chippendale by N. I. Bienenstock, editor of *Furniture World*. Reproduced from the 1762 edition. 200 plates, plus 19 photographic plates. vi + 249pp. 9⅛ x 12¼. 21601-2 Paperbound $3.50

AMERICAN ANTIQUE FURNITURE: A BOOK FOR AMATEURS,
Edgar G. Miller, Jr.
Standard introduction and practical guide to identification of valuable American antique furniture. 2115 illustrations, mostly photographs taken by the author in 148 private homes, are arranged in chronological order in extensive chapters on chairs, sofas, chests, desks, bedsteads, mirrors, tables, clocks, and other articles. Focus is on furniture accessible to the collector, including simpler pieces and a larger than usual coverage of Empire style. Introductory chapters identify structural elements, characteristics of various styles, how to avoid fakes, etc. "We are frequently asked to name some book on American furniture that will meet the requirements of the novice collector, the beginning dealer, and . . . the general public. . . . We believe Mr. Miller's two volumes more completely satisfy this specification than any other work," *Antiques*. Appendix. Index. Total of vi + 1106pp. 7⅞ x 10¾.
21599-7, 21600-4 Two volume set, paperbound $7.50

IT'S FUN TO MAKE THINGS FROM SCRAP MATERIALS, *Evelyn Glantz Hershoff*
What use are empty spools, tin cans, bottle tops? What can be made from rubber bands, clothes pins, paper clips, and buttons? This book provides simply worded instructions and large diagrams showing you how to make cookie cutters, toy trucks, paper turkeys, Halloween masks, telephone sets, aprons, linoleum block- and spatter prints — in all 399 projects! Many are easy enough for young children to figure out for themselves; some challenging enough to entertain adults; all are remarkably ingenious ways to make things from materials that cost pennies or less! Formerly "Scrap Fun for Everyone." Index. 214 illustrations. 373pp. 5⅜ x 8½. 21251-3 Paperbound $1.75

SYMBOLIC LOGIC and THE GAME OF LOGIC, *Lewis Carroll*
"Symbolic Logic" is not concerned with modern symbolic logic, but is instead a collection of over 380 problems posed with charm and imagination, using the syllogism and a fascinating diagrammatic method of drawing conclusions. In "The Game of Logic" Carroll's whimsical imagination devises a logical game played with 2 diagrams and counters (included) to manipulate hundreds of tricky syllogisms. The final section, "Hit or Miss" is a lagniappe of 101 additional puzzles in the delightful Carroll manner. Until this reprint edition, both of these books were rarities costing up to $15 each. Symbolic Logic: Index. xxxi + 199pp. The Game of Logic: 96pp. 2 vols. bound as one. 5⅜ x 8. 20492-8 Paperbound $2.50

MATHEMATICAL PUZZLES OF SAM LOYD, PART I
selected and edited by M. Gardner
Choice puzzles by the greatest American puzzle creator and innovator. Selected from his famous collection, "Cyclopedia of Puzzles," they retain the unique style and historical flavor of the originals. There are posers based on arithmetic, algebra, probability, game theory, route tracing, topology, counter and sliding block, operations research, geometrical dissection. Includes the famous "14-15" puzzle which was a national craze, and his "Horse of a Different Color" which sold millions of copies. 117 of his most ingenious puzzles in all. 120 line drawings and diagrams. Solutions. Selected references. xx + 167pp. 5⅜ x 8. 20498-7 Paperbound $1.35

STRING FIGURES AND HOW TO MAKE THEM, *Caroline Furness Jayne*
107 string figures plus variations selected from the best primitive and modern examples developed by Navajo, Apache, pygmies of Africa, Eskimo, in Europe, Australia, China, etc. The most readily understandable, easy-to-follow book in English on perennially popular recreation. Crystal-clear exposition; step-by-step diagrams. Everyone from kindergarten children to adults looking for unusual diversion will be endlessly amused. Index. Bibliography. Introduction by A. C. Haddon. 17 full-page plates, 960 illustrations. xxiii + 401pp. 5⅜ x 8½. 20152-X Paperbound $2.25

PAPER FOLDING FOR BEGINNERS, *W. D. Murray and F. J. Rigney*
A delightful introduction to the varied and entertaining Japanese art of origami (paper folding), with a full, crystal-clear text that anticipates every difficulty; over 275 clearly labeled diagrams of all important stages in creation. You get results at each stage, since complex figures are logically developed from simpler ones. 43 different pieces are explained: sailboats, frogs, roosters, etc. 6 photographic plates. 279 diagrams. 95pp. 5⅝ x 8⅜. 20713-7 Paperbound $1.00

THE MUSIC OF THE SPHERES: THE MATERIAL UNIVERSE—FROM ATOM TO QUASAR, SIMPLY EXPLAINED, *Guy Murchie*
Vast compendium of fact, modern concept and theory, observed and calculated data, historical background guides intelligent layman through the material universe. Brilliant exposition of earth's construction, explanations for moon's craters, atmospheric components of Venus and Mars (with data from recent fly-by's), sun spots, sequences of star birth and death, neighboring galaxies, contributions of Galileo, Tycho Brahe, Kepler, etc.; and (Vol. 2) construction of the atom (describing newly discovered sigma and xi subatomic particles), theories of sound, color and light, space and time, including relativity theory, quantum theory, wave theory, probability theory, work of Newton, Maxwell, Faraday, Einstein, de Broglie, etc. "Best presentation yet offered to the intelligent general reader," *Saturday Review*. Revised (1967). Index. 319 illustrations by the author. Total of xx + 644pp. 5⅜ x 8½.
21809-0, 21810-4 Two volume set, paperbound $5.00

FOUR LECTURES ON RELATIVITY AND SPACE, *Charles Proteus Steinmetz*
Lecture series, given by great mathematician and electrical engineer, generally considered one of the best popular-level expositions of special and general relativity theories and related questions. Steinmetz translates complex mathematical reasoning into language accessible to laymen through analogy, example and comparison. Among topics covered are relativity of motion, location, time; of mass; acceleration; 4-dimensional time-space; geometry of the gravitational field; curvature and bending of space; non-Euclidean geometry. Index. 40 illustrations. x + 142pp. 5⅜ x 8½. 61771-8 Paperbound $1.35

HOW TO KNOW THE WILD FLOWERS, *Mrs. William Starr Dana*
Classic nature book that has introduced thousands to wonders of American wild flowers. Color-season principle of organization is easy to use, even by those with no botanical training, and the genial, refreshing discussions of history, folklore, uses of over 1,000 native and escape flowers, foliage plants are informative as well as fun to read. Over 170 full-page plates, collected from several editions, may be colored in to make permanent records of finds. Revised to conform with 1950 edition of Gray's Manual of Botany. xlii + 438pp. 5⅜ x 8½. 20332-8 Paperbound $2.50

MANUAL OF THE TREES OF NORTH AMERICA, *Charles Sprague Sargent*
Still unsurpassed as most comprehensive, reliable study of North American tree characteristics, precise locations and distribution. By dean of American dendrologists. Every tree native to U.S., Canada, Alaska; 185 genera, 717 species, described in detail—leaves, flowers, fruit, winterbuds, bark, wood, growth habits, etc. plus discussion of varieties and local variants, immaturity variations. Over 100 keys, including unusual 11-page analytical key to genera, aid in identification. 783 clear illustrations of flowers, fruit, leaves. An unmatched permanent reference work for all nature lovers. Second enlarged (1926) edition. Synopsis of families. Analytical key to genera. Glossary of technical terms. Index. 783 illustrations, 1 map. Total of 982pp. 5⅜ x 8.
20277-1, 20278-X Two volume set, paperbound $6.00